D0983891

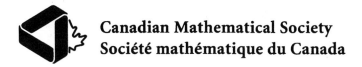

Canadian Mathematical Society
Société mathématique du Canada

Editors-in-Chief
Rédacteurs-en-chef
Jonathan M. Borwein
Peter Borwein

Springer
New York
Berlin
Heidelberg
Barcelona
Hong Kong
London
Milan
Paris
Singapore
Tokyo

CMS Books in Mathematics
Ouvrages de mathématiques de la SMC

Jiří Herman   Radan Kučera
Jaromír Šimša

Equations and Inequalities

Elementary Problems and Theorems in Algebra and Number Theory

Translated by Karl Dilcher

 Springer

Jiří Herman
Brno Grammar School
tř. Kpt. Jaroše 14
658 70 Brno
Czech Republic

Jaromír Šimša
Mathematical Institute
Academy of Sciences of the Czech
  Republic
Žižkova 22
616 62 Brno
Czech Republic

Radan Kučera
Department of Mathematics
Faculty of Science, Masaryk University
Janáčkovo nam. 2a
662 95 Brno
Czech Republic

Translator
Karl Dilcher
Department of Mathematics and Statistics
Dalhousie University
Halifax, Nova Scotia B3H 3J5
Canada

Editors-in-Chief
Rédacteurs-en-chef
Jonathan Borwein
Peter Borwein
Centre for Experimental and Constructive Mathematics
Department of Mathematics and Statistics
Simon Fraser University
Burnaby, British Columbia V5A 1S6
Canada

Mathematics Subject Classification (2000): 16G20, 20Kxx

Library of Congress Cataloging-in-Publication Data
Herman, Jiří.
    [Metody řešení matematických úloh I. English]
    Equations and inequalities : elementary problems and theorems in algebra and number
theory / Jiří Herman, Radan Kučera, Jaromír Šimša ; translated by Karl Dilcher.
        p. cm. — (CMS books in mathematics)
    Includes bibliographical references and index.
    ISBN 0-387-98942-0 (alk. paper)
    1. Equations—Numerical solutions.   2. Inequalities.   3. Problem solving.   I. Kučera
Radan.   II. Šimša, Jaromír.   III. Title.   IV. Series.
    QA218.H4713   2000
    512.9′4—dc21   99-047384

Printed on acid-free paper.

Production managed by Allan Abrams; manufacturing supervised by Jerome Basma.
Photocomposed copy prepared from the translator's LaTeX files.
Printed and bound by R.R. Donnelley and Sons, Harrisonburg, VA.
Printed in the United States of America.

9 8 7 6 5 4 3 2 1

ISBN 0-387-98942-0 Springer-Verlag New York Berlin Heidelberg   SPIN 10745911

Preface

This book is intended as a text for a problem-solving course at the first- or second-year university level, as a text for enrichment classes for talented high-school students, or for mathematics competition training. It can also be used as a source of supplementary material for any course dealing with algebraic equations or inequalities, or to supplement a standard elementary number theory course.

There are already many excellent books on the market that can be used for a problem-solving course. However, some are merely collections of problems from a variety of fields and lack cohesion. Others present problems according to topic, but provide little or no theoretical background. Most problem books have a limited number of rather challenging problems. While these problems tend to be quite beautiful, they can appear forbidding and discouraging to a beginning student, even with well-motivated and carefully written solutions. As a consequence, students may decide that problem solving is only for the few high performers in their class, and abandon this important part of their mathematical, and indeed overall, education.

One of the reasons why problem solving is often found to be difficult is the fact that in recent decades there has been less emphasis on technical skills in North American high-school mathematics. Furthermore, such skills are rarely taught at university, where most courses are quite theoretical or structure-oriented. As a result, a lack of "mathematical fluency" is often evident even in upper years at university; this reduces the enjoyment of the subject and impairs progress and success.

A second reason is that most students are not used to more complex or multi-layered problems. It would certainly be wrong to give the impression that all mathematical problems should succumb to a straightforward approach. Indeed, much of the attractiveness of mathematics lies in the satisfaction derived from solving difficult problems after much effort and several futile attempts. On the other hand, being "stuck" too often and for too long can be very discouraging and counterproductive, and is ultimately a waste of time that could be better spent learning and practicing new techniques.

This book attempts to address these issues and offers a partial remedy. This is done by emphasizing basic algebraic operations and other technical skills that are reinforced in numerous examples and exercises. However, even the easiest ones require a small twist. We therefore hope that this process of practicing does not become purely rote, but will retain the reader's interest, raise his or her level of confidence, and encourage attempts to solve some of the more challenging problems.

Another aim of this book is to familiarize the reader with methods for solving problems in elementary mathematics, accessible to beginning university and advanced high-school students. This can be done in different ways; for instance, the authors of some books introduce several general methods (e.g., induction, analogy, or the pigeonhole principle) and illustrate each one with concrete problems from different areas of mathematics and with varying degrees of difficulty.

Our approach, however, is different. We present a relatively self-contained overview over some parts of elementary mathematics that do not receive much attention in high-school and university education. We give only enough theoretical background to make these topics self-contained and rigorous, and concentrate on solving particular problems in the three main areas corresponding to the first three chapters of this book.

These chapters are fairly independent of each other, with only a limited number of cross-references. Within each chapter, clusters of sections and subsections are tied together either by topic, or by the methods needed to solve the examples and exercises. The problem-book character of this text is underlined by the large number of exercises; they can be solved by using a method or methods previously introduced. We suggest that the reader first carefully study any relevant examples, before attempting to solve any of the exercises.

The individual problems (divided into approximately 330 examples and 760 exercises) are of varying degrees of difficulty, from completely straightforward, where the use of a method under consideration will immediately lead to a solution, to much more difficult problems whose solutions will sometimes require considerable effort. The more demanding exercises are marked with an asterisk (*). Answers to all exercises can be found in the final chapter, where additional hints and instructions to the more difficult ones are given as well.

The problems were taken from a variety of mainly Eastern European sources, such as Mathematical Olympiads and other competitions. Many of them are therefore not otherwise easily accessible to the English-speaking reader. An important criterion for the selection of problems was that their solutions should, in principle, be accessible to high-school students. We believe that even with this limitation one can successfully stimulate creative work in mathematics and illustrate its diversity and richness.

One of our objectives has been to stress alternative approaches to solving a given problem. Sometimes we provide several different solutions in one place, while at other times we return to a problem in later parts of the book.

This book is a translation of the second Czech edition of *Metody řešení matematických úloh* I ("Methods for solving mathematical problems I") published in 1996 at Masaryk University in Brno. Apart from the correction of some minor misprints, the main material has been left unchanged. Three short sections with "competition-type" problems at the ends of the chapters were deleted, but some of these problems were incorporated into the main body of the book. An alphabetical index was added.

The Czech editions of this book have been used by the authors in special enrichment classes for secondary-school students and for Mathematical Olympiad training in the Czech Republic. It has also been used on a regular basis at Masaryk University in courses for future mathematics teachers at secondary schools. A preliminary English version of the first two chapters was successfully tried out by the translator in a newly created problem-solving course in the Fall of 1998 at Dalhousie University, in Halifax, Canada. It was felt that a course partly based on this book worked quite well even in a class with a wide range of mathematical backgrounds and abilities. The book's structure, the worked examples, and the range in level of difficulty of the exercises make it particularly well suited as a source for assigned readings and homework exercises.

The translator also feels that, following the Czech example, this book may be suitable for a course in a teachers' education program for high-school mathematics teachers. Finally, we hope that the reader will find this book a rich source of useful identities, equations, and inequalities.

Brno and Halifax   Jiří Herman
July 1999   Radan Kučera
                                                              Jaromír Šimša
                                                              Karl Dilcher

Contents

Symbols

In addition to usual mathematical notation, the following symbols for sets of numbers will be used:

$\mathbb{N}$ the set $\{1, 2, 3, \dots\}$ of all natural numbers, or positive integers

$\mathbb{N}_0$ the set of all nonnegative integers

$\mathbb{Z}$ the set of all integers

$\mathbb{Q}$ the set of all rational numbers

$\mathbb{Q}^+$ the set of all positive rational numbers

$\mathbb{R}$ the set of all real numbers

$\mathbb{R}^+$ the set of all positive real numbers

$\mathbb{R}^-$ the set of all negative real numbers

$\mathbb{R}_0^+$ the set of all nonnegative real numbers

$\mathbb{R}_0^-$ the set of all nonpositive real numbers

$\mathbb{C}$ the set of all complex numbers

1

Algebraic Identities and Equations

In contrast to the other chapters of this book, this first chapter will not be restricted to one topic; rather, it is devoted to several relatively disparate matters. The emphasis lies on *finite sums*, *polynomials*, and solutions of *systems of equations*. The final section deals with several types of problems that can be solved by using *complex numbers*.

1 Formulas for Powers

This is a short preparatory section. We introduce several identities that will be used throughout this text.

1.1 Combinatorial Numbers

Definition. For an arbitrary number $n \in \mathbb{N}_0$ we define the number $n!$ (read n factorial) as follows: $0! = 1$, and for $n > 0$ we let $n!$ be the product of all natural numbers not exceeding the number n:

$$n! = 1 \cdot 2 \cdots (n-1) \cdot n.$$

Factorials help us define *combinatorial numbers* $\binom{n}{k}$ (read n choose k) by the relation

$$\binom{n}{k} = \frac{n!}{k!(n-k)!}, \quad \text{where } n, k \in \mathbb{N}_0, k \leq n.$$

From this definition we obtain directly the identities

$$\binom{n}{0} = \binom{n}{n} = 1 \quad \text{and} \quad \binom{n}{k} = \binom{n}{n-k}, \quad \text{where } 0 \le k \le n; \quad (1)$$

in the case where $0 < k \le n$, we have furthermore

$$\binom{n}{k} + \binom{n}{k-1} = \frac{n!}{k!(n-k)!} + \frac{n!}{(k-1)!(n-k+1)!}$$

$$= \frac{n!}{k!(n-k+1)!}[(n-k+1) + k] = \binom{n+1}{k}. \quad (2)$$

These identities enable us to show that combinatorial numbers are the coefficients in the expansion of the nth power of a binomial expression. Therefore they are often called *binomial coefficients*. In the following theorem, which generalizes the known formula $(A+B)^2 = A^2 + 2AB + B^2$, we can take A, B to be numbers or expressions.

1.2   The Binomial Theorem

Theorem. *For an arbitrary natural number n we have*

$$(A+B)^n = \binom{n}{0}A^n + \binom{n}{1}A^{n-1}B + \binom{n}{2}A^{n-2}B^2 + \cdots + \binom{n}{n}B^n.$$

PROOF.   Since $\binom{1}{0} = \binom{1}{1} = 1$, the theorem is certainly true for $n = 1$. We assume that it is true for some fixed $n \ge 1$; we will prove that it holds for $n + 1$. If we add the two identities

$$A \cdot (A+B)^n = \binom{n}{0}A^{n+1} + \binom{n}{1}A^nB + \cdots + \binom{n}{n}AB^n,$$
$$B \cdot (A+B)^n = \qquad\quad \binom{n}{0}A^nB + \cdots + \binom{n}{n-1}AB^n + \binom{n}{n}B^{n+1},$$

we obtain on the left-hand side the power $(A+B)^{n+1}$, while the sum on the right is, according to (2), equal to

$$\binom{n}{0}A^{n+1} + \binom{n+1}{1}A^nB + \binom{n+1}{2}A^{n-1}B^2 + \cdots + \binom{n+1}{n}AB^n + \binom{n}{n}B^{n+1}.$$

Since $\binom{n}{0} = 1 = \binom{n+1}{0}$ and $\binom{n}{n} = 1 = \binom{n+1}{n+1}$, the statement of the theorem holds also for $n + 1$, which completes the proof by induction.   □

1.3   Examples

(i) If we set $A = B = 1$, resp. $A = 1$, $B = -1$, we obtain interesting identities between combinatorial numbers, valid for all $n \in \mathbb{N}$:

$$2^n = \binom{n}{0} + \binom{n}{1} + \cdots + \binom{n}{n-1} + \binom{n}{n}, \quad (3)$$

$$0 = \binom{n}{0} - \binom{n}{1} + \cdots + (-1)^{n-1}\binom{n}{n-1} + (-1)^n\binom{n}{n}. \tag{4}$$

(ii) We will determine the largest summand in the expansion of $(1+\sqrt{2})^{100}$. To do this, we determine the quotient of the $(i+1)$th and the ith summand $(0 < i \leq 100)$ of the expansion, namely

$$\frac{\binom{100}{i}(\sqrt{2})^i}{\binom{100}{i-1}(\sqrt{2})^{i-1}} = \frac{100!}{i!(100-i)!} \cdot \frac{(i-1)!(101-i)!}{100!} \cdot \sqrt{2}$$

$$= \frac{101-i}{i} \cdot \sqrt{2}.$$

This quotient is bigger than 1 exactly when $(101-i)\sqrt{2} > i$, that is, exactly when

$$i < \frac{101\sqrt{2}}{1+\sqrt{2}} = 101(2 - \sqrt{2}) \approx 59.1644,$$

and so for $i \leq 59$ the $(i+1)$th summand is bigger than the ith. Similarly, we convince ourselves that for $i \geq 60$ the $(i+1)$th summand is less than the ith. Therefore, the largest summand is $\binom{100}{59} \cdot (\sqrt{2})^{59}$.

1.4 Exercises

Show that (i)–(iii) hold for arbitrary natural numbers n:

(i) $\binom{n}{0} + \binom{n}{2} + \binom{n}{4} + \cdots = 2^{n-1}$.

(ii) $\binom{n}{1} + \binom{n}{3} + \binom{n}{5} + \cdots = 2^{n-1}$
(the sums in (i), (ii) end with $\binom{n}{n}$ or $\binom{n}{n-1}$).

(iii) $\binom{2n+1}{0} + \binom{2n+1}{1} + \binom{2n+1}{2} + \cdots + \binom{2n+1}{n} = 4^n$.

(iv) Find the largest summand in the expansion of $(\sqrt{2} + \sqrt{3})^{50}$.

(v) Determine all natural numbers m, n satisfying

$$\binom{n+1}{m+1} = \binom{n+1}{m} = \tfrac{5}{3}\binom{n+1}{m-1}.$$

The binomial theorem is going to be used extensively in what follows, in particular in Sections 2.9 and 2.10, where we find identities similar to (3) and (4). Also, in Section 3.7 of Chapter 2 we show how the binomial theorem can be used to prove inequalities. Now, however, we introduce a generalization of the known identity $A^2 - B^2 = (A - B) \cdot (A + B)$. As in 1.2, the letters A, B may denote numbers or expressions.

1.5 A Factorization

Theorem. *For an arbitrary natural number n we have*

$$A^n - B^n = (A - B) \cdot (A^{n-1} + A^{n-2}B + \cdots + AB^{n-2} + B^{n-1}). \tag{5}$$

PROOF. By direct calculation: If we subtract the identities

$$A \cdot (A^{n-1} + A^{n-2}B + \cdots + B^{n-1}) = A^n + A^{n-1}B + \cdots + AB^{n-1},$$
$$B \cdot (A^{n-1} + A^{n-2}B + \cdots + B^{n-1}) = \qquad A^{n-1}B + \cdots + AB^{n-1} + B^n,$$

we immediately obtain (5).   □

1.6 Example

Find the coefficients of the polynomial

$$F(x) = (1+x)^r + (1+x)^{r+1} + \cdots + (1+x)^s,$$

where r, s are integers, $0 \le r < s$.

SOLUTION. If we factor out $(1+x)^r$, we obtain

$$F(x) = (1+x)^r \cdot [1 + (1+x) + \cdots + (1+x)^{s-r}].$$

Replacing $A = 1 + x$, $B = 1$, $n = s - r + 1$ in (5), we get

$$(1+x)^{s-r+1} - 1 = x \cdot [(1+x)^{s-r} + (1+x)^{s-r-1} + \cdots + (1+x) + 1],$$

hence

$$F(x) = \frac{1}{x} \cdot (1+x)^r [(1+x)^{s-r+1} - 1] = \frac{1}{x} \cdot [(1+x)^{s+1} - (1+x)^r].$$

According to the binomial theorem 1.2,

$$F(x) = \sum_{i=1}^{s+1} \binom{s+1}{i} x^{i-1} - \sum_{i=1}^{r} \binom{r}{i} x^{i-1},$$

and therefore the coefficient of x^j in $F(x)$ is $\binom{s+1}{j+1} - \binom{r}{j+1}$ for $0 \le j < r$ and $\binom{s+1}{j+1}$ for $r \le j \le s$.   □

1.7 Exercises

Show that for an arbitrary natural number m we have

(i) $A^{2m} - B^{2m} = (A+B)(A^{2m-1} - A^{2m-2}B + \cdots - B^{2m-1}),$ (6)

(ii) $A^{2m-1} + B^{2m-1} = (A+B)(A^{2m-2} - A^{2m-3}B + \cdots + B^{2m-2}).$ (7)

2   Finite Sums

In this section we give several methods for obtaining formulas for some interesting *finite sums*, that is, sums with finitely many summands. We point out that we avoid on purpose the use of mathematical induction, since in many cases it is quite difficult to obtain the necessary hypothesis on how the result should look. This does not mean, however, that this method could not be used for certain problems. Indeed, the many examples in this section would be very useful for practicing this important method of proof.

For the remainder of this section, let k, m, n be natural numbers.

2.1   Example

Determine $S_1(n) = 1 + 2 + 3 + \cdots + n$.

SOLUTION.   We write out the sum $S_1(n)$ twice, with the second one written in opposite order; then we add the two expressions:

$$
\begin{array}{cccccccc}
S_1(n) = & 1 & + & 2 & + & 3 & + \cdots + (n-1) + & n \\
S_1(n) = & n & + (n-1) & + (n-2) & + \cdots + & & 2 & + & 1 \\
\hline
2S_1(n) = & (n+1) & + (n+1) & + (n+1) & + \cdots + & (n+1) & + (n+1).
\end{array}
$$

That is, $2S_1(n) = n(n+1)$, and therefore

$$
S_1(n) = 1 + 2 + 3 + \cdots + n = \frac{n(n+1)}{2}. \tag{8}
$$

We were able to determine this sum very easily, simply by *changing the order* of the summands. It is now clear that this method can be used for summing finitely many terms of any *arithmetic progression*. If an arithmetic progression is determined by its first term a_1 and difference d, then we get for the sum S_A of its first n terms,

$$
\begin{array}{cccc}
S_A = & a_1 & + & [a_1 + d] & + \cdots + & [a_1 + (n-1)d] \\
S_A = & [a_1 + (n-1)d] & + & [a_1 + (n-2)d] & + \cdots + & a_1 \\
\hline
2S_A = & [2a_1 + (n-1)d] & + & [2a_1 + (n-1)d] & + \cdots + & [2a_1 + (n-1)d],
\end{array}
$$

that is,

$$
S_A = \frac{n}{2}[2a_1 + (n-1)d] = \frac{n}{2}[a_1 + a_1 + (n-1)d] = \frac{n}{2}(a_1 + a_n), \tag{9}
$$

where we have used the fact that $a_n = a_1 + (n-1)d$.   □

2.2   Remark

If we change the above procedure a little, we can also determine the sum of the first n terms of a *geometric progression*. If a geometric progression is determined by its first term a_1 and quotient $q \neq 1$, we obtain a formula for the sum S_G of its first n terms as follows:

$$
\begin{aligned}
S_G &= a_1 + a_1 q + a_1 q^2 + \cdots + a_1 q^{n-1} \\
q \cdot S_G &= a_1 q + a_1 q^2 + \cdots + a_1 q^{n-1} + a_1 q^n \\
\hline
q \cdot S_G - S_G &= -a_1 + a_1 q^n.
\end{aligned}
$$

Since $q \neq 1$, we obtain

$$
S_G = a_1 \cdot \frac{q^n - 1}{q - 1}. \tag{10}
$$

The trick that led to the expression (10) is based on an appropriate *multiplication* of the original expression for S_G, followed by a *subtraction* of the two equations, so that on the resulting right-hand side all but two terms disappear. We note that if we begin by factoring out a_1 in the sum S_G, then formula (10) can be obtained directly from (5) by setting $A = 1$, $B = q$.

2.3   Exercises

Evaluate the following sums:

(i) $S = 1 + 3 + 5 + \cdots + (2n - 1)$.

(ii) $S = 1 - 2 + 3 - 4 + \cdots + (-1)^{n+1} n$.

(iii) $S = n + (n + 3) + (n + 6) + \cdots + 4n$.

(iv) $S = (-31) + (-27) + (-23) + \cdots + 29 + 33$.

(v) $S = 2 + 2^2 + 2^3 + \cdots + 2^n$.

(vi) $S = 1 - \frac{1}{2} + \frac{1}{2^2} - \frac{1}{2^3} + \cdots + (-1)^{n-1} \cdot \frac{1}{2^{n-1}}$.

(vii) $S = 2 + 5 + 11 + \cdots + (3 \cdot 2^{n-1} - 1)$.

(viii) $S = \left(x + \frac{1}{x}\right)^2 + \left(x^2 + \frac{1}{x^2}\right)^2 + \cdots + \left(x^n + \frac{1}{x^n}\right)^2$, where $x \in \mathbb{R}$, $x \neq 0$.

2.4   Some Combinatorial Identities

The methods of *changing the order* of summands or appropriate *multiplication* of the sum can also be applied to certain combinatorial sums, that is, sums that contain factorials or binomial coefficients. In some cases both methods can be combined in an appropriate way.

2.5 Example

Determine
$$S = 1 \cdot \binom{n}{1} + 2 \cdot \binom{n}{2} + 3 \cdot \binom{n}{3} + \cdots + n \cdot \binom{n}{n}.$$

(i) FIRST SOLUTION. If we use the fact that $k \cdot \binom{n}{k} = k \cdot \binom{n}{n-k}$, we can write the given sum in a different way, as
$$S = n \cdot \binom{n}{0} + (n-1) \cdot \binom{n}{1} + (n-2) \cdot \binom{n}{2} + \cdots + 1 \cdot \binom{n}{n-1}.$$

We then add both expressions of the sum S:

$$
\begin{array}{llllll}
S = & 1 \cdot \binom{n}{1} & + & 2 \cdot \binom{n}{2} & + \cdots + (n-1) \cdot \binom{n}{n-1} & + n \cdot \binom{n}{n} \\
S = & n \cdot \binom{n}{0} + (n-1) \cdot \binom{n}{1} & + & (n-2) \cdot \binom{n}{2} & + \cdots + \quad 1 \cdot \binom{n}{n-1} \\
\hline
2S = & n \cdot \binom{n}{0} + & n \cdot \binom{n}{1} & + & n \cdot \binom{n}{2} + \cdots + \quad n \cdot \binom{n}{n-1} & + n \cdot \binom{n}{n}.
\end{array}
$$

Hence, according to (3),
$$S = \tfrac{1}{2} n \cdot \left[\binom{n}{0} + \binom{n}{1} + \binom{n}{2} + \cdots + \binom{n}{n} \right] = \tfrac{1}{2} \cdot n \cdot 2^n = n \cdot 2^{n-1}. \qquad \square$$

(ii) SECOND SOLUTION. Each term in the sum S is of the form $k \cdot \binom{n}{k}$ (for $k = 1, 2, \ldots, n$) and we have
$$k \cdot \frac{n!}{k!(n-k)!} = k \cdot \frac{n(n-1)!}{k \cdot (k-1)!(n-k)!} = n \cdot \binom{n-1}{k-1}.$$

Then
$$
\begin{aligned}
S &= 1 \cdot \binom{n}{1} + 2 \cdot \binom{n}{2} + \cdots + (n-1) \cdot \binom{n}{n-1} + n \cdot \binom{n}{n} \\
&= n \cdot \binom{n-1}{0} + n \cdot \binom{n-1}{1} + \cdots + n \cdot \binom{n-1}{n-1} \\
&= n \cdot \left[\binom{n-1}{0} + \binom{n-1}{1} + \cdots + \binom{n-1}{n-2} + \binom{n-1}{n-1} \right] = n \cdot 2^{n-1}. \qquad \square
\end{aligned}
$$

2.6 Example

Determine
$$S = 1 \cdot \binom{n}{1} - 2 \cdot \binom{n}{2} + 3 \cdot \binom{n}{3} - 4 \cdot \binom{n}{4} + \cdots + (-1)^{n+1} \cdot n \cdot \binom{n}{n}.$$

SOLUTION. If we try the method of 2.5.(i), we notice that it works only for even n, in which case according to (4) we obtain $S = 0$. Let us instead rewrite each term in the sum S exactly as in 2.5.(ii); then we get
$$
\begin{aligned}
S &= 1 \cdot \binom{n}{1} - 2 \cdot \binom{n}{2} + 3 \cdot \binom{n}{3} - 4 \cdot \binom{n}{4} + \cdots + (-1)^{n+1} \cdot n \cdot \binom{n}{n} \\
&= n \cdot \left[\binom{n-1}{0} - \binom{n-1}{1} + \binom{n-1}{2} - \binom{n-1}{3} + \cdots + (-1)^{n+1} \binom{n-1}{n-1} \right],
\end{aligned}
$$

and so, according to (4), $S = 0$ for $n \geq 2$ and also $S = 1$ for $n = 1$. $\qquad \square$

2.7 Example

Determine
$$S = \tfrac{1}{1}\binom{n}{0} + \tfrac{1}{2}\binom{n}{1} + \tfrac{1}{3}\binom{n}{2} + \cdots + \tfrac{1}{n+1}\binom{n}{n}.$$

SOLUTION.   Here each summand is of the form $\frac{1}{k+1} \cdot \binom{n}{k}$. We immediately reach our goal if we multiply each summand (and thus the whole sum) by the number $n + 1$, since

$$(n + 1) \cdot \tfrac{1}{k+1}\binom{n}{k} = \tfrac{n+1}{k+1} \cdot \tfrac{n!}{k!(n-k)!} = \tfrac{(n+1)!}{(k+1)!(n-k)!} = \binom{n+1}{k+1}.$$

Then we have

$$
\begin{aligned}
(n + 1) \cdot S &= \tfrac{n+1}{1}\binom{n}{0} + \tfrac{n+1}{2}\binom{n}{1} + \tfrac{n+1}{3}\binom{n}{2} + \cdots + \tfrac{n+1}{n}\binom{n}{n-1} + \tfrac{n+1}{n+1}\binom{n}{n} \\
&= \binom{n+1}{1} + \binom{n+1}{2} + \binom{n+1}{3} + \cdots + \binom{n+1}{n} + \binom{n+1}{n+1} \\
&= \left[\binom{n+1}{0} + \binom{n+1}{1} + \binom{n+1}{2} + \cdots + \binom{n+1}{n+1}\right] - \binom{n+1}{0} \\
&= 2^{n+1} - 1.
\end{aligned}
$$

From this we obtain $S = \frac{2^{n+1}-1}{n+1}$. □

2.8 Exercises

Determine the following sums:

(i) $S = 1 \cdot \binom{n}{0} + 2 \cdot \binom{n}{1} + 3 \cdot \binom{n}{2} + \cdots + (n + 1) \cdot \binom{n}{n}.$

(ii) $S = 1 \cdot \binom{n}{2} - 2 \cdot \binom{n}{3} + 3 \cdot \binom{n}{4} - 4 \cdot \binom{n}{5} + \cdots + (-1)^n(n - 1) \cdot \binom{n}{n}$
for $n > 2.$

(iii) $S = 2 \cdot \binom{n}{2} + 6 \cdot \binom{n}{3} + 12 \cdot \binom{n}{4} + \cdots + n(n - 1) \cdot \binom{n}{n}$ for $n > 2.$

*(iv) $S = 1^2 \cdot \binom{n}{1} + 2^2 \cdot \binom{n}{2} + 3^2 \cdot \binom{n}{3} + \cdots + n^2 \cdot \binom{n}{n}.$

(v) $S = \tfrac{1}{2}\binom{n}{1} - \tfrac{1}{3}\binom{n}{2} + \tfrac{1}{4}\binom{n}{3} - \tfrac{1}{5}\binom{n}{4} + \cdots + \tfrac{(-1)^{n+1}}{n+1}\binom{n}{n}.$

(vi) $S = \tfrac{1}{2}\binom{n}{0} + \tfrac{1}{3}\binom{n}{1} + \tfrac{1}{4}\binom{n}{2} + \cdots + \tfrac{1}{n+2}\binom{n}{n}.$

*(vii) $S = \tfrac{2^2}{2}\binom{n}{1} + \tfrac{2^3}{3}\binom{n}{2} + \tfrac{2^4}{4}\binom{n}{3} + \cdots + \tfrac{2^{n+1}}{n+1}\binom{n}{n}.$

(viii) $S = \tfrac{1}{1!(2n-1)!} + \tfrac{1}{3!(2n-3)!} + \tfrac{1}{5!(2n-5)!} + \cdots + \tfrac{1}{(2n-1)!1!}.$

2.9 Applications of the Binomial Theorem

The *binomial theorem* 1.2 can be used successfully for deriving several combinatorial identities. We consider the binomial expansion of the expression

$$(1 + x)^k = \sum_{i=0}^{k}\binom{k}{i}x^i,$$

which is a polynomial of degree k in the variable x. If we succeed in expressing this polynomial in a different form for certain k, then by equating the coefficients of equal powers x^i we can obtain interesting formulas between binomial coefficients. For example, let us prove the identity

$$\binom{n}{0}^2 + \binom{n}{1}^2 + \binom{n}{2}^2 + \cdots + \binom{n}{n}^2 = \binom{2n}{n}.$$

Here the appropriate exponent is $k = 2n$ since clearly $(1+x)^{2n} = (1+x)^n \cdot (1+x)^n$. After expanding both sides of this equation according to the binomial theorem, we obtain the required identity by equating the coefficients of x^n:

$$(1+x)^{2n} = \binom{2n}{0} + \binom{2n}{1}x + \cdots + \binom{2n}{n}x^n + \cdots + \binom{2n}{2n}x^{2n},$$

$$(1+x)^n \cdot (1+x)^n = \left(\binom{n}{0} + \binom{n}{1}x + \cdots + \binom{n}{n-1}x^{n-1} + \binom{n}{n}x^n\right)$$

$$\times \left(\binom{n}{0} + \binom{n}{1}x + \cdots + \binom{n}{n-1}x^{n-1} + \binom{n}{n}x^n\right)$$

$$= \binom{n}{0}\binom{n}{0} + \left[\binom{n}{0}\binom{n}{1} + \binom{n}{1}\binom{n}{0}\right]x + \cdots$$

$$+ \left[\binom{n}{0}\binom{n}{n} + \binom{n}{1}\binom{n}{n-1} + \cdots + \binom{n}{n}\binom{n}{0}\right]x^n + \cdots$$

$$+ \binom{n}{n}\binom{n}{n} \cdot x^{2n}.$$

Therefore, we have

$$\binom{2n}{n} = \binom{n}{0}\binom{n}{n} + \binom{n}{1}\binom{n}{n-1} + \cdots + \binom{n}{n-1}\binom{n}{1} + \binom{n}{n}\binom{n}{0}$$

$$= \binom{n}{0}^2 + \binom{n}{1}^2 + \cdots + \binom{n}{n-1}^2 + \binom{n}{n}^2,$$

where in the last equality we have again used (1).

In general, the two expressions of this polynomial form the basis of the method of *generating functions*. With the help of this method one can obtain a large number of interesting identities. The interested reader is encouraged to consult, e.g., the books [9] and [6].

2.10 *Exercises*

Using the method of 2.9, prove (i) and (ii), where $p, r \in \mathbb{N}$, $p \le m$, $p \le n$, and $r \le n$:

(i) $\binom{n}{0}\binom{m}{p} + \binom{n}{1}\binom{m}{p-1} + \binom{n}{2}\binom{m}{p-2} + \cdots + \binom{n}{p}\binom{m}{0} = \binom{m+n}{p}$,

(ii) $\binom{n}{0}\binom{n}{r} + \binom{n}{1}\binom{n}{r+1} + \binom{n}{2}\binom{n}{r+2} + \cdots + \binom{n}{n-r}\binom{n}{n} = \dfrac{(2n)!}{(n-r)!(n+r)!}$.

(iii) Determine the sum $S = 1 \cdot \binom{n}{1}^2 + 2 \cdot \binom{n}{2}^2 + 3 \cdot \binom{n}{3}^2 + \cdots + n \cdot \binom{n}{n}^2$.

2.11 Example

Let us determine the sum

$$\binom{m}{0} + \binom{m+1}{1} + \binom{m+2}{2} + \cdots + \binom{m+n-1}{n-1}.$$

Using (1), we rewrite this sum as

$$\binom{m}{m} + \binom{m+1}{m} + \binom{m+2}{m} + \cdots + \binom{m+n-1}{m},$$

and we see that it is the coefficient of x^m in the polynomial

$$(1+x)^m + (1+x)^{m+1} + (1+x)^{m+2} + \cdots + (1+x)^{m+n-1},$$

which, according to Example 1.6 with $r = m$ and $s = m+n-1$, is equal to $\binom{m+n}{m+1}$. This solves the problem. For later use, we write the result as

$$\binom{m}{m} + \binom{m+1}{m} + \binom{m+2}{m} + \cdots + \binom{m+n-1}{m} = \binom{m+n}{m+1}. \tag{11}$$

2.12 Remark

We can convince ourselves that the sum in Example 2.11 in its original form can be found as the coefficient of x^n in the polynomial

$$x^n(1+x)^m + x^{n-1}(1+x)^{m+1} + x^{n-2}(1+x)^{m+2} + \cdots + x^1(1+x)^{m+n-1}.$$

This polynomial is equal to

$$x(1+x)^{m+n} - x^{n+1}(1+x)^m.$$

Formula (11), by the way, is easy to prove by induction with respect to n; it suffices to use formulas (1) and (2). It also has a nice combinatorial interpretation, but we will not pursue this further in this book. In Section 7 of this chapter we will return to some problems concerning sums that involve binomial coefficients.

2.13 The Sums $B(k, n)$

With the help of relation (11) we now derive a formula for the sum

$$B(k, n) = 1 \cdot 2 \cdot 3 \cdots k + 2 \cdot 3 \cdot 4 \cdots (k+1) + \cdots$$
$$+ n(n+1)(n+2) \cdots (n+k-1),$$

which will be used in Sections 2.14 and 2.15. If we put $m = k$ in (11), then upon multiplying by $k!$ we obtain

$$B(k, n) = k! \binom{k+n}{k+1} = \frac{n(n+1)(n+2)\cdots(n+k)}{k+1}. \tag{12}$$

Let us note that according to the definition we have

$$B(1, n) = S_1(n) = \frac{n(n+1)}{2}.$$

Therefore, the derivation of (12) for $k = 1$ constitutes another proof of formula (8).

2.14   The Sums $S_k(n)$

Now we turn our attention to the sum $S_k(n)$ of *kth powers of the first n natural numbers*:

$$S_k(n) = 1^k + 2^k + 3^k + \cdots + n^k,$$

where $k \in \mathbb{N}_0$. This notation is consistent with that in Section 2.1, where we showed that

$$S_1(n) = 1^1 + 2^1 + 3^1 + \cdots + n^1 = \frac{n(n+1)}{2}.$$

Furthermore, it is clear that

$$S_0(n) = 1^0 + 2^0 + 3^0 + \cdots + n^0 = n.$$

Let us now determine $S_2(n) = 1^2 + 2^2 + 3^2 + \cdots + n^2$. According to (12) we have for $k = 2$,

$$
\begin{aligned}
B(2,n) &= 1 \cdot 2 + 2 \cdot 3 + 3 \cdot 4 + \cdots + n(n+1) \\
&= (1 \cdot 1 + 1) + (2 \cdot 2 + 2) + (3 \cdot 3 + 3) + \cdots + (n \cdot n + n) \\
&= (1^2 + 2^2 + 3^2 + \cdots + n^2) + (1 + 2 + 3 + \cdots + n) \\
&= S_2(n) + S_1(n) = \frac{n(n+1)(n+2)}{3},
\end{aligned}
$$

from which we get

$$S_2(n) = B(2,n) - S_1(n) = \tfrac{n(n+1)(n+2)}{3} - \tfrac{n(n+1)}{2} = \tfrac{n(n+1)(2n+1)}{6}. \tag{13}$$

2.15   Exercises

Using the method in 2.14, show:

(i) $S_3(n) = 1^3 + 2^3 + 3^3 + \cdots + n^3 = \frac{n^2(n+1)^2}{4}.$ $\hspace{2cm}$ (14)

(ii) $S_4(n) = 1^4 + 2^4 + 3^4 + \cdots + n^4 = \frac{n(n+1)(2n+1)(3n^2+3n-1)}{30}.$ $\hspace{1cm}$ (15)

(iii) $S_5(n) = 1^5 + 2^5 + 3^5 + \cdots + n^5 = \frac{n^2(n+1)^2(2n^2+2n-1)}{12}.$ $\hspace{1cm}$ (16)

2.16   Alternative Derivation of Expressions for $S_k(n)$

Expressions for $S_k(n)$ can also be obtained by using a binomial expansion for the term $(i + 1)^{k+1}$, where we replace i successively by $1, 2, 3, \ldots, n$.

Thus, for computing $S_2(n)$ we add the following n equations:

$$
\begin{aligned}
(1+1)^3 = \quad & 2^3 \quad = \quad 1^3 \;+\; 3 \cdot 1^2 \;+\; 3 \cdot 1 \;+\; 1 \\
(2+1)^3 = \quad & 3^3 \quad = \quad 2^3 \;+\; 3 \cdot 2^2 \;+\; 3 \cdot 2 \;+\; 1 \\
(3+1)^3 = \quad & 4^3 \quad = \quad 3^3 \;+\; 3 \cdot 3^2 \;+\; 3 \cdot 3 \;+\; 1 \\
\vdots \qquad & \qquad \vdots \\
(n+1)^3 = \quad & (n+1)^3 \quad = \quad n^3 \;+\; 3 \cdot n^2 \;+\; 3 \cdot n \;+\; 1
\end{aligned}
$$

$$S_3(n+1) - 1^3 = S_3(n) + 3 \cdot S_2(n) + 3 \cdot S_1(n) + S_0(n),$$

from which, in view of the equation $S_3(n+1) - S_3(n) = (n+1)^3$, it follows that

$$
\begin{aligned}
3 \cdot S_2(n) &= (n+1)^3 - 1^3 - 3S_1(n) - S_0(n) \\
&= n^3 + 3n^2 + 3n + 1 - 1 - \frac{3}{2}n(n+1) - n.
\end{aligned}
$$

From this we easily obtain formula (13) for the sum $S_2(n)$.

2.17   Exercise

Use the method of 2.16 to determine $S_3(n)$ and $S_4(n)$.

2.18   Remark

Both methods for determining $S_k(n)$ that we have used in 2.14 and 2.16 are *recursive*: The sum $S_k(n)$ can be found if we have already determined $S_0(n), S_1(n), \ldots, S_{k-1}(n)$. In general, one can use the method of paragraph 2.16 to show that the recurrence relation

$$
\binom{k+1}{1} S_k(n) + \binom{k+1}{2} S_{k-1}(n) + \cdots + \binom{k+1}{k} S_1(n) + \binom{k+1}{k+1} S_0(n)
$$
$$
= (n+1)^{k+1} - 1
$$

holds. Looking at formulas (8) and (13)–(16) one might expect that $S_k(n)$ for each k is a polynomial in the variable n of degree $k+1$. That this is indeed true follows from the above recurrence relation by induction on k. Finally, we have the following theorem, which we state without proof (see, e.g., [2], Section 5.8, Theorem 3).

2.19   Bernoulli's Formula

Theorem. *If $k \geq 2$, then for the sum $S_k(n) = 1^k + 2^k + \cdots + n^k$ the following formula holds:*

$$S_k(n) = \frac{n^{k+1}}{k+1} + \frac{n^k}{2} + \binom{k+1}{2} \cdot B_2 \cdot \frac{n^{k-1}}{k+1} + \binom{k+1}{3} \cdot B_3 \cdot \frac{n^{k-2}}{k+1} + \cdots$$
$$+ \binom{k+1}{k} \cdot B_k \cdot \frac{n}{k+1} \tag{17}$$
$$= \frac{n^{k+1}}{k+1} + \frac{n^k}{2} + \sum_{i=2}^{k} \binom{k+1}{i} \cdot B_i \cdot \frac{n^{k+1-i}}{k+1},$$

where B_i is the so-called ith Bernoulli number.

2.20 Remark on Bernoulli Numbers

Bernoulli numbers are rational numbers; they occur in many fields of mathematics. The first few Bernoulli numbers are

$$B_0 = 1, \quad B_1 = -\frac{1}{2}, \quad B_2 = \frac{1}{6}, \quad B_3 = 0, \qquad B_4 = -\frac{1}{30},$$
$$B_5 = 0, \quad B_6 = \frac{1}{42}, \quad B_7 = 0, \quad B_8 = -\frac{1}{30}, \quad B_9 = 0.$$

We notice that starting with B_3, all Bernoulli numbers in the above list with odd index are zero. This is no coincidence. In general, it is true that for an arbitrary natural number k we have $B_{2k+1} = 0$, $B_{4k-2} > 0$, and $B_{4k} < 0$. The Bernoulli numbers can be defined by the following recurrence relations: We set $B_0 = 1$ and require that for an arbitrary natural number k we have

$$1 + \binom{k+1}{1} B_1 + \binom{k+1}{2} B_2 + \cdots + \binom{k+1}{k} B_k = 0$$

(compare this with (17) for $n = 1$, keeping in mind that $S_k(1) = 1$ and $B_1 = -\frac{1}{2}$). We can use this to compute Bernoulli numbers with even index that are not in the above list. For instance,

$$B_{10} = -\frac{1}{11}(B_0 + 11B_1 + 55B_2 + 330B_4 + 462B_6 + 165B_8)$$
$$= -\left(\frac{1}{11} - \frac{1}{2} + \frac{5}{6} - 1 + 1 - \frac{1}{2}\right) = \frac{5}{66}.$$

2.21 Remark

If we know the first $(k+1)$ formulas for $S_0(n)$, $S_1(n)$, ..., $S_k(n)$, we can easily determine the value of the sum $F(1) + F(2) + \cdots + F(n)$, where $F(x)$ is an arbitrary polynomial of degree k. If $F(x) = a_0 + a_1 x + \cdots + a_k x^k$, then

$$\sum_{i=1}^{n} F(i) = \sum_{i=1}^{n}(a_0 + a_1 i + \cdots + a_k i^k)$$
$$= a_0 \sum_{i=1}^{n} i^0 + a_1 \sum_{i=1}^{n} i + \cdots + a_k \sum_{i=1}^{n} i^k$$
$$= a_0 S_0(n) + a_1 S_1(n) + \cdots + a_k S_k(n).$$

Thus, for example,

$$\sum_{i=1}^{n}(2i-1) = 2 \cdot S_1(n) - S_0(n) = n(n+1) - n = n^2$$

(compare with 2.3.(i)).

2.22 *Exercises*

Following Remark 2.21, determine the following sums:

(i) $S = 1 + 3 + 7 + \cdots + (n^2 - n + 1)$.

(ii) $S = 1 + 15 + 65 + \cdots + (4n^3 - 6n^2 + 4n - 1)$.

(iii) $S = 5 + 35 + 113 + \cdots + (4n^3 + 2n - 1)$.

2.23 *Example*

Determine $S = 1^2 - 2^2 + 3^2 - 4^2 + \cdots + (2n-1)^2 - (2n)^2$.

(i) FIRST SOLUTION. Using appropriate rules of summation, we get

$$
\begin{aligned}
S &= 1^2 - 2^2 + 3^2 - 4^2 + \cdots + (2n-1)^2 - (2n)^2 \\
&= [1^2 + 2^2 + 3^2 + \cdots + (2n-1)^2 + (2n)^2] - 2 \cdot [2^2 + 4^2 + \cdots + (2n)^2] \\
&= [1^2 + 2^2 + 3^2 + \cdots + (2n-1)^2 + (2n)^2] - 8 \cdot [1^2 + 2^2 + \cdots + n^2] \\
&= S_2(2n) - 8 \cdot S_2(n) \\
&= \frac{2n(2n+1)(4n+1)}{6} - 8 \cdot \frac{n(n+1)(2n+1)}{6} = -n(2n+1). \qquad \square
\end{aligned}
$$

(ii) SECOND SOLUTION. We divide the $2n$ terms into n pairs. First we find the values of each of these pairs and add the results:

$$
\begin{array}{ccccc}
1^2 - 2^2 & = & -(2-1) \cdot (2+1) & = & -3, \\
3^2 - 4^2 & = & -(4-3) \cdot (4+3) & = & -7, \\
5^2 - 6^2 & = & -(6-5) \cdot (6+5) & = & -11, \\
& \vdots & & & \vdots \\
(2n-1)^2 - (2n)^2 & = & -[2n - (2n-1)] \cdot [2n + (2n-1)] & = & -(4n-1).
\end{array}
$$

Then, adding the n equations by way of identity (9), we obtain

$$S = -[3 + 7 + 11 + \cdots + (4n-1)] = -\frac{n}{2}[4n - 1 + 3] = -n(2n+1). \qquad \square$$

2.24 Remark

Both methods in the previous paragraph can be used to find expressions for *alternating sums* $\sum_{i=1}^{2n}(-1)^{i+1}F(i)$ with an even number of terms, where $F(x)$ is a polynomial (in 2.23 we had $F(x) = x^2$). In 2.23.(i) we used the fact that

$$\sum_{i=1}^{2n}(-1)^{i+1}F(i) = \sum_{i=1}^{2n}F(i) - 2 \cdot \sum_{i=1}^{n}F(2i),$$

and in 2.23.(ii) also the fact that

$$\sum_{i=1}^{2n}(-1)^{i+1}F(i) = \sum_{i=1}^{n}G(i),$$

with $G(i) = F(2i-1) - F(2i)$, where in both cases the sums on the right sides were determined according to Section 2.21.

If the number of terms in the alternating sum is odd, we separate out the last term and thus reduce the problem to the above case.

2.25 Exercises

Find the following sums:

(i) $S = 1^2 + 3^2 + 5^2 + \cdots + (2n-1)^2.$

(ii) $S = 2 \cdot 1^2 + 3 \cdot 2^2 + 4 \cdot 3^2 + \cdots + (k+1) \cdot k^2.$

(iii) $S = 1^3 + 3^3 + 5^3 + \cdots + (2n-1)^3.$

(iv) $S = 1^3 - 2^3 + 3^3 - 4^3 + \cdots + (-1)^{n+1} \cdot n^3.$

*(v) $S = \frac{1^2}{1} + \frac{1^2+2^2}{2} + \frac{1^2+2^2+3^2}{3} + \cdots + \frac{1^2+2^2+3^2+\cdots+n^2}{n}.$

2.26 The Sums $R_k(q,n)$

We now study the sums $R_k(q,n)$ defined by

$$R_k(q,n) = 1^k \cdot q + 2^k \cdot q^2 + 3^k \cdot q^3 + \cdots + n^k \cdot q^n.$$

We will further assume that $q \neq 1$, since for $q = 1$ we have $R_k(1,n) = S_k(n)$; Sections 2.14–2.20 were devoted to this case. Furthermore, for $k = 0$ we obviously have

$$R_0(q,n) = 1^0q + 2^0q^2 + \cdots + n^0q^n = q + q^2 + \cdots + q^n = q \cdot \frac{q^n - 1}{q - 1}. \tag{18}$$

2.27 Example

Determine $R_1(q, n) = 1 \cdot q + 2 \cdot q^2 + 3 \cdot q^3 + \cdots + n \cdot q^n$ $(q \neq 1)$.

(i) FIRST SOLUTION. We use *partial summation*, where step by step we will sum the terms of n geometric progressions (see 2.2). Consider the following n equations:

$$
\begin{aligned}
q + q^2 + q^3 + \cdots + q^{n-1} + q^n &= q \cdot \tfrac{q^n - 1}{q - 1} &= \tfrac{q^{n+1} - q}{q - 1}, \\
q^2 + q^3 + \cdots + q^{n-1} + q^n &= q^2 \cdot \tfrac{q^{n-1} - 1}{q - 1} &= \tfrac{q^{n+1} - q^2}{q - 1}, \\
q^3 + \cdots + q^{n-1} + q^n &= q^3 \cdot \tfrac{q^{n-2} - 1}{q - 1} &= \tfrac{q^{n+1} - q^3}{q - 1}, \\
&\ \ \vdots &\ \ \vdots \\
q^{n-1} + q^n &= q^{n-1} \cdot \tfrac{q^2 - 1}{q - 1} &= \tfrac{q^{n+1} - q^{n-1}}{q - 1}, \\
q^n &= q^n \cdot \tfrac{q - 1}{q - 1} &= \tfrac{q^{n+1} - q^n}{q - 1}.
\end{aligned}
$$

This sum is

$$
\begin{aligned}
R_1(q, n) &= \tfrac{q^{n+1} - q}{q - 1} + \tfrac{q^{n+1} - q^2}{q - 1} + \cdots + \tfrac{q^{n+1} - q^n}{q - 1} \\
&= \tfrac{1}{q - 1} \cdot [n \cdot q^{n+1} - (q + q^2 + \cdots + q^n)] \\
&= \tfrac{1}{q - 1} \cdot \left[n \cdot q^{n+1} - \tfrac{q^{n+1} - q}{q - 1} \right],
\end{aligned}
$$

from which we obtain

$$
R_1(q, n) = \frac{q}{(q - 1)^2} \cdot [n \cdot q^{n+1} - (n + 1) \cdot q^n + 1]. \tag{19}
$$

$\square$

(ii) SECOND SOLUTION. We determine the *difference* between the sum $R_1(q, n)$ and its product with q (this is the same trick that led to the sum of S_G in Section 2.2). If we write equal powers q^i underneath each other,

$$
\begin{aligned}
R_1(q, n) &= q + 2q^2 + 3q^3 + \cdots + & n \cdot q^n, \\
q \cdot R_1(q, n) &= q^2 + 2q^3 + \cdots + (n - 1) \cdot q^n & + n \cdot q^{n+1},
\end{aligned}
$$

then by subtracting the first equation from the second, we obtain

$$
\begin{aligned}
(q - 1) \cdot R_1(q, n) &= -(q + q^2 + q^3 + \cdots + q^n) + n \cdot q^{n+1} \\
&= n \cdot q^{n+1} - \frac{q^{n+1} - q}{q - 1} \\
&= \frac{q}{q - 1} \cdot [n \cdot q^{n+1} - (n + 1) \cdot q^n + 1],
\end{aligned}
$$

which leads to (19).   □

(iii) THIRD SOLUTION. We use differential calculus: We have

$$R_1(q, n) = q + 2q^2 + 3q^3 + \cdots + n \cdot q^n = q \cdot (1 + 2q + 3q^2 + \cdots + n \cdot q^{n-1}),$$

where the sum $C(q, n) = 1 + 2q + 3q^2 + \cdots + nq^{n-1}$ is clearly the derivative of the sum $R_0(q, n)$ with respect to the variable q. Since $R_0(q, n) = \frac{q^{n+1}-q}{q-1}$ by (18), we have

$$C(q, n) = \frac{d}{dq}[R_0(q, n)] = \frac{[(n+1)q^n - 1](q-1) - (q^{n+1} - q)}{(q-1)^2}$$
$$= \frac{n \cdot q^{n+1} - (n+1) \cdot q^n + 1}{(q-1)^2}.$$

Since $R_1(q, n) = q \cdot C(q, n)$, we again obtain relation (19).   □

2.28 Example

Determine

$$R_2(q, n) = 1^2 q^1 + 2^2 q^2 + 3^2 q^3 + \cdots + n^2 \cdot q^n.$$

To solve this problem, we follow the ideas of the second solution in 2.27, where we derived the recurrence relation

$$(q-1) \cdot R_1(q, n) = n \cdot q^{n+1} - R_0(q, n).$$

We determine the difference between the sum $R_2(q, n)$ and its product by q; in a similar fashion as in solution (ii) in Example 2.27 we obtain

$$(q-1) \cdot R_2(q, n) = -\left[1^2 \cdot q + (2^2 - 1^2)q^2 + (3^2 - 2^2)q^3 + \cdots \right.$$
$$\left. + \left(n^2 - (n-1)^2\right) q^n\right] + n^2 \cdot q^{n+1}.$$

Here each of the n summands between the square brackets is of the form

$$\left(k^2 - (k-1)^2\right) \cdot q^k = (2k-1) \cdot q^k = 2k \cdot q^k - q^k.$$

Therefore,

$$(q-1) \cdot R_2(q, n) = -[(2q - q) + (2 \cdot 2q^2 - q^2) + (2 \cdot 3q^3 - q^3) + \cdots$$
$$+ (2nq^n - q^n)] + n^2 \cdot q^{n+1}$$
$$= n^2 q^{n+1} - [2(q + 2q^2 + 3q^3 + \cdots + n \cdot q^n)$$
$$- (q + q^2 + q^3 + \cdots + q^n)]$$
$$= n^2 \cdot q^{n+1} - 2R_1(q, n) + R_0(q, n).$$

If we now substitute $R_0(q, n)$ and $R_1(q, n)$ according to (18) and (19), we obtain

$$R_2(q, n) = \frac{q}{(q-1)^3} \cdot [n^2 \cdot q^{n+2} - (2n^2 + 2n - 1) \cdot q^{n+1} + (n+1)^2 \cdot q^n - q - 1].$$
$$\tag{20}$$

2.29   Remark

The method of solution in the previous example can also be used for determining $R_k(q,n)$ for $k \geq 3$. This will be a recursive method, similar to the method for finding $S_k(n)$ that was explained in 2.18.

2.30   Exercises

In problems (i) and (ii) find the sums by using the method of 2.27.(iii); use any method to find the sums in (iii)–(iv):

(i) $R_2(q,n)$.

(ii) $S(q) = 2 + 6q + 12q^2 + \cdots + n(n-1)q^{n-2}$, where $n \geq 2$ and $q \neq 1$.

(iii) $S(t) = 1 + 2t + 3t^2 + \cdots + (n-1)t^{n-2}$.

(iv) $S(v) = 1^2 + 2^2 \cdot v + 3^2 \cdot v^2 + \cdots + n^2 \cdot v^{n-1}$.

(v) For $x \neq 0$ determine

$$S(x) = \left(x^{n-1} + \frac{1}{x^{n-1}}\right) + 2 \cdot \left(x^{n-2} + \frac{1}{x^{n-2}}\right) + \cdots$$
$$+ (n-1) \cdot \left(x + \frac{1}{x}\right) + n.$$

(vi) Determine the sum $R_3(q,n) = 1^3 \cdot q^1 + 2^3 \cdot q^2 + 3^3 \cdot q^3 + \cdots + n^3 \cdot q^n$.

*(vii) Express $R_k(q,n)$ recursively in terms of $R_0(q,n), R_1(q,n), \ldots$, $R_{k-1}(q,n)$.

2.31   The Sums $S(q,n)$

Let us now consider the sums

$$S(q,n) = P(1)q + P(2)q^2 + P(3)q^3 + \cdots + P(n)q^n,$$

where $q \neq 1$ and $P(x) = a_0 + a_1 x + a_2 x^2 + \cdots + a_m x^m$ is a given polynomial of degree n. Clearly,

$$S(q,n) = \sum_{i=1}^{n} P(i)q^i = \sum_{i=1}^{n}(a_0 + a_1 i + a_2 i^2 + \cdots + a_m i^m)q^i$$

$$= a_0 \sum_{i=1}^{n} q^i + a_1 \sum_{i=1}^{n} iq^i + a_2 \sum_{i=1}^{n} i^2 q^i + \cdots + a_m \sum_{i=1}^{n} i^m q^i$$

$$= a_0 R_0(q,n) + a_1 R_1(q,n) + a_2 R_2(q,n) + \cdots + a_m R_m(q,n).$$

We can therefore express each sum $S(q,n)$ as a linear combination of the sums $R_k(q,n)$ $(k = 0,1,2,\ldots,m)$ which we already investigated in 2.26–2.30.

2.32 *Example*

Evaluate

$$S = \frac{3}{2} + \frac{4}{2^2} + \frac{5}{2^3} + \cdots + \frac{n+2}{2^n}.$$

SOLUTION. According to 2.31 we obtain

$$S = \sum_{i=1}^{n} \frac{i+2}{2^i} = \sum_{i=1}^{n} \frac{i}{2^i} + 2 \cdot \sum_{i=1}^{n} \frac{1}{2^i} = R_1(\tfrac{1}{2}, n) + 2R_0(\tfrac{1}{2}, n),$$

and therefore, using (18) and (19) with $q = \frac{1}{2}$,

$$S = \frac{\frac{1}{2}}{\left(\frac{1}{2} - 1\right)^2} \cdot \left[n \cdot \left(\frac{1}{2}\right)^{n+1} - (n+1) \cdot \left(\frac{1}{2}\right)^n + 1 \right] + 2 \cdot \frac{1}{2} \cdot \frac{\left(\frac{1}{2}\right)^n - 1}{\frac{1}{2} - 1}$$

$$= \frac{2^{n+2} - n - 4}{2^n}. \qquad \qquad \square$$

2.33 *Exercises*

Determine the following sums:

(i) $S = 2^1 + 2 \cdot 2^2 + 3 \cdot 2^3 + \cdots + n \cdot 2^n$.

(ii) $S = 2 \cdot 2^1 + 6 \cdot 2^2 + 12 \cdot 2^3 + \cdots + n(n+1) \cdot 2^n$.

(iii) $S = 1 + \frac{3}{2} + \frac{5}{4} + \cdots + \frac{2n-1}{2^{n-1}}$.

(iv) $S = 1 - \frac{3}{2} + \frac{5}{4} - \frac{7}{8} + \cdots + (-1)^{n-1} \cdot \frac{2n-1}{2^{n-1}}$.

2.34 *Remark*

The evaluation of $S(q, n)$ by way of the sums $R_k(q, n)$ may require some work because of the form of the expression $a_0 R_0(q, n) + a_1 R_1(q, n) + \cdots + a_m R_m(q, n)$. If, in addition, we do not have tables of the necessary formulas for $R_k(q, n)$, we can use the *method of undetermined coefficients* to find the sum of $S(q, n)$. This method is based on the following theorem, which we state without proof.

2.35 *Transforming $S(q, n)$*

Theorem. *Let $P(x)$ be a polynomial of degree m, and $q \neq 1$ a real number. Then there exist a polynomial $Q(x)$ of degree m and a number $d \in \mathbb{R}$ such that for each n we have*

$$S(q, n) = \sum_{i=1}^{n} P(i)q^i = d + Q(n)q^n.$$

2.36   The Method of Undetermined Coefficients

We assume that in Theorem 2.35 we have $P(x) = a_0 + a_1 x + a_2 x^2 + \cdots + a_m x^m$. Then the polynomial $Q(x)$ is of degree m; we will therefore try to determine it in the form $Q(x) = b_0 + b_1 x + b_2 x^2 + \cdots + b_m x^m$. To evaluate the sum $S(q, n)$, we must find the constant d and $m+1$ coefficients $b_0, b_1, \ldots, b_m$. First we find the coefficients b_i from the following pairs of expressions for the differences $S(q, n+1) - S(q, n)$:

$$S(q, n+1) - S(q, n) = \sum_{i=1}^{n+1} P(i)q^i - \sum_{i=1}^{n} P(i)q^i = P(n+1)q^{n+1},$$

$$S(q, n+1) - S(q, n) = d + Q(n+1)q^{n+1} - (d + Q(n)q^n)$$
$$= [Q(n+1)q - Q(n)]q^n.$$

Therefore, $P(n+1)q = Q(n+1)q - Q(n)$. From the polynomial equation $P(x+1)q = Q(x+1)q - Q(x)$ we determine the coefficients $b_0, b_1, \ldots, b_m$. Finally, we obtain the constant d from the equation $S(q, 1) = d + Q(1)q$. We illustrate the use of this method in the following example.

2.37   Example

Determine the sum $S = \frac{1}{2} + \frac{2}{2^2} + \frac{3}{2^3} + \cdots + \frac{n}{2^n}$.

SOLUTION.   We have

$$S = S(\tfrac{1}{2}, n) = \sum_{i=1}^{n} \frac{i}{2^i} = \sum_{i=1}^{n} i(\tfrac{1}{2})^i.$$

We will therefore attempt to find the sum in the form

$$S(\tfrac{1}{2}, n) = d + \frac{a + bn}{2^n}.$$

Since

$$S(\tfrac{1}{2}, n+1) - S(\tfrac{1}{2}, n) = \frac{n+1}{2^{n+1}},$$

for each n we must have

$$\frac{n+1}{2^{n+1}} = \frac{a + b(n+1)}{2^{n+1}} - \frac{a + bn}{2^n},$$

that is, $n + 1 = -bn + (b - a)$. The polynomials $x + 1 = -bx + (b - a)$ are equal if and only if $1 = -b$, $1 = b - a$, that is, $b = -1$, $a = -2$. From the formula for $n = 1$, namely $\frac{1}{2} = d + \frac{a+b}{2}$, we finally obtain $d = 2$. Therefore, $S(\tfrac{1}{2}, n) = 2 - \frac{n+2}{2^n}$. Let us add that the above can also be considered to be a proof by induction.   □

2.38 Exercises

Use the method of undetermined coefficients to find the sums in Exercise 2.33.

For the remainder of this section we will be interested in sums of the form $\sum_{i=1}^{n} \frac{P(i)}{Q(i)}$, where $P(x)$, $Q(x)$ are polynomials and $Q(x)$ has degree at least 1. In some (rather exceptional) cases one can write the result in the form $\frac{F(n)}{G(n)}$, where $F(x)$ and $G(x)$ are appropriate polynomials. Such expressions, however, may not exist even for simple sums, such as $\sum_{i=1}^{n} \frac{1}{i}$ or $\sum_{i=1}^{n} \frac{1}{i^2}$. In the case where such an expression exists, we use the method of *partial fraction decomposition* to actually find it.

2.39 Example

Find the sum $S = \frac{1}{1 \cdot 2} + \frac{1}{2 \cdot 3} + \frac{1}{3 \cdot 4} + \cdots + \frac{1}{n(n+1)}$.

SOLUTION. In this example we have $P(x) = 1$, $Q(x) = x(x+1)$. We decompose the fraction $\frac{1}{x(x+1)}$ into partial fractions. If

$$\frac{1}{x(x+1)} = \frac{a}{x} + \frac{b}{x+1},$$

then $1 = (a+b)x + a$, that is, $a = 1$ and $b = -1$, and therefore

$$\frac{1}{x(x+1)} = \frac{1}{x} - \frac{1}{x+1}.$$

(This expression is, of course, easy to "guess" by way of the calculation $\frac{1}{x(x+1)} = \frac{(x+1)-x}{x(x+1)} = \frac{1}{x} - \frac{1}{x+1}$.) We have therefore

$$S = \frac{1}{1 \cdot 2} + \frac{1}{2 \cdot 3} + \frac{1}{3 \cdot 4} + \cdots + \frac{1}{n(n+1)}$$
$$= \frac{1}{1} - \frac{1}{2} + \frac{1}{2} - \frac{1}{3} + \frac{1}{3} - \frac{1}{4} + \cdots + \frac{1}{n} - \frac{1}{n+1}.$$

In this sum of $2n$ summands we split off the first and the last term and collect the remaining $2n - 2$ summands into $n - 1$ pairs of neighboring terms, with each pair adding up to zero. We obtain therefore

$$S = 1 - \frac{1}{n+1} = \frac{n}{n+1}. \qquad \square$$

2.40 Exercises

Determine the following sums:

(i) $S = \frac{1}{k(k+1)} + \frac{1}{(k+1)(k+2)} + \frac{1}{(k+2)(k+3)} + \cdots + \frac{1}{n(n+1)}$, where $k \leq n$.

(ii) $S = \frac{1}{1\cdot3} + \frac{1}{3\cdot5} + \frac{1}{5\cdot7} + \cdots + \frac{1}{(2n-1)(2n+1)}$.

(iii) $S = \frac{1}{1\cdot3} + \frac{1}{2\cdot4} + \frac{1}{3\cdot5} + \cdots + \frac{1}{n(n+2)}$.

(iv) $S = \frac{1}{1\cdot4} + \frac{1}{4\cdot7} + \frac{1}{7\cdot10} + \cdots + \frac{1}{(3n-2)(3n+1)}$.

(v) $S = \frac{1}{1\cdot2\cdot3} + \frac{1}{2\cdot3\cdot4} + \frac{1}{3\cdot4\cdot5} + \cdots + \frac{1}{n(n+1)(n+2)}$.

(vi) $S = \frac{1}{1\cdot3\cdot5} + \frac{1}{3\cdot5\cdot7} + \frac{1}{5\cdot7\cdot9} + \cdots + \frac{1}{(2n-1)(2n+1)(2n+3)}$.

2.41 Example

Determine the sum

$$S = \frac{3}{4} + \frac{5}{36} + \frac{7}{144} + \cdots + \frac{2n+1}{n^2(n+1)^2}.$$

SOLUTION.   Here each summand is of the form $\frac{2i+1}{i^2(i+1)^2}$ $(i = 1, 2, \ldots, n)$, and we have

$$\frac{2i+1}{i^2(i+1)^2} = \frac{i^2 + 2i + 1 - i^2}{i^2(i+1)^2} = \frac{(i+1)^2 - i^2}{i^2(i+1)^2} = \frac{1}{i^2} - \frac{1}{(i+1)^2},$$

so that

$$S = \frac{3}{4} + \frac{5}{36} + \frac{7}{144} + \cdots + \frac{2n+1}{n^2(n+1)^2}$$

$$= \frac{2^2 - 1^2}{1^2 \cdot 2^2} + \frac{3^2 - 2^2}{2^2 \cdot 3^2} + \frac{4^2 - 3^2}{3^2 \cdot 4^2} + \cdots + \frac{(n+1)^2 - n^2}{n^2(n+1)^2}$$

$$= \frac{1}{1^2} - \frac{1}{2^2} + \frac{1}{2^2} - \frac{1}{3^2} + \frac{1}{3^2} - \frac{1}{4^2} + \cdots + \frac{1}{n^2} - \frac{1}{(n+1)^2}$$

$$= 1 - \frac{1}{(n+1)^2} = \frac{n(n+2)}{(n+1)^2}. \qquad \square$$

2.42 Exercises

Evaluate:

(i) $S = \frac{3}{1\cdot2} + \frac{7}{2\cdot3} + \frac{13}{3\cdot4} + \cdots + \frac{n^2+n+1}{n(n+1)}$.

(ii) $S = \frac{3}{1\cdot2} + \frac{13}{2\cdot3} + \frac{37}{3\cdot4} + \cdots + \frac{n^3+n^2+1}{n(n+1)}$.

(iii) $S = \frac{5}{1\cdot2} + \frac{13}{2\cdot3} + \frac{25}{3\cdot4} + \cdots + \frac{2n^2+2n+1}{n(n+1)}$.

2.43 Example

Let us determine

$$S = \frac{1^2}{1 \cdot 3} + \frac{2^2}{3 \cdot 5} + \frac{3^2}{5 \cdot 7} + \cdots + \frac{n^2}{(2n-1)(2n+1)}.$$

SOLUTION. In this sum each term is of the form $\frac{i^2}{(2i-1)(2i+1)} = \frac{i^2}{4i^2-1}$. If we consider the sum $4S$, we can write each term in the form

$$\frac{4i^2}{4i^2-1} = \frac{4i^2-1+1}{4i^2-1} = 1 + \frac{1}{4i^2-1} = 1 + \frac{1}{(2i-1)(2i+1)}.$$

Therefore, we have, by Exercise 2.40.(ii),

$$4S = \left(1 + \tfrac{1}{1\cdot3}\right) + \left(1 + \tfrac{1}{3\cdot5}\right) + \cdots + \left(1 + \tfrac{1}{(2n-1)(2n+1)}\right) = n + \tfrac{n}{2n+1}.$$

So finally, $S = \frac{n(n+1)}{2(2n+1)}$. $\qquad\qquad\qquad\qquad\qquad\qquad\qquad\qquad\square$

2.44 Exercises

Evaluate

*(i) $S = \dfrac{1}{1\cdot3\cdot5} + \dfrac{2}{3\cdot5\cdot7} + \dfrac{3}{5\cdot7\cdot9} + \cdots + \dfrac{n}{(2n-1)(2n+1)(2n+3)}.$

(ii) $S = \dfrac{1^4}{1\cdot3} + \dfrac{2^4}{3\cdot5} + \dfrac{3^4}{5\cdot7} + \cdots + \dfrac{n^4}{(2n-1)\cdot(2n+1)}.$

3   Polynomials

The search for zeros of polynomials was for many centuries the basic problem of all of algebra. Even when algebra in its development departed to a great extent from this problem, polynomials continued to play an important role in algebra as well as in a number of other areas (among the more remarkable occurrences are the *Taylor polynomial* for the approximate evaluation of functions, and the use of polynomials in interpolation questions). In this section we recall some basic terms connected with polynomials, and also several methods for finding their zeros. However, we are not concerned with providing a well-rounded theory of polynomials; instead, we limit ourselves to some properties that receive relatively little attention in basic algebra courses but that are useful for solving certain problems.

Although we have already used polynomials in the preceding sections, we will first recall some basic terms from their theory.

3.1 Definitions

A *polynomial* in the variable x is an expression that can be written in the form

$$F(x) = a_n x^n + a_{n-1} x^{n-1} + \cdots + a_1 x + a_0, \tag{21}$$

where $n \in \mathbb{N}_0$ and a_i $(i = 0, 1, \ldots, n)$, called *coefficients*, are numbers from some number domain $(\mathbb{Z}, \mathbb{Q}, \mathbb{R}, \mathbb{C})$. The expressions $a_i x^i$ are called terms of the polynomial $F(x)$, and a_0 is called the constant term. Those terms $a_i x^i$ whose coefficients a_i are zero can be disregarded in (21). A polynomial all of whose coefficients are zero $(F(x) = 0)$ is called the *zero polynomial*. After dropping possible zero terms, any nonzero polynomial can be written in the form (21) with the additional condition $a_n \neq 0$. In this case the number n is called *degree of the polynomial $F(x)$*, denoted by $n = \deg F(x)$. The degree of the zero polynomial will remain undefined. Polynomials of degree 1 are called *linear*, those of degree 2 *quadratic*, and of degree 3 *cubic*.

We remark that the polynomial $F(x)$ is uniquely determined by its coefficients, although it can be expressed in a variety of different forms.

3.2 Example

Show that the polynomial
$$F(x) = (1 - x + x^2 - \cdots - x^{99} + x^{100}) \cdot (1 + x + x^2 + \cdots + x^{99} + x^{100})$$
has zero coefficients for all odd powers of x.

SOLUTION.   Using (5) and (7) with $A = 1$ and $B = x$, we get
$$F(x) = \frac{1 + x^{101}}{1 + x} \cdot \frac{1 - x^{101}}{1 - x} = \frac{1 - x^{202}}{1 - x^2},$$
from which, using (5) with $A = 1$, $B = x^2$ and $n = 101$, we get
$$F(x) = 1 + x^2 + x^4 + \cdots + x^{198} + x^{200}.$$

Thus we have proved the statement of this example, and at the same time we also found the coefficients of the even powers of x in $F(x)$. Let us add that in general, the following is true: The coefficients of the odd powers of x in $F(x)$ are zero exactly when $F(-x)$ is identical with $F(x)$.   □

3.3 Exercises

For arbitrary $k \in \mathbb{N}$ find the coefficients of the polynomials

(i) $F(x) = (1 + x)^{2k} + (1 - x)^{2k} - (1 + x^2)^k$,

(ii) $G(x) = (1 + x) \cdot (1 + x^2) \cdot (1 + x^4) \cdots (1 + x^{2^{k-1}})$.

*(iii) Decide which of the two polynomials
$$F(x) = (1 + x^2 - x^3)^{1000}, \qquad G(x) = (1 - x^2 + x^3)^{1000}$$
has the larger coefficient of x^{20}.

(iv) For each $k \in \mathbb{N}$ consider the polynomial
$$F_k(x) = (x^2 - x + 1)(x^4 - x^2 + 1) \cdots (x^{2^k} - x^{2^{k-1}} + 1).$$
Determine the coefficients of all polynomials $(x^2 + x + 1)F_k(x)$.

3.4 Zeros of Polynomials

We can substitute any number for the variable x and carry out the required operations. If $F(x) = a_n x^n + a_{n-1} x^{n-1} + \cdots + a_1 x + a_0$ is a polynomial and c a number, then the number

$$F(c) = a_n \cdot c^n + a_{n-1} \cdot c^{n-1} + \cdots + a_1 c + a_0$$

is called the *value* of the polynomial $F(x)$ at the number c. If $F(c) = 0$, then the number c is called a *zero* of the polynomial $F(x)$ (or a *root* of the polynomial equation $F(x) = 0$).

The following paragraphs will be devoted to the problem of finding zeros of polynomials, which is, as we already mentioned, one of the basic problems of algebra.

Finding the (single) zero of a linear polynomial is very simple. We also have a formula for determining the zeros of a quadratic polynomial: The polynomial $ax^2 + bx + c$ with real coefficients a, b, c and *discriminant* $D = b^2 - 4ac$ has the zeros $(-b \pm \sqrt{D})/2a$, which are real if and only if $D \geq 0$. Formulas for determining the zeros of polynomials of degree 3 and 4 do exist, but their application is so cumbersome that they are almost useless. For $n \geq 5$ the young Norwegian mathematician Niels Henrik Abel (1802–1829) showed in 1823 that there does not exist a formula that would allow one to evaluate the zeros of an arbitrary polynomial of degree n from its coefficients with the help of algebraic operations, including roots of arbitrary degree. This is why the approximate evaluation of zeros of polynomials belongs to the basic problems of numerical mathematics.

3.5 Examples

(i) Which conditions will the real numbers a_1, a_2, a_3, b_1, b_2, b_3 have to satisfy so that the polynomial

$$(a_1 + b_1 x)^2 + (a_2 + b_2 x)^2 + (a_3 + b_3 x)^2$$

is the square of some linear polynomial with real coefficients?

SOLUTION. We assume that there exist real numbers c, d with $d \neq 0$ such that

$$(a_1 + b_1 x)^2 + (a_2 + b_2 x)^2 + (a_3 + b_3 x)^2 = (c + dx)^2. \qquad (22)$$

If the two polynomials are equal, then their values for any number must also be equal, in particular for $-\frac{c}{d}$. Therefore,

$$\left(a_1 - b_1 \cdot \frac{c}{d}\right)^2 + \left(a_2 - b_2 \cdot \frac{c}{d}\right)^2 + \left(a_3 - b_3 \cdot \frac{c}{d}\right)^2 = 0,$$

from which in view of the fact that the numbers in parentheses are real, it follows that

$$a_1 = q \cdot b_1, \quad a_2 = q \cdot b_2, \quad a_3 = q \cdot b_3, \tag{23}$$

where $q = \frac{c}{d}$. Furthermore, if we equate the coefficients of x^2 in (22), we obtain

$$b_1^2 + b_2^2 + b_3^2 = d^2 \neq 0. \tag{24}$$

Conversely, if (24) and (23) are satisfied, then (22) holds for $c = q \cdot d$. The required condition for the numbers $a_1, a_2, a_3, b_1, b_2, b_3$ is therefore $b_1^2 + b_2^2 + b_3^2 \neq 0$ and the existence of a real number q satisfying (23). □

(ii) Find a real number p such that the difference between the real zeros of $x^2 - px + p - 9$ is equal to 6.

SOLUTION.   The zeros x_1, x_2 of the polynomial are given by $x_1 = \frac{p + \sqrt{D}}{2}$ and $x_2 = \frac{p - \sqrt{D}}{2}$, where $D = p^2 - 4p + 36$. Since in this notation we have $x_1 \geq x_2$, the condition of the problem can be written as

$$6 = x_1 - x_2 = \frac{(p + \sqrt{D}) - (p - \sqrt{D})}{2} = \sqrt{D},$$

which holds if and only if $D = 36$. From the equation $p^2 - 4p + 36 = 36$ we find the values $p = 0$ and $p = 4$, and therefore obtain the polynomials $x^2 - 9 = (x + 3)(x - 3)$ and $x^2 - 4x + 5 = (x + 1)(x - 5)$ with the sets of zeros $\{-3, 3\}$ and $\{-1, 5\}$. □

3.6   Exercises

(i) Find the sum S of the coefficients of the polynomial

$$F(x) = (1 - 3x + 3x^2)^{1988} \cdot (1 + 3x - 3x^2)^{1989}.$$

(ii) Show that for arbitrary real numbers a, p, q with $a \neq 0$ and $p \neq q$, the equation

$$\frac{1}{x - p} + \frac{1}{x - q} = \frac{1}{a^2}$$

has two distinct real roots.

*(iii) Decide whether there exist real numbers a, b, c such that for each $\lambda \in \mathbb{R}^+$ the polynomial $ax^2 + bx + c + \lambda$ has exactly two distinct positive real zeros.

*(iv) Determine the numbers $a, b, p, q \in \mathbb{R}$ such that

$$(2x - 1)^{20} - (ax + b)^{20} = (x^2 + px + q)^{10}.$$

In analogy to Section 3.4, where we substituted a number for the variable x in $F(x) = a_n x^n + a_{n-1}x^{n-1} + \cdots + a_1 x + a_0$, we can also substitute a polynomial $G(x)$ for x. We thus obtain the polynomial

$$F(G(x)) = a_n(G(x))^n + a_{n-1}(G(x))^{n-1} + \cdots + a_1 G(x) + a_0.$$

3.7 Example

Find those nonzero polynomials $F(x)$ for which $F(x^2)$ is equal to $(F(x))^2$.

SOLUTION. We will try to find the polynomial $F(x)$ in the form (21), where $a_n \neq 0$. We assume that at least one of the coefficients $a_{n-1}, a_{n-2}, \ldots, a_1, a_0$ is different from 0 and we denote by $k < n$ the largest k for which $a_k \neq 0$. Then the condition $F(x^2) = (F(x))^2$ can be written as

$$a_n x^{2n} + a_k x^{2k} + \cdots + a_1 x^2 + a_0 = (a_n x^n + a_k x^k + \cdots + a_1 x + a_0)^2.$$

If we compare the coefficients of x^{n+k}, we obtain $0 = 2a_n a_k$, which contradicts the assumption. Therefore, $a_{n-1} = \cdots = a_1 = a_0 = 0$ and $F(x) = a_n x^n$. The condition $F(x) = (F(x))^2$ thus takes the form $a_n x^{2n} = a_n^2 x^{2n}$, which holds if and only if $a_n = a_n^2$, that is, $a_n = 1$. The required polynomials are therefore exactly the polynomials x^n, where $n \in \mathbb{N}_0$. □

3.8 Exercises

Find all polynomials $F(x)$ that satisfy:

(i) $F(x + 3) = x^2 + 7x + 12$.

(ii) $F(x^2 + 1) = x^{11} - 8x^7 + 6x^5 - 4$.

(iii) $F(x - 1) + 2F(x + 1) = 3x^2 - 7x$.

*(iv) $F(x^2 - 2x) = (F(x - 2))^2$.

*(v) $F(F(x)) = (F(x))^n$ for a given $n \in \mathbb{N}$ (deg $F(x) > 0$).

3.9 Polynomial Division with Remainder

Theorem. *Let $F(x), G(x)$ be polynomials, with $G(x)$ nonzero. Then there exists a unique pair of polynomials $H(x), R(x)$ such that*

$$F(x) = H(x) \cdot G(x) + R(x), \qquad (25)$$

where $R(x) = 0$ or deg $R(x) < $ deg $G(x)$. The coefficients of the polynomials $H(x)$, $R(x)$ belong to the same number domain $\mathbb{Q}$, $\mathbb{R}$, or $\mathbb{C}$ as the coefficients of $F(x)$, $G(x)$.

A proof can be found in [4]; the polynomials $H(x), R(x)$ can be determined by way of the well-known division algorithm, which we will not introduce here.

We remark that the polynomial $H(x)$ is called the (incomplete) *quotient* and $R(x)$ the *remainder* from the division of the polynomial $F(x)$ by the polynomial $G(x)$.

If $R(x) = 0$ in (25), we say that the polynomial $G(x)$ *divides* the polynomial $F(x)$, which we denote by $G(x) \mid F(x)$. If furthermore $F(x)$ is nonzero, then the inequality $\deg G(x) \le \deg F(x)$ is true.

If, using Theorem 3.9, we divide $F(x)$ by $(x - c)$ with remainder, then $R(x) = 0$ or $\deg R(x) = 0$; hence $R(x) = r$ for some real number r. If in (25) we substitute the number c for x, then

$$F(c) = H(c) \cdot (c - c) + r;$$

hence $F(c) = r$. We have thus derived the following theorem.

3.10 Bézout's Theorem

Theorem. *The remainder in the division of the polynomial $F(x)$ by $(x-c)$ is equal to $F(c)$. The polynomial $F(x)$ is divisible by $(x - c)$ if and only if the number c is a zero of $F(x)$.*

3.11 Multiple Roots

It can happen that the number c is a zero not only of the nonzero polynomial $F(x)$, but also of the polynomial $F(x)/(x - c)$. Then $(x - c)^2 \mid F(x)$, and the number c is called a *multiple zero* of the polynomial $F(x)$. Obviously, there exists a largest $k \le \deg F(x)$ such that $(x - c)^k \mid F(x)$; the number c is then called a *k-fold zero* (or *zero with multiplicity k*) of the polynomial $F(x)$. In the case where $(x - c)^k \mid F(x)$, it is appropriate to call the number c an *at least k-fold zero* of the polynomial $F(x)$. We also remark that a zero that is not multiple, that is, a zero with multiplicity 1, is called a *simple zero*.

3.12 Example

Determine the values a, b, n with $a, b \in \mathbb{R}$, $n \in \mathbb{N}$, for which the number 1 is at least a double zero of the polynomial

$$F(x) = x^n - ax^{n-1} + bx - 1.$$

SOLUTION. $F(x)$ has 1 as a zero if and only if $F(1) = b - a = 0$, hence $a = b$. For $F(x)$ to have a double zero, we need $\deg F(x) = n \ge 2$. For

$n = 2$ we have $F(x) = x^2 - 1 = (x - 1)(x + 1)$, and therefore 1 is only a simple zero of $F(x)$. So, let $n \geq 3$. By (5) we have

$$F(x) = x^n - 1 - ax(x^{n-2} - 1)$$
$$= (x - 1)[(x^{n-1} + x^{n-2} + \cdots + x + 1) - ax(x^{n-3} + \cdots + x + 1)],$$

which means that 1 is at least a double zero if and only if it is a zero of the polynomial in square brackets, that is, if and only if

$$n - a(n - 2) = 0,$$

and thus, $a = b = n/(n - 2)$. We have shown that there exists for each $n \geq 3$ a unique polynomial $F(x)$ with the required properties.   □

3.13 Exercises

(i) Find the remainder of division of the polynomial

$$x + x^3 + x^9 + x^{27} + x^{81} + x^{243}$$

by (a) $x - 1$. (b) $x^2 - 1$.

(ii) For which natural numbers n is 0 a zero of $(x + 1)^n + (x - 1)^n - 2$? For which n is it a multiple zero?

(iii) Determine for which natural numbers n the polynomial

$$1 + x^2 + x^4 + \cdots + x^{2n}$$

is divisible by

$$1 + x + x^2 + \cdots + x^n.$$

(iv) Determine the polynomial $F(x)$ of smallest possible degree that upon division by $(x - 1)^2$ and $(x - 2)^3$ leaves remainder $2x$, resp. $3x$.

(v) Let $F(x) = 1 + x + x^2 + \cdots + x^{1999}$. Find the remainder when $F(x^5)$ is divided by $1 + x + x^2 + x^3 + x^4$.

Certain polynomials (for instance, $2x^2 + 1$, $x^4 + x^2 + 1$, $x^2 - x + 1$) have no zeros in the field of real numbers. The situation changes if we extend the domain in which we look for zeros to include the field of complex numbers.

3.14 The Fundamental Theorem of Algebra

Theorem. *Each nonzero polynomial $F(x)$ of degree n has exactly n complex zeros $x_1, x_2, \ldots, x_n$ if we count their multiplicities, and we have*

$$F(x) = a_n x^n + a_{n-1} x^{n-1} + \cdots + a_1 x + a_0$$
$$= a_n(x - x_1)(x - x_2) \cdots (x - x_n). \tag{26}$$

A proof can be found in [8], p. 184.

3.15 Consequences

(i) If $F(x)$, $G(x)$ are polynomials of degree not exceeding n and such that for $n + 1$ distinct numbers $c_0, c_1, \ldots, c_n$ we have $F(c_i) = G(c_i)$ $(i = 0, 1, \ldots, n)$, then $F(x) = G(x)$.

PROOF. If we assume that the polynomials $F(x)$ and $G(x)$ are distinct, then $F(x) - G(x)$ is a nonzero polynomial of degree not exceeding n and has $n + 1$ zeros $c_0, c_1, \ldots, c_n$; this is a contradiction. □

(ii) Each (complex) zero of $F(x)$ is a zero of $G(x)$ of the same or of higher multiplicity if and only if $F(x)$ divides $G(x)$.

PROOF. This follows by writing $F(x)$ and $G(x)$ in the form (26). □

3.16 Examples

(i) Determine the natural number n for which the polynomial

$$F(x) = x^n + (x - 1)^n + (2x - 1)^n - (3^n + 2^n + 1)$$

is divisible by $G(x) = x^2 - x - 2$.

SOLUTION. Since $G(x) = x^2 - x - 2 = (x + 1)(x - 2)$, by 3.15.(ii) the polynomial $F(x)$ is divisible by $G(x)$ if and only if $F(-1) = F(2) = 0$. Now

$$F(-1) = (-1)^n + (-2)^n + (-3)^n - (3^n + 2^n + 1^n)$$

$$= \begin{cases} 0 & \text{if } n \text{ is even,} \\ -2(3^n + 2^n + 1^n) \neq 0 & \text{if } n \text{ is odd,} \end{cases}$$

and $F(2) = 2^n + 1^n + 3^n - (3^n + 2^n + 1^n) = 0$. Hence $G(x)$ divides $F(x)$ if and only if n is even. □

(ii) Let a, b, c be distinct real numbers. Show that

$$\frac{(a + b)(a + c)}{(a - b)(a - c)} + \frac{(b + c)(b + a)}{(b - c)(b - a)} + \frac{(c + a)(c + b)}{(c - a)(c - b)} = 1.$$

SOLUTION. By multiplying with the common denominator we obtain the equation

$$(a + b)(a + c)(b - c) + (b + c)(b + a)(c - a) + (c + a)(c + b)(a - b)$$
$$= (a - b)(a - c)(b - c), \tag{27}$$

which can be verified by multiplying out all the parentheses on both sides. Instead of this laborious approach, a different method of proof can be used for identities of this kind: We substitute one of the given numbers (a, for example) by the variable x, and rather than verifying (27), we show with the help of 3.15.(i) that the polynomials

$$F(x) = (x+b)(x+c)(b-c) + (b+c)(b+x)(c-x)$$
$$+ (c+x)(c+b)(x-b),$$
$$G(x) = (x-b)(x-c)(b-c)$$

are equal. Since we are dealing with polynomials of degree at most 2, it suffices to verify the equation $F(t) = G(t)$ for three distinct values of t, which we choose such that the evaluation of $F(t)$, $G(t)$ will be as easy as possible. Since $G(b) = G(c) = 0$, we determine the values $F(b)$ and $F(c)$:

$$F(b) = 2b(b^2 - c^2) - 2b(b^2 - c^2) + (b+c)^2 \cdot 0 = 0,$$

and similarly $F(c) = 0$. The third appropriate value is $x = 0$:

$$F(0) = bc(b-c) + bc(b+c) - bc(b+c) = bc(b-c) = G(0).$$

Thus the proof is *almost* complete, but we must not forget that the numbers $b, c, 0$ are distinct only when $bc \neq 0$. However, we have already verified equation (27) for $a = 0$, and it is similarly easy to verify it in the cases $b = 0$ or $c = 0$. $\square$

(iii) Let the numbers $a, b, c \in \mathbb{R}$ be such that $a \neq b$ and $c^3 = a^3 + b^3 + 3abc$. Show that $c = a + b$.

SOLUTION. Set $F(x) = x^3 - 3abx - a^3 - b^3$. Clearly, $F(a+b) = 0$, and thus $a + b$ is a zero of $F(x)$. Since $F(c) = 0$, it suffices to show that $F(x)$ has a unique real zero. Since upon division we obtain

$$F(x) = [x - (a+b)][x^2 + (a+b)x + a^2 - ab + b^2],$$

it suffices to show that the trinomial $x^2 + (a+b)x + a^2 - ab + b^2$ has a negative discriminant in the case $a \neq b$. But this is clear, since $D = -3(a-b)^2$. $\square$

3.17 Exercises

Show that (i)–(iv) hold for arbitrary distinct real numbers a, b, c:

(i) $a \cdot \frac{(a+b)(a+c)}{(a-b)(a-c)} + b \cdot \frac{(b+c)(b+a)}{(b-c)(b-a)} + c \cdot \frac{(c+a)(c+b)}{(c-a)(c-b)} = a + b + c.$

(ii) $a^2 \cdot \frac{(a+b)(a+c)}{(a-b)(a-c)} + b^2 \cdot \frac{(b+c)(b+a)}{(b-c)(b-a)} + c^2 \cdot \frac{(c+a)(c+b)}{(c-a)(c-b)} = (a+b+c)^2.$

(iii) $\frac{(x-b)(x-c)}{(a-b)(a-c)} + \frac{(x-c)(x-a)}{(b-c)(b-a)} + \frac{(x-a)(x-b)}{(c-a)(c-b)} = 1.$

(iv) $\frac{bc(x-a)^2}{(a-b)(a-c)} + \frac{ca(x-b)^2}{(b-c)(b-a)} + \frac{ab(x-c)^2}{(c-a)(c-b)} = x^2.$

(v) Show that for arbitrary $n \in \mathbb{N}$ the polynomial

$$(x+1)^{2n} - x^{2n} - 2x - 1$$

is divisible by $2x^3 + 3x^2 + x$.

*(vi) Determine all polynomials $F(x)$ for which

$$x \cdot F(x-1) = (x-26) \cdot F(x).$$

3.18   Vieta's Relations

By multiplying out the parentheses on the right-hand side of (26) we obtain relations between the coefficients of a polynomial and its zeros. Thus, the numbers x_1, x_2 are zeros of the quadratic trinomial $a_2 x^2 + a_1 x + a_0$ exactly when

$$-\frac{a_1}{a_2} = x_1 + x_2, \qquad \frac{a_0}{a_2} = x_1 \cdot x_2. \tag{28}$$

Similarly, x_1, x_2, x_3 are zeros of the cubic polynomial $a_3 x^3 + a_2 x^2 + a_1 x + a_0$ if and only if

$$-\frac{a_2}{a_3} = x_1 + x_2 + x_3, \quad \frac{a_1}{a_3} = x_1 x_2 + x_1 x_3 + x_2 x_3, \quad -\frac{a_0}{a_3} = x_1 x_2 x_3. \tag{29}$$

Analogous formulas can be written down for higher-degree polynomials. Finally, we note that in relations (28) (resp. (29)) we find the so-called *elementary symmetric polynomials* with two (resp. three) variables. Here they are evaluated at the roots of the quadratic (resp. cubic) polynomials considered above. We will study these in detail in Section 4.

3.19   Example

Show that if the sum of any two of the zeros of the polynomial

$$F(x) = x^4 + ax^3 + bx^2 + cx + d \tag{30}$$

is equal to the sum of the remaining two, then $a^3 - 4ab + 8c = 0$.

SOLUTION.   We denote by x_1, x_2, x_3, x_4 the zeros of the polynomial (30) and assume that $x_1 + x_2 = x_3 + x_4$. By 3.14 and by (28) we have

$$F(x) = (x - x_1)(x - x_2)(x - x_3)(x - x_4) = (x^2 + px + q)(x^2 + px + r),$$

where $p = -(x_1 + x_2) = -(x_3 + x_4)$, $q = x_1 x_2$ a $r = x_3 x_4$. Since

$$(x^2 + px + q)(x^2 + px + r) = x^4 + 2px^3 + (p^2 + q + r)x^2 + p(q + r)x + qr,$$

we obtain the following four equations from (30) by comparing coefficients:

$$a = 2p, \qquad b = p^2 + q + r, \qquad c = p(q + r), \qquad d = qr,$$

from which, upon substitution,

$$a^3 - 4ab + 8c = 8p^3 - 8p(p^2 + q + r) + 8p(q + r) = 0. \qquad \square$$

3.20 Exercises

(i) Suppose that the polynomial $x^2 + px + q$ has the following property: The sum of its zeros is equal to $p+q$, and the product is pq. Find the coefficients p, q.

(ii) Show that if the product of two zeros of the polynomial (30) is equal to the product of the remaining two, then $a^2d = c^2$.

(iii) Let s_k be the sum of the kth powers of the two roots of the equation $ax^2 + bx + c = 0$. Show that for each integer $k \geq 0$ the equation $as_{k+2} + bs_{k+1} + cs_k = 0$ is satisfied.

3.21 Example

Show that if the polynomial $F(x) = x^3 + px + q$, where $p, q \in \mathbb{R}$, $q \neq 0$, has three real zeros, then $p < 0$.

SOLUTION.   By (29) the zeros x_1, x_2, x_3 of the polynomial $F(x)$ satisfy

$$0 = (x_1 + x_2 + x_3)^2 = x_1^2 + x_2^2 + x_3^2 + 2(x_1x_2 + x_1x_3 + x_2x_3)$$
$$= x_1^2 + x_2^2 + x_3^2 + 2p.$$

Since $q \neq 0$, we have $x_1x_2x_3 \neq 0$ and therefore $x_1^2 + x_2^2 + x_3^2 > 0$; hence $p < 0$. □

3.22 Exercises

(i) Suppose that the sum of two zeros of the polynomial

$$F(x) = 2x^3 - x^2 - 7x + \lambda$$

is equal to 1. Determine λ and the zeros of $F(x)$.

(ii) Suppose that the zeros x_1, x_2, x_3 of the polynomial $x^3 + px + q$ satisfy

$$x_3 = \frac{1}{x_1} + \frac{1}{x_2}.$$

Which condition do the coefficients p, q satisfy?

(iii) Find a relation satisfied by the coefficients of the polynomial $x^3 + px^2 + qx + r$ if one of its zeros is equal to the sum of the other two.

(iv) Let the polynomial $F(x) = x^3 + px^2 + qx + r$ have the zeros x_1, x_2, x_3. Find a polynomial $G(x)$ that has zeros x_1x_2, x_1x_3, x_2x_3, and a polynomial $H(x)$ which has zeros $x_1 + x_2, x_1 + x_3, x_2 + x_3$.

(v) Express the coefficient r of the polynomial $x^3 + px^2 + qx + r$ in terms of the remaining coefficients if it is known that one of the zeros of this polynomials is the arithmetic mean of the remaining two.

(vi) Show that for arbitrary nonzero numbers $a, b \in \mathbb{R}$ the zeros x_1, x_2, x_3 of the polynomial $ax^3 - ax^2 + bx + b$ satisfy the condition

$$(x_1 + x_2 + x_3) \cdot \left(\frac{1}{x_1} + \frac{1}{x_2} + \frac{1}{x_3} \right) = -1.$$

*(vii) Let p, q, r, b, c be five different nonzero real numbers. Determine the sum $x + y + z$, provided that the numbers x, y, z satisfy the system

$$\frac{x}{p} + \frac{y}{p-b} + \frac{z}{p-c} = 1,$$

$$\frac{x}{q} + \frac{y}{q-b} + \frac{z}{q-c} = 1,$$

$$\frac{x}{r} + \frac{y}{r-b} + \frac{z}{r-c} = 1.$$

3.23 Reduction of the Degree by Way of a Known Zero

If we know a zero x_1 of a polynomial $F(x)$, we can use this fact for finding its remaining zeros $x_2, x_3, \ldots, x_n$. By 3.10, $F(x)$ is divisible by the linear polynomial $(x - x_1)$, and according to 3.14, $F(x)/(x - x_1)$ has exactly the zeros $x_2, x_3, \ldots, x_n$.

3.24 Example

Show that

$$\sqrt[3]{3 \cdot \sqrt{21} + 8} - \sqrt[3]{3 \cdot \sqrt{21} - 8} = 1. \tag{31}$$

SOLUTION.   Let A denote the left-hand side of (31). By the binomial theorem 1.2 we have

$$A^3 = 16 - 3 \cdot A \cdot \sqrt[3]{(3\sqrt{21} + 8)(3\sqrt{21} - 8)} = 16 - 15A,$$

and therefore A is a zero of the polynomial $F(x) = x^3 + 15x - 16$. Since $F(1) = 0$, the number 1 is a zero of $F(x)$. By dividing, we obtain the identity $F(x)/(x - 1) = x^2 + x + 16$. This last trinomial has no real zeros, since its discriminant is -63. The only real zero of $F(x)$ is therefore the number 1. Since A is a real zero of $F(x)$, we have $A = 1$. $\square$

3.25 Exercises

Prove:

(i) $\sqrt[3]{20 - 14\sqrt{2}} + \sqrt[3]{20 + 14\sqrt{2}} = 4.$

(ii) $\sqrt[3]{\sqrt{5} + 2} + \sqrt[3]{\sqrt{5} - 2} = \sqrt{5}.$

3.26   Rational Zeros of Polynomials with Integer Coefficients

There exists an easy algorithm for finding rational zeros of a polynomial with integer coefficients (compare this with the general case described in 3.4). Suppose we are given the polynomial

$$F(x) = a_n x^n + a_{n-1} x^{n-1} + \cdots + a_1 x + a_0 \tag{32}$$

with integer coefficients a_i, where $a_n \neq 0$. We wish to find its rational zeros. If $a_0 = 0$, then one of the zeros is 0, and using the method described in 3.23, we may reduce the degree of $F(x)$ and search for the zeros of $F(x)/x$. After a finite number of steps we always obtain a polynomial with nonzero constant term. We may therefore assume in addition that $F(x)$ in (32) satisfies $a_0 \neq 0$. If a rational number α is a zero of $F(x)$, then we have $F(\alpha) = 0$, and if we express $\alpha = r/s$, where $r \in \mathbb{Z}$, $s \in \mathbb{N}$ and r, s are relatively prime, then

$$F\left(\frac{r}{s}\right) = a_n \cdot \left(\frac{r}{s}\right)^n + a_{n-1} \left(\frac{r}{s}\right)^{n-1} + \cdots + a_1 \cdot \frac{r}{s} + a_0 = 0,$$

from which upon multiplying by s^n we obtain

$$a_n r^n + a_{n-1} r^{n-1} s + \cdots + a_1 \cdot r \cdot s^{n-1} + a_0 s^n = 0.$$

Since the integer r divides the first n terms on the left of this last equation, it also divides the last term $a_0 s^n$, which implies that r divides a_0, since r and s are relatively prime. In complete analogy we find that s divides a_n. Since a_0 and a_n have only finitely many divisors, the rational zero α can take on only finitely many values. By successively substituting these values into $F(x)$, we find all rational zeros of $F(x)$.

We remark that it is not necessary to substitute all possible values of α into $F(x)$. Indeed, for each integer k it is true that if r/s is a zero of the polynomial (32) and r and s are relatively prime, then the integer $F(k)$ has to be divisible by $(r - ks)$. This result will be proved in Section 10.11 in the third chapter. Using this criterion will save work with the substitutions: To begin with, it suffices to calculate $F(k)$ for suitably chosen $k \in \mathbb{Z}$ (the usual choice is $k = \pm 1$), then for each "suspected" value $\alpha = r/s$ to check whether $F(k)$ is divisible by $(r - ks)$. An example for this approach is given in 3.27.(ii).

3.27   Examples

(i) Find the rational zeros of the polynomial

$$F(x) = x^4 - 4x^2 + x + 2.$$

SOLUTION.   By 3.26 we expect the rational zeros of $F(x)$ to be of the form r/s, where $r \in \{-2, -1, 1, 2\}$ and $s = 1$. Therefore, all rational zeros are in fact integers (this is true for any polynomial with integer coefficients that has 1 as leading coefficient). Substituting, we get $F(-2) = 0$, $F(-1) = -2$, $F(1) = 0$, $F(2) = 4$, and therefore -2 and 1 are the only rational zeros of $F(x)$. We can now apply 3.23 and determine all zeros of $F(x)$: We divide $F(x)$ by the polynomial $(x + 2)(x - 1)$ and get $F(x) = (x+2)(x-1)(x^2 - x - 1)$. The quadratic polynomial $x^2 - x - 1$ has zeros $(1 \pm \sqrt{5})/2$, and from this we have that the zeros of $F(x)$ are -2, 1, $(1 + \sqrt{5})/2$, $(1 - \sqrt{5})/2$. $\square$

(ii) Find the rational zeros of

$$G(x) = 3x^4 + 5x^3 + x^2 + 5x - 2.$$

SOLUTION.   If the rational number r/s, with r, s relatively prime integers and $s > 0$, is a zero of $G(x)$, then $r \in \{1, -1, 2, -2\}$, $s \in \{1, 3\}$, and therefore $r/s \in \{1, -1, \frac{1}{3}, -\frac{1}{3}, 2, -2, \frac{2}{3}, -\frac{2}{3}\}$. To avoid having to check whether each of these eight values is a zero of $G(x)$, we use the criterion from the end of Section 3.26: We set $k = 1$ (resp. $k = -1$), so $G(1) = 12$ (resp. $G(-1) = -8$); therefore, 1 and -1 are not zeros of $G(x)$. Now we set up a table with all six remaining triples of the values of r/s, $r - s$ and $r + s$.

$\frac{r}{s}$	$\frac{1}{3}$	$-\frac{1}{3}$	2	-2	$\frac{2}{3}$	$-\frac{2}{3}$	
$r - s$	-2	-4	1	-3	-1	-5	$G(1) = 12$
$r + s$	4	2	3	-1	5	1	$G(-1) = -8$

Since -5 is not a divisor of 12, and 3, 5 are not divisors of -8, it remains only to decide for $r/s \in \{\frac{1}{3}, -\frac{1}{3}, -2\}$ whether $G(r/s) = 0$. We verify by direct calculation that $G(\frac{1}{3}) = G(-2) = 0$, $G(-\frac{1}{3}) = -100/27$, which means that $G(x)$ has exactly two rational zeros: $\frac{1}{3}$ and -2. $\square$

(iii) Decide whether there exist rational numbers a, b such that the polynomials $x^2 + ax + b$, $x^5 - x - 1$ have a common zero.

SOLUTION.   We assume that some number α is a zero of both polynomials. Then

$$\alpha^5 = \alpha + 1 \quad \text{and} \quad \alpha^2 = -a\alpha - b.$$

From the equation $\alpha^2 = -a\alpha - b$ we first find an expression $\alpha^5 = c\alpha + d$ with coefficients c, d depending on a, b:

$$\begin{aligned}
\alpha^5 &= \alpha \cdot (\alpha^2)^2 = \alpha \cdot (a\alpha + b)^2 = \alpha[a^2(-a\alpha - b) + 2ab\alpha + b^2] \\
&= (-a\alpha - b)(2ab - a^3) + (b^2 - a^2 b)\alpha \\
&= (a^4 - 3a^2 b + b^2)\alpha + (a^3 b - 2ab^2).
\end{aligned}$$

From the equation

$$\alpha^5 = \alpha + 1 = (a^4 - 3a^2b + b^2)\alpha + (a^3b - 2ab^2)$$

it follows that

$$\alpha(a^4 - 3a^2b + b^2 - 1) = -a^3b + 2ab^2 + 1.$$

Since the polynomial $x^5 - x - 1$ has no rational zeros (according to 3.26 the only rational zeros would be ± 1, which obviously does not occur), α is not a rational number. If $a^4 - 3a^2b + b^2 - 1 \neq 0$, we would have

$$\alpha = \frac{-a^3b + 2ab^2 + 1}{a^4 - 3a^2b + b^2 - 1} \in \mathbb{Q},$$

which is a contradiction. Therefore,

$$a^4 - 3a^2b + b^2 - 1 = 0, \tag{33}$$
$$-a^3b + 2ab^2 + 1 = 0. \tag{34}$$

We will now eliminate b from equations (33) and (34). If we subtract the identity (34) from (33) multiplied by $2a$, we get

$$2a^5 - 5a^3b - 2a - 1 = 0, \tag{35}$$

and thus

$$b = \frac{2a^5 - 2a - 1}{5a^3} \quad \text{and} \quad b^2 = \frac{4a^{10} - 8a^6 - 4a^5 + 4a^2 + 4a + 1}{25a^6}.$$

If we substitute these expressions for b and b^2 into (34), we obtain after simplification the following equation for the coefficient a:

$$a^{10} + 3a^6 - 11a^5 - 4a^2 - 4a - 1 = 0.$$

This equation, however, has no rational roots (by 3.26 the only candidates would be ± 1 which, however, is impossible). This means that there are no rational numbers a, b satisfying the required condition. $\qquad\square$

3.28 Exercises

Find the zeros of the polynomials in (i)–(iii):

(i) $F(x) = x^4 + 2x^3 - 2x^2 - 6x + 5$.

(ii) $F(x) = 2x^4 + 7x^3 - 12x^2 - 38x + 21$.

(iii) $F(x) = 3x^6 + 17x^5 + 22x^4 - 4x^3 - 23x^2 - 13x - 2$.

*(iv) Let $a, b, p, q \in \mathbb{Q}$ and suppose that the two polynomials $F(x) = x^3 + px + q$, $G(x) = x^2 + ax + b$ have a common irrational zero. Show that $G(x)$ divides $F(x)$.

(v) The parentheses in the equation $(\ )x^4 + (\ )x^3 + (\ )x^2 + (\ )x + (\ ) = 0$ are replaced in an arbitrary order by the numbers 1, -2, 3, 4, and -6. Show that each such equation has at least one rational root.

4   Symmetric Polynomials

For the solution of systems of algebraic equations in Section 5 we will require the concept of a polynomial in several variables. In this section we will study such polynomials and will be especially interested in symmetric polynomials and their connection with Vieta's relations from Section 3.18.

4.1   Definitions

A *polynomial* in n variables $x_1, x_2, \ldots, x_n$ is a sum of finitely many expressions, called *monomials*, of the form

$$a \cdot x_1^{k_1} \cdot x_2^{k_2} \cdots x_n^{k_n} \qquad (k_1, k_2, \ldots, k_n \in \mathbb{N}_0), \qquad (36)$$

where a is a number (depending on the n-tuple $k_1, k_2, \ldots, k_n$), called a *coefficient*. If some k_i is zero, we just leave out x_i^0. Similarly, we drop the terms with zero coefficients. Since monomials with the same exponent n-tuple can always be added together (by adding their coefficients), we will for the remainder of this section assume that a given polynomial is a sum of terms (36) with distinct n-tuples of exponents $k_1, k_2, \ldots, k_n$. A polynomial all of whose coefficients are zero is called the *zero polynomial*. The sum $k_1 + k_2 + \cdots + k_n$ is called the degree of the monomial (36). The *degree of a nonzero polynomial* will be defined as the largest degree of the monomials occurring in the given polynomial. If in a polynomial all monomials with nonzero coefficients have degree r, then we talk about a *homogeneous polynomial of degree r*.

A polynomial $F(x_1, x_2, \ldots, x_n)$ is called *symmetric* if for an arbitrary order $y_1, y_2, \ldots, y_n$ of the variables $x_1, x_2, \ldots, x_n$ we have

$$F(y_1, y_2, \ldots, y_n) = F(x_1, x_2, \ldots, x_n).$$

It is clear that in order to check whether $F(x_1, x_2, \ldots, x_n)$ is symmetric, it suffices to verify that $F(x_1, x_2, \ldots, x_n)$ does not change when x_1 is interchanged with x_i for arbitrary $i = 2, 3, \ldots, n$.

For example, the polynomial $x_1^2 + 2x_1x_2 + 3x_2^2$ is homogeneous of second degree in the two variables x_1, x_2, but is not symmetric (by interchanging x_1 and x_2 we obtain $3x_1^2 + 2x_1x_2 + x_2^2$, which is a different polynomial). On the other hand, the fourth-degree polynomial $x_1^3 + x_1^2x_2^2 + x_2^3$ in the two variables x_1, x_2 is not homogeneous, but it is symmetric. However, this last polynomial can also be considered as a polynomial in three variables x_1, x_2, x_3 (where x_3 occurs with exponent zero in all terms); then this polynomial is not symmetric (for example, by interchanging x_1 and x_3 we obtain $x_2^3 + x_2^2x_3^2 + x_3^2$).

Now, we could begin constructing a general theory of symmetric polynomials in n variables. However, since in applications we most often come across polynomials in two or three variables, we will basically restrict ourselves to these cases.

4.2 Elementary Symmetric Polynomials in Two Variables

It is easy to convince ourselves that the polynomials

$$\sigma_1(x_1, x_2) = x_1 + x_2, \qquad\qquad (37)$$
$$\sigma_2(x_1, x_2) = x_1 \cdot x_2$$

are symmetric in the two variables x_1, x_2. More generally, the following is true: If we choose an arbitrary polynomial $F(y_1, y_2)$ and substitute $y_1 = \sigma_1(x_1, x_2)$ and $y_2 = \sigma_2(x_1, x_2)$, then we obtain the polynomial $F(x_1 + x_2, x_1 x_2)$, which is symmetric in the variables x_1, x_2. The question arises whether this simple method of constructing symmetric polynomials is sufficiently general, that is, whether one can obtain any symmetric polynomial in this way.

4.3 Representation of Symmetric Polynomials in Two Variables

Theorem. *For an arbitrary symmetric polynomial $F(x_1, x_2)$ in two variables there exists a unique polynomial $H(y_1, y_2)$ in two variables such that*

$$F(x_1, x_2) = H(x_1 + x_2, \, x_1 \cdot x_2).$$

A proof of a more general statement can be found in [13].

By Theorem 4.3 it is therefore possible to use the polynomials (37) to construct arbitrary symmetric polynomials. Therefore, σ_1 and σ_2 are called *elementary symmetric polynomials*. We will now describe a method for expressing a given symmetric polynomial $F(x_1, x_2)$ in terms of these. The terms of $F(x_1, x_2)$ of the form $a_{ii} x_1^i x_2^i$ will be changed to $a_{ii} \sigma_2^i$, and we collect the remaining terms in pairs of the form $a_{ij} x_1^i x_2^j + a_{ji} x_1^j x_2^i$, where $i > j$. The symmetry of $F(x_1, x_2)$ implies the equation $a_{ij} = a_{ji}$, and thus we can factor out $a_{ij}(x_1 x_2)^j = a_{ij} \sigma_2^j$. It therefore remains to find expressions for the sums of powers $s_{i-j} = x_1^{i-j} + x_2^{i-j}$.

4.4 Sums of Powers

Using the binomial theorem 1.2, we obtain the following expressions for $s_k = x_1^k + x_2^k$:

$$
\begin{aligned}
s_1 &= x_1 + x_2 = \sigma_1, \\
s_2 &= x_1^2 + x_2^2 = (x_1 + x_2)^2 - 2x_1 x_2 = \sigma_1^2 - 2\sigma_2, \\
s_3 &= x_1^3 + x_2^3 = (x_1 + x_2)^3 - 3x_1 x_2(x_1 + x_2) = \sigma_1^3 - 3\sigma_1\sigma_2, \\
s_4 &= x_1^4 + x_2^4 = (x_1 + x_2)^4 - 4x_1 x_2(x_1^2 + x_2^2) - 6x_1^2 x_2^2 \\
&= \sigma_1^4 - 4\sigma_2 s_2 - 6\sigma_2^2 = \sigma_1^4 - 4\sigma_2(\sigma_1^2 - 2\sigma_2) - 6\sigma_2^2 \\
&= \sigma_1^4 - 4\sigma_1^2\sigma_2 + 2\sigma_2^2,
\end{aligned}
$$

$$s_5 = x_1^5 + x_2^5 = (x_1 + x_2)^5 - 5x_1x_2(x_1^3 + x_2^3) - 10x_1^2x_2^2(x_1 + x_2)$$
$$= \sigma_1^5 - 5\sigma_2 s_3 - 10\sigma_2^2\sigma_1$$
$$= \sigma_1^5 - 5\sigma_2(\sigma_1^3 - 3\sigma_1\sigma_2) - 10\sigma_2^2\sigma_1$$
$$= \sigma_1^5 - 5\sigma_1^3\sigma_2 + 5\sigma_1\sigma_2^2.$$

We could continue in this way; however, in this text we will not require formulas for s_i ($i > 5$). Another way of deriving these formulas is based on the recurrence relation $s_{i+2} = \sigma_1 s_{i+1} - \sigma_2 s_i$, of whose validity we can easily convince ourselves.

4.5 Examples

To illustrate the method described above we will express the given polynomials $F(x_1, x_2)$, $G(x_1, x_2)$, and $H(x_1, x_2)$ by way of the elementary symmetric polynomials σ_1 and σ_2 introduced in (37).

(i) $F(x_1, x_2) = x_1^5 x_2 + x_1^4 x_2^2 + x_1^2 x_2^4 + x_1 x_2^5 + x_1^2 x_2 + x_1 x_2^2.$

SOLUTION. We divide the terms of the polynomials $F(x_1, x_2)$ into pairs of the form $x_1^i x_2^j + x_1^j x_2^i$, and factor out and substitute the sums of powers from Section 4.4:

$$F(x_1, x_2) = (x_1^5 x_2 + x_1 x_2^5) + (x_1^4 x_2^2 + x_1^2 x_2^4) + (x_1^2 x_2 + x_1 x_2^2)$$
$$= x_1 x_2(x_1^4 + x_2^4) + (x_1 x_2)^2(x_1^2 + x_2^2) + x_1 x_2(x_1 + x_2)$$
$$= \sigma_2 s_4 + \sigma_2^2 s_2 + \sigma_2 \sigma_1$$
$$= \sigma_2(\sigma_1^4 - 4\sigma_1^2\sigma_2 + 2\sigma_2^2) + \sigma_2^2(\sigma_1^2 - 2\sigma_2) + \sigma_2\sigma_1$$
$$= \sigma_1^4\sigma_2 - 3\sigma_1^2\sigma_2^2 + \sigma_1\sigma_2. \qquad \Box$$

(ii) $G(x_1, x_2) = (x_1 - x_2)^{2000}.$

SOLUTION. We have

$$G(x_1, x_2) = (x_1 - x_2)^{2000} = [(x_1 - x_2)^2]^{1000}$$
$$= (x_1^2 + x_2^2 - 2x_1x_2)^{1000} = [(x_1 + x_2)^2 - 4x_1x_2]^{1000}$$
$$= (\sigma_1^2 - 4\sigma_2)^{1000}. \qquad \Box$$

(iii) $H(x_1, x_2) = (x_1 + 4x_2)(2x_1 + 3x_2)(3x_1 + 2x_2)(4x_1 + x_2).$

SOLUTION. By combining factors in an appropriate way, we obtain

$$H(x_1, x_2) = (x_1 + 4x_2)(4x_1 + x_2)(2x_1 + 3x_2)(3x_1 + 2x_2)$$
$$= [4(x_1^2 + x_2^2) + 17x_1x_2][6(x_1^2 + x_2^2) + 13x_1x_2]$$
$$= [4(\sigma_1^2 - 2\sigma_2) + 17\sigma_2][6(\sigma_1^2 - 2\sigma_2) + 13\sigma_2]$$
$$= (4\sigma_1^2 + 9\sigma_2)(6\sigma_1^2 + \sigma_2). \qquad \Box$$

If we compare (37) and Vieta's relations (28), we notice that the right-hand sides are identical. More exactly, the right-hand sides in (28) are the values of the elementary symmetric polynomials σ_1 and σ_2 evaluated at the zeros of the single-variable polynomial $a_2x^2 + a_1x + a_0$. Theorem 4.3 has therefore the following important consequence: *An arbitrary symmetric polynomial in two variables evaluated at the zeros of a given quadratic polynomial $F(x)$ in one variable x can be expressed in terms of the coefficients of $F(x)$.* (A similar statement is true for polynomials of higher degrees.)

4.6 Examples

(i) Compute the number $c = x_1^5 + x_2^5$, where x_1, x_2 are the zeros of $F(x) = x^2 + x - 19$.

SOLUTION.   One possibility would be to find the zeros of $F(x)$ and simply substitute. However, it will be easier to express c in terms of the coefficients of $F(x)$. Since by (28), $\sigma_1(x_1, x_2) = x_1 + x_2 = -1$ and $\sigma_2(x_1, x_2) = x_1 x_2 = -19$, we have

$$c = s_5 = \sigma_1^5 - 5\sigma_1^3\sigma_2 + 5\sigma_1\sigma_2^2$$
$$= (-1)^5 - 5 \cdot (-1)^3 \cdot (-19) + 5 \cdot (-1) \cdot (-19)^2$$
$$= -1901. \qquad \square$$

(ii) Determine the coefficient c in the polynomial $x^2 + x + c$ if its zeros x_1, x_2 satisfy the equation

$$\frac{2x_1^3}{2 + x_2} + \frac{2x_2^3}{2 + x_1} = -1.$$

SOLUTION.   We rewrite this equation into the equivalent form $(x_1 \neq -2, x_2 \neq -2)$

$$2x_1^3(2 + x_1) + 2x_2^3(2 + x_2) + (2 + x_1)(2 + x_2) = 0,$$
$$4(x_1^3 + x_2^3) + 2(x_1^4 + x_2^4) + 4 + 2(x_1 + x_2) + x_1 x_2 = 0.$$

Using the relations for sums of powers from the table in 4.4 we express the left-hand side of this last equation in terms of $\sigma_1 = x_1 + x_2$ and $\sigma_2 = x_1 x_2$, and get

$$4\sigma_1^3 - 12\sigma_1\sigma_2 + 2\sigma_1^4 - 8\sigma_1^2\sigma_2 + 4\sigma_2^2 + 4 + 2\sigma_1 + \sigma_2 = 0.$$

Since by (28) we have $\sigma_1 = x_1 + x_2 = -1$, $\sigma_2 = x_1 x_2 = c$, we obtain by substitution a quadratic equation for the unknown coefficient c, namely

$$4c^2 + 5c = 0,$$

whose roots $c_1 = 0$, $c_2 = -\frac{5}{4}$ are the solutions to our problem. $\square$

4.7 Exercises

Let x_1, x_2 be the zeros of the polynomial $x^2 + px + q$. In (i), (ii) determine the quadratic polynomial with the zeros

(i) x_1^3, x_2^3.

(ii) $x_1^2 + x_2, x_1 + x_2^2$.

(iii) Find the real coefficient b of the polynomial $x^2 + bx - 1$ if its zeros x_1, x_2 satisfy the equation

$$\frac{x_1^3}{x_2^2 - 3} + \frac{x_2^3}{x_1^2 - 3} = 0.$$

4.8 Example

Find the quadratic polynomial with zeros x_1, x_2, given $x_1^5 + x_2^5 = 31$ and $x_1 + x_2 = 1$.

SOLUTION.   Let the polynomial in question be of the form $x^2 + px + q$. Then

$$-p = x_1 + x_2 = \sigma_1,$$
$$q = x_1 x_2 = \sigma_2.$$

By the identity for the power sum s_5 in 4.4 we have

$$31 = x_1^5 + x_2^5 = s_5 = \sigma_1^5 - 5\sigma_1^3\sigma_2 + 5\sigma_1\sigma_2^2 = -p^5 + 5p^3q - 5pq^2.$$

Since $p = -\sigma_1 = -(x_1 + x_2) = -1$, we obtain for q the quadratic equation $5q^2 - 5q - 30 = 0$ which has the roots $q_1 = 3$, $q_2 = -2$. The problem is therefore solved by the two polynomials $x^2 - x + 3$ and $x^2 - x - 2$ (note that one of them has complex zeros). □

4.9 Exercises

In problems (i)–(iii) find a quadratic polynomial with zeros x_1, x_2 satisfying

(i) $4(x_1 + x_2) = 3x_1 x_2$, $x_1 + x_2 + x_1^2 + x_2^2 = 26$.

(ii) $x_1^2 + x_1 x_2 + x_2^2 = 49$, $x_1^4 + x_1^2 x_2^2 + x_2^4 = 931$.

(iii) $x_1^2 + x_2^2 = 7 + x_1 x_2$, $x_1^3 + x_2^3 = 6x_1 x_2 - 1$.

*(iv) Let a, b be two distinct roots of the equation $x^4 + x^3 = 1$. Show that their product ab is a root of the equation $x^6 + x^4 + x^3 - x^2 - 1 = 0$.

4.10 Elementary Symmetric Polynomials in Three Variables

To express symmetric polynomials in three variables we need the three elementary symmetric polynomials

$$\sigma_1(x_1, x_2, x_3) = x_1 + x_2 + x_3,$$
$$\sigma_2(x_1, x_2, x_3) = x_1 x_2 + x_1 x_3 + x_2 x_3, \qquad (38)$$
$$\sigma_3(x_1, x_2, x_3) = x_1 x_2 x_3.$$

If we substitute them into an arbitrary (not necessarily symmetric) polynomial in three variables σ_1, σ_2, and σ_3, we obtain a symmetric polynomial in the variables x_1, x_2, and x_3.

4.11 Representation of Symmetric Polynomials in Three Variables

Theorem. *For an arbitrary symmetric polynomial $F(x_1, x_2, x_3)$ in three variables there exists a unique polynomial $H(y_1, y_2, y_3)$ in three variables such that*

$$F(x_1, x_2, x_3) = H(x_1 + x_2 + x_3, \ x_1 x_2 + x_1 x_3 + x_2 x_3, \ x_1 x_2 x_3).$$

Once again, a proof of a more general result, which also includes Theorem 4.3, can be found in [13].

4.12 Table

Finding the polynomial $H(y_1, y_2, y_3)$ of the previous theorem without deeper knowledge of the theory can be difficult. Therefore, we give at least a few such expressions in the following table; with its help we can continue our work in this and the following section.

$$s_2 = x_1^2 + x_2^2 + x_3^2 = \sigma_1^2 - 2\sigma_2,$$
$$s_3 = x_1^3 + x_2^3 + x_3^3 = \sigma_1^3 - 3\sigma_1\sigma_2 + 3\sigma_3,$$
$$s_4 = x_1^4 + x_2^4 + x_3^4 = \sigma_1^4 - 4\sigma_1^2\sigma_2 + 2\sigma_2^2 + 4\sigma_1\sigma_3,$$
$$s_5 = x_1^5 + x_2^5 + x_3^5 = \sigma_1^5 - 5\sigma_1^3\sigma_2 + 5\sigma_1\sigma_2^2 + 5\sigma_1^2\sigma_3 - 5\sigma_2\sigma_3,$$
$$s_6 = x_1^6 + x_2^6 + x_3^6$$
$$= \sigma_1^6 - 6\sigma_1^4\sigma_2 + 9\sigma_1^2\sigma_2^2 - 2\sigma_2^3 + 6\sigma_1^3\sigma_3 - 12\sigma_1\sigma_2\sigma_3 + 3\sigma_3^2,$$
$$x_1^2 x_2^2 + x_1^2 x_3^2 + x_2^2 x_3^2 = \sigma_2^2 - 2\sigma_1\sigma_3,$$
$$x_1^3 x_2^3 + x_1^3 x_3^3 + x_2^3 x_3^3 = \sigma_2^3 + 3\sigma_3^2 - 3\sigma_1\sigma_2\sigma_3,$$
$$x_1^2 x_2 + x_1 x_2^2 + x_1^2 x_3 + x_1 x_3^2 + x_2^2 x_3 + x_2 x_3^2 = \sigma_1\sigma_2 - 3\sigma_3,$$
$$x_1^3 x_2 + x_1 x_2^3 + x_1^3 x_3 + x_1 x_3^3 + x_2^3 x_3 + x_2 x_3^3 = \sigma_1^2\sigma_2 - 2\sigma_2^2 - \sigma_1\sigma_3,$$
$$x_1^4 x_2^2 + x_1^2 x_2^4 + x_1^4 x_3^2 + x_1^2 x_3^4 + x_2^4 x_3^2 + x_2^2 x_3^4$$
$$= \sigma_1^2\sigma_2^2 - 2\sigma_2^3 - 2\sigma_1^3\sigma_3 + 4\sigma_1\sigma_2\sigma_3 - 3\sigma_3^2.$$

4.13 Examples

(i) Find all values of the parameter $a \in \mathbb{R}$ such that the zeros x_1, x_2, x_3 of $x^3 - 6x^2 + ax + a$ satisfy

$$(x_1 - 3)^3 + (x_2 - 3)^3 + (x_3 - 3)^3 = 0.$$

SOLUTION.   We use the formulas in 4.12 for $s_k = x_1^k + x_2^k + x_3^k$ ($k = 2, 3$). Since x_1, x_2, x_3 are zeros of the given polynomial, we have $\sigma_1 = 6$, $\sigma_2 = a$, $\sigma_3 = -a$, and therefore

$$s_1 = \sigma_1 = 6, \qquad s_2 = \sigma_1^2 - 2\sigma_2 = 36 - 2a,$$
$$s_3 = \sigma_1^3 - 3\sigma_1\sigma_2 + 3\sigma_3 = 216 - 21a.$$

Using the binomial theorem we then obtain

$$0 = (x_1 - 3)^3 + (x_2 - 3)^3 + (x_3 - 3)^3 = s_3 - 3s_2 \cdot 3 + 3s_1 \cdot 3^2 - 3 \cdot 3^3$$
$$= 216 - 21a - 9(36 - 2a) + 27 \cdot 6 - 81 = -27 - 3a,$$

and therefore $a = -9$.   □

(ii) Prove that the number

$$c = \sqrt[3]{\tfrac{1}{9}} + \sqrt[3]{-\tfrac{2}{9}} + \sqrt[3]{\tfrac{4}{9}}.$$

is a zero of $F(x) = x^3 + \sqrt[3]{6}\,x^2 - 1$.

SOLUTION.   We set $a_1 = \sqrt[3]{\tfrac{1}{9}}$, $a_2 = \sqrt[3]{-\tfrac{2}{9}}$, $a_3 = \sqrt[3]{\tfrac{4}{9}}$. Clearly,

$$a_1^3 + a_2^3 + a_3^3 = \tfrac{1}{9} - \tfrac{2}{9} + \tfrac{4}{9} = \tfrac{1}{3},$$
$$a_1 \cdot a_2 \cdot a_3 = \sqrt[3]{\tfrac{1}{9} \cdot \left(-\tfrac{2}{9}\right) \cdot \left(\tfrac{4}{9}\right)} = -\tfrac{2}{9},$$
$$a_1 a_2 + a_1 a_3 + a_2 a_3 = \sqrt[3]{-\tfrac{2}{81}} + \sqrt[3]{\tfrac{4}{81}} + \sqrt[3]{-\tfrac{8}{81}}$$
$$= \sqrt[3]{-\tfrac{2}{9}} \left(\sqrt[3]{\tfrac{1}{9}} + \sqrt[3]{-\tfrac{2}{9}} + \sqrt[3]{\tfrac{4}{9}} \right) = \sqrt[3]{-\tfrac{2}{9}} \cdot c.$$

By 4.12, for $x_1 = a_1$, $x_2 = a_2$, $x_3 = a_3$ we have

$$a_1^3 + a_2^3 + a_3^3 = (a_1 + a_2 + a_3)^3 - 3(a_1 + a_2 + a_3)(a_1 a_2 + a_1 a_3 + a_2 a_3)$$
$$+ 3a_1 a_2 a_3,$$

and therefore

$$\tfrac{1}{3} = c^3 - 3c \cdot \sqrt[3]{-\tfrac{2}{9}} \cdot c - \tfrac{2}{3},$$

which implies that the number c is a zero of $F(x) = x^3 + \sqrt[3]{6}\,x^2 - 1$.   □

4.14 Exercises

(i) Construct a cubic polynomial whose zeros are squares of the zeros of $x^3 - 2x^2 + x - 12$.

*(ii) Suppose that the three distinct numbers $a, b, c \in \mathbb{R}$ satisfy

$$a(a^2 + p) = b(b^2 + p) = c(c^2 + p)$$

for an appropriate $p \in \mathbb{R}$. Show that in this case $a + b + c = 0$.

(iii) Show that if $a, b, c \in \mathbb{R}$ and $a + b + c = 0$, then

$$a^4 + b^4 + c^4 = 2(ab + ac + bc)^2.$$

(iv) Construct a cubic polynomial with zeros $x_1 + x_2$, $x_1 + x_3$, $x_2 + x_3$, where x_1, x_2, x_3 are the zeros of $x^3 + px^2 + qx + r$.

(v) Prove that if the numbers $a, b, c \in \mathbb{R}$ satisfy

$$\frac{1}{a} + \frac{1}{b} + \frac{1}{c} = \frac{1}{a + b + c},$$

then for each odd $n \in \mathbb{N}$,

$$\left(\frac{1}{a} + \frac{1}{b} + \frac{1}{c} \right)^n = \frac{1}{a^n + b^n + c^n} = \frac{1}{(a + b + c)^n}.$$

Elementary symmetric polynomials can also be successfully used to find decompositions of symmetric polynomials (into symmetric factors).

4.15 Example

Decompose the polynomial

$$F(x_1, x_2, x_3) = x_1^3 + x_2^3 + x_3^3 - 3x_1 x_2 x_3$$

into a product of symmetric factors.

SOLUTION.   By 4.12 we have

$$F(x_1, x_2, x_3) = \sigma_1^3 - 3\sigma_1 \sigma_2 + 3\sigma_3 - 3\sigma_3 = \sigma_1^3 - 3\sigma_1 \sigma_2 = \sigma_1(\sigma_1^2 - 3\sigma_2),$$

and thus

$$\begin{aligned} F(x_1, x_2, x_3) &= (x_1 + x_2 + x_3)[(x_1 + x_2 + x_3)^2 - 3(x_1 x_2 + x_1 x_3 + x_2 x_3)] \\ &= (x_1 + x_2 + x_3) \cdot (x_1^2 + x_2^2 + x_3^2 - x_1 x_2 - x_1 x_3 - x_2 x_3). \quad \square \end{aligned}$$

4.16 Exercises

Write the polynomials (i)–(iv) as products of symmetric factors:

(i) $(x_1 + x_2)(x_1 + x_3)(x_2 + x_3) + x_1 x_2 x_3$.

(ii) $2(x_1^3 + x_2^3 + x_3^3) + x_1^2 x_2 + x_1 x_2^2 + x_1^2 x_3 + x_1 x_3^2 + x_2^2 x_3 + x_2 x_3^2 - 3x_1 x_2 x_3$.

(iii) $(x_1^2 + x_2^2 + x_3^2 + x_1 x_2 + x_1 x_3 + x_2 x_3)^2 - (x_1 + x_2 + x_3)^2 (x_1^2 + x_2^2 + x_3^2)$.

*(iv) $(x_1 + x_2 + x_3)^3 - (x_1 + x_2 - x_3)^3 - (x_1 - x_2 + x_3)^3 - (-x_1 + x_2 + x_3)^3$.

5 Systems of Equations

The problem of solving systems of equations belongs to the basic questions of classical algebra; it receives considerable attention both in secondary-school mathematics and in introductory university algebra courses. Unfortunately, attention is mainly concentrated on systems of *linear* equations, while as a rule much less attention is paid to systems of equations of *higher degree*, and to methods of their solutions. We will therefore try to partially fill this gap in what follows. The main emphasis in this section will be on Sections 5.7–5.14 (especially the method of symmetric polynomials). In the initial parts we will introduce several nonstandard approaches to solving systems of linear equations (5.1–5.6).

5.1 Systems of Linear Equations

As we already mentioned, we assume that the reader is acquainted with algorithmic methods for solving systems of linear equations (Gaussian elimination, Cramer's rule). We will therefore restrict ourselves to those problems for which it may be advantageous to solve them differently, especially by introducing new *auxiliary variables*.

We point out that we will solve systems of equations with *real coefficients* and look for their *real solutions*.

5.2 Example

Solve the following system of n equations in n variables:

$$
\begin{aligned}
x_1 + 2x_2 + 3x_3 + \cdots + (n-1)x_{n-1} + nx_n &= 1, \\
x_2 + 2x_3 + 3x_4 + \cdots + (n-1)x_n + nx_1 &= 2, \\
x_3 + 2x_4 + 3x_5 + \cdots + (n-1)x_1 + nx_2 &= 3, \\
&\ \vdots \\
x_n + 2x_1 + 3x_2 + \cdots + (n-1)x_{n-2} + nx_{n-1} &= n.
\end{aligned}
$$

SOLUTION.   It turns out to be very useful to introduce the new variable $s = x_1 + x_2 + \cdots + x_n$, since by adding all equations we now obtain

$$s + 2s + 3s + \cdots + (n-1)s + ns = 1 + 2 + 3 + \cdots + (n-1) + n,$$

and thus $s = 1$.

Next we subtract the second equation from the first, the third from the second, etc., and finally the first equation from the nth, thus obtaining the system

$$
\begin{aligned}
(1-n)x_1 + & \quad x_2 + \cdots + & \quad x_{n-1} + & \quad x_n = -1, \\
x_1 + (1-n)x_2 + \cdots + & & x_{n-1} + & \quad x_n = -1, \\
& & & \vdots \\
x_1 + & \quad x_2 + \cdots + (1-n)x_{n-1} + & & x_n = -1, \\
x_1 + & \quad x_2 + \cdots + & \quad x_{n-1} + (1-n)x_n = n-1,
\end{aligned}
$$

which can be rewritten as

$$
\begin{aligned}
s - nx_1 &= -1, \\
s - nx_2 &= -1, \\
&\vdots \\
s - nx_{n-1} &= -1, \\
s - nx_n &= n-1.
\end{aligned}
$$

From this we get $x_1 = x_2 = \cdots = x_{n-1} = \frac{2}{n}$, $x_n = \frac{2-n}{n}$.

We realize that the last two systems are only consequences of (as opposed to being equivalent to) the original system; we convince ourselves by a *test* that the n-tuple obtained above is indeed a solution of that system: For $k = 1, 2, \ldots, n$ the kth equation of the original system has the form

$$x_k + 2 \cdot x_{k+1} + \cdots + (n-k+1)x_n + (n-k+2)x_1 + \cdots + n \cdot x_{k-1} = k.$$

By substituting the left-hand sides we then obtain

$$
\begin{aligned}
&\tfrac{2}{n} + 2 \cdot \tfrac{2}{n} + \cdots + (n-k) \cdot \tfrac{2}{n} + (n-k+1) \cdot \tfrac{2-n}{n} \\
&+ (n-k+2) \cdot \frac{2}{n} + \cdots + n \cdot \frac{2}{n} \\
&= \tfrac{2}{n} \cdot (1 + 2 + \cdots + (n-k+1) + \cdots + n) + (n-k+1) \cdot \left(-\tfrac{n}{n}\right) \\
&= \tfrac{2}{n} \cdot \tfrac{n(n+1)}{2} - (n-k+1) = n+1 - n+k-1 = k.
\end{aligned}
$$

Therefore, the system has a unique solution.   □

5.3 *Exercises*

Solve the following systems of equations:

(i) $2x_1 + x_2 + x_3 + \cdots + x_n = 1,$
$x_1 + 2x_2 + x_3 + \cdots + x_n = 2,$
$x_1 + x_2 + 2x_3 + \cdots + x_n = 3,$

$$\vdots$$

$x_1 + x_2 + x_3 + \cdots + 2x_n = n.$

(ii) $x_2 + x_3 + \cdots + x_{n-1} + x_n = 0,$
$x_1 \qquad + x_3 + \cdots + x_{n-1} + x_n = 1,$
$x_1 + x_2 \qquad + \cdots + x_{n-1} + x_n = 2,$

$$\vdots$$

$x_1 + x_2 + x_3 + \cdots + x_{n-1} \qquad = n - 1.$

(iii) $x_1 + 2x_2 + 3x_3 + \cdots + (n-1)x_{n-1} + nx_n = n,$
$x_1 \qquad + 3x_3 + \cdots + (n-1)x_{n-1} + nx_n = n - 2,$
$x_1 + 2x_2 \qquad + \cdots + (n-1)x_{n-1} + nx_n = n - 3,$

$$\vdots$$

$x_1 + 2x_2 + 3x_3 + \cdots + (n-1)x_{n-1} \qquad = 0.$

5.4 *Systems of Linear Equations with Parameters*

These systems actually represent a whole (normally infinite) set of systems, one each for a fixed choice of values of the parameters. For solving them, it is often necessary to discuss the form of the solutions (possibly also their number) depending on the values of the parameters. In the following examples we will explain several methods of solution of systems of linear equations with parameters. Here, x_i are variables and a_i are parameters.

(i) $x_1 + x_2 + x_3 \qquad = a_1,$
$x_1 + x_2 \qquad + x_4 = a_2,$
$x_1 \qquad + x_3 + x_4 = a_3,$
$x_2 + x_3 + x_4 = a_4.$

$$(39)$$

Introducing the auxiliary variable $s = x_1 + x_2 + x_3 + x_4$, we can rewrite the system (39) as

$$x_1 = s - a_4,$$
$$x_2 = s - a_3,$$
$$x_3 = s - a_2,$$
$$x_4 = s - a_1.$$

Summing the equations of the last system, we obtain $3s = a_1 + a_2 + a_3 + a_4$, hence $s = (a_1 + a_2 + a_3 + a_4)/3$, and the system (39) has the unique solution

$$x_1 = (a_1 + a_2 + a_3 - 2a_4)/3, \qquad x_2 = (a_1 + a_2 - 2a_3 + a_4)/3,$$
$$x_3 = (a_1 - 2a_2 + a_3 + a_4)/3, \qquad x_4 = (-2a_1 + a_2 + a_3 + a_4)/3,$$

which *must* be verified by substituting.

(ii)
$$
\begin{aligned}
x_1 - \ \ x_2 - \ \ x_3 \ - \cdots - x_{n-1} - \ \ & \ x_n = 2a, \\
-x_1 + 3x_2 - \ \ x_3 \ - \cdots - x_{n-1} - \ \ & \ x_n = 2^2 a, \\
-x_1 - \ \ x_2 + 7x_3 \ - \cdots - x_{n-1} - \ \ & \ x_n = 2^3 a, \\
& \ \ \vdots \\
-x_1 - \ \ x_2 - \ \ x_3 \ - \cdots - x_{n-1} + (2^n - 1)x_n = 2^n a.
\end{aligned}
\tag{40}
$$

We rewrite the system (40) in the equivalent form

$$
\begin{aligned}
2x_1 - (x_1 + x_2 + x_3 + \cdots + x_n) &= 2a, \\
2^2 x_2 - (x_1 + x_2 + x_3 + \cdots + x_n) &= 2^2 a, \\
2^3 x_3 - (x_1 + x_2 + x_3 + \cdots + x_n) &= 2^3 a, \\
&\vdots \\
2^n x_n - (x_1 + x_2 + x_3 + \cdots + x_n) &= 2^n a;
\end{aligned}
$$

then, again setting $s = x_1 + x_2 + x_3 + \cdots + x_n$, we have

$$
\begin{aligned}
x_1 - \tfrac{s}{2} &= a, \\
x_2 - \tfrac{s}{2^2} &= a, \\
x_3 - \tfrac{s}{2^3} &= a, \\
&\vdots \\
x_n - \tfrac{s}{2^n} &= a,
\end{aligned}
$$

and by adding the equations of the last system we get

$$s - s\left(\tfrac{1}{2} + \tfrac{1}{2^2} + \tfrac{1}{2^3} + \cdots + \tfrac{1}{2^n}\right) = n \cdot a,$$

from which by formula (10), $s = n \cdot 2^n \cdot a$. The system (40) has therefore a unique solution: $x_k = a + s/2^k = a(1 + n \cdot 2^{n-k})$ for $k = 1, 2, \ldots, n$, which has to be verified by substituting.

(iii)
$$
\begin{aligned}
2x_1 - \ \ x_2 \qquad\qquad\qquad\qquad &= a_1, \\
-x_1 + 2x_2 - \ \ x_3 \qquad\qquad &= 0, \\
- \ \ x_2 + 2x_3 - x_4 \qquad &= 0, \\
&\ \ \vdots \\
-x_{n-2} + 2x_{n-1} - \ \ x_n &= 0, \\
- \ \ x_{n-1} + 2x_n &= a_2.
\end{aligned}
\tag{41}
$$

We rewrite the equations of the system (41) from the second to the second-to-last one in the form $x_1 - x_2 = x_2 - x_3$, $x_2 - x_3 = x_3 - x_4$, ..., $x_{n-2} - x_{n-1} = x_{n-1} - x_n$. The "trick" of elimination relies on the introduction of the new variable $t = x_1 - x_2$, and in view of the preceding equations we have

$$t = x_1 - x_2 = x_2 - x_3 = x_3 - x_4 = \cdots = x_{n-1} - x_n.$$

The first and the last equations in the system (41) can also be expressed in terms of the variable t:

$$x_1 = a_1 - t, \quad x_n = a_2 + t.$$

Thus we obtain the system

$$
\begin{aligned}
x_1 &= & & a_1 - t, \\
x_2 &= & x_1 - t &= a_1 - 2t, \\
x_3 &= & x_2 - t &= a_1 - 3t, \\
&\vdots & \vdots & \\
x_{n-1} &= & x_{n-2} - t &= a_1 - (n-1)t, \\
x_n &= & x_{n-1} - t &= a_1 - nt, \\
x_n &= & & a_2 + t.
\end{aligned}
\tag{42}
$$

From the last two equations of (42) we then determine $t = (a_1 - a_2)/(n+1)$. Substituting this into the first n equations of (42), we find that $x_k = a_1 - k \cdot (a_1 - a_2)/(n+1)$ for each $k = 1, 2, \ldots, n$. To complete the solution it remains only to substitute into (41).

5.5 Example

Solve the system with parameters $a_1, a_2, \ldots, a_n \in \mathbb{R}$:

$$
\begin{aligned}
x_1 + x_2 &= a_1, \\
x_2 + x_3 &= a_2, \\
x_3 + x_4 &= a_3, \\
&\vdots \\
x_{n-1} + x_n &= a_{n-1}, \\
x_n + x_1 &= a_n.
\end{aligned}
\tag{43}
$$

(i) FIRST SOLUTION. We begin with step-by-step elimination of the variables in the individual equations in (43):

$$x_2 = a_1 - x_1,$$
$$x_3 = a_2 - x_2 = a_2 - a_1 + x_1,$$
$$x_4 = a_3 - x_3 = a_3 - a_2 + a_1 - x_1,$$
$$\vdots$$
$$x_i = a_{i-1} - x_{i-1} = a_{i-1} - a_{i-2} + a_{i-3} - \cdots + (-1)^{i+1}x_1, \qquad (44)$$
$$\vdots$$
$$x_n = a_{n-1} - x_{n-1} = a_{n-1} - a_{n-2} + \cdots + (-1)^{n+1}x_1,$$
$$x_1 = a_n - x_n = a_n - a_{n-1} + \cdots + (-1)^n x_1.$$

From the last two equations in (44) it is apparent that to proceed further we have to distinguish between the cases n even and n odd.

(a) $n = 2k$. From the last equation in (44) it is clear that this system, and therefore also (43), has a solution if and only if $a_1 + a_3 + \cdots + a_{2k-1} = a_2 + a_4 + \cdots + a_{2k}$. In this case, (43) has infinitely many solutions, which can be expressed, by (44), in terms of the free variable x_1.

(b) $n = 2k + 1$. In this case it follows from the last equation in (44) that $2x_1 = a_{2k+1} - a_{2k} + \cdots + a_1$; hence $x_1 = (a_1 - a_2 + \cdots - a_{2k} + a_{2k+1})/2$. If we write the original system as

$$x_2 + x_3 = a_2,$$
$$x_3 + x_4 = a_3,$$
$$\vdots$$
$$x_n + x_1 = a_n,$$
$$x_1 + x_2 = a_1,$$

then by repeating the above approach we obtain

$$x_2 = \frac{a_2 - a_3 + \cdots - a_{2k+1} + a_1}{2},$$

and in complete analogy,

$$x_i = \frac{a_i - a_{i+1} + \cdots - a_{i-2} + a_{i-1}}{2}. \qquad \square$$

(ii) SECOND SOLUTION. We distinguish between the parities of n from the beginning, and introduce the auxiliary variable $s = x_1 + x_2 + \cdots + x_n$.

(a) $n = 2k$. By adding the odd equations in the system (43) we find that

$$s = x_1 + x_2 + \cdots + x_{2k-1} + x_{2k} = a_1 + a_3 + \cdots + a_{2k-1},$$

and adding the even ones,

$$s = x_2 + x_3 + \cdots + x_{2k} + x_1 = a_2 + a_4 + \cdots + a_{2k};$$

therefore, a necessary condition for the solvability of the system is the equation $a_1 + a_3 + \cdots + a_{2k-1} = a_2 + a_4 + \cdots + a_{2k}$. If this condition is satisfied, then (43) has infinitely many solutions, which can be expressed, as in 5.5.(i)(a), in terms of the free variable x_1; in the opposite case (43) has no solutions.

(b) $n = 2k + 1$. If we add all the equations in (43), then upon dividing by 2 we get

$$s = x_1 + x_2 + \cdots + x_{2k+1} = (a_1 + a_2 + \cdots + a_{2k+1})/2.$$

To evaluate the variable x_1, we add all odd equations in (43), and obtain

$$2x_1 + x_2 + \cdots + x_{2k} + x_{2k+1} = x_1 + s = a_1 + a_3 + \cdots + a_{2k+1},$$

hence $x_1 = (a_1 - a_2 + a_3 - \cdots - a_{2k} + a_{2k+1})/2$. This method can also be used to find x_i, $i = 2, 3, \ldots, 2k+1$; we only have to change the order of the equations in (43) and rewrite that system in the form

$$x_i + x_{i+1} = a_i,$$
$$x_{i+1} + x_{i+2} = a_{i+1},$$
$$\vdots \qquad\qquad (45)$$
$$x_{i-1} + x_i = a_{i-1}.$$

Then by adding all odd equations in (45) we get

$$x_i + x_{i+1} + x_{i+2} + \cdots + x_{i-1} + x_i = x_i + s = a_i + a_{i+2} + \cdots + a_{i-1};$$

thus $x_i = (a_i - a_{i+1} + \cdots - a_{i-2} + a_{i-1})/2$. □

5.6 Exercises

Solve the systems (i)–(v), where p and a_i are real parameters:

(i) $x_1 + x_2 + x_3 + x_4 = 2a_1,$
  $x_1 + x_2 - x_3 - x_4 = 2a_2,$
  $x_1 - x_2 + x_3 - x_4 = 2a_3,$
  $x_1 - x_2 - x_3 + x_4 = 2a_4.$

(ii) $(1+p)x_1 + \qquad x_2 + \qquad x_3 = 1,$
  $x_1 + (1+p)x_2 + \qquad x_3 = 1,$
  $x_1 + \qquad x_2 + (1+p)x_3 = 1.$

(iii) $x_1 + x_2 = a_1 a_2,$

   $x_1 + x_3 = a_1 a_3,$

   $x_1 + x_4 = a_1 a_4,$

   $x_2 + x_3 = a_2 a_3,$

   $x_2 + x_4 = a_2 a_4,$

   $x_3 + x_4 = a_3 a_4.$

In (iv) and (v), assume that $a_1 + a_2 + a_3 \neq 0$:

(iv) $(a_2 + a_3)(x_2 + x_3) - a_1 x_1 = a_2 - a_3,$

   $(a_1 + a_3)(x_1 + x_3) - a_2 x_2 = a_3 - a_1,$

   $(a_1 + a_2)(x_1 + x_2) - a_3 x_3 = a_1 - a_2.$

(v) $a_1 x_1 + a_2 x_2 + a_3 x_3 = a_1 + a_2 + a_3,$

   $a_2 x_1 + a_3 x_2 + a_1 x_3 = a_1 + a_2 + a_3,$

   $a_3 x_1 + a_1 x_2 + a_2 x_3 = a_1 + a_2 + a_3.$

(vi) Solve the following system with the parameter $a \in \mathbb{R}$:

$$\frac{3}{2} x_1 + \quad x_2 + \quad x_3 + \cdots + \qquad x_n = \frac{a}{2},$$

$$x_1 + \frac{5}{4} x_2 + \quad x_3 + \cdots + \qquad x_n = \frac{a}{4},$$

$$x_1 + \quad x_2 + \frac{9}{8} x_3 + \cdots + \qquad x_n = \frac{a}{8},$$

$$\vdots$$

$$x_1 + \quad x_2 + \quad x_3 + \cdots + \frac{2^n + 1}{2^n} x_n = \frac{a}{2^n}.$$

(vii) Solve the following system with parameters a_1, a_2, a_3, a_4:

$$2x_1 - \quad x_2 \qquad\qquad = a_1,$$

$$-x_1 + 2x_2 - \quad x_3 \qquad = a_2,$$

$$-x_2 + 2x_3 - \quad x_4 = a_3,$$

$$-x_3 + 2x_4 = a_4.$$

(viii) Show that if

$$x_1 + x_2 + x_3 = 0,$$

$$x_2 + x_3 + x_4 = 0,$$

$$\vdots$$

$$x_{99} + x_{100} + x_1 = 0,$$

$$x_{100} + x_1 + x_2 = 0,$$

then $x_1 = x_2 = x_3 = \cdots = x_{99} = x_{100} = 0.$

5.7 Systems of Equations of Higher Degree

As opposed to systems of linear equations, there is no universal method for solving systems of higher-degree equations. However, there are several methods that can be used in special cases. The simplest one is the *elimination method*, which with the help of one equation of the system eliminates several variables in the remaining equations. The disadvantage, however, is that if we use a nonlinear equation, then often the degree of the remaining equations will increase.

5.8 Examples

We will illustrate the advantages and shortcomings of the elimination method by solving three systems of nonlinear equations:

(i) $x_1 + 2x_2 + x_3 = 8,$

$\quad x_1 + 3x_2 + 2x_3 = 12,$

$\quad x_1^2 + x_2^2 + x_3^2 = 19.$

SOLUTION.   If we subtract the first equation from the second, we obtain $x_2 = 4 - x_3$. Substituting this into the first and third equations, we obtain a system of two equations in two variables:

$$x_1 - x_3 = 0,$$
$$x_1^2 + 2x_3^2 - 8x_3 = 3.$$

If we now substitute the first equation into the second, we get $3x_3^2 - 8x_3 - 3 = 0$, which has the two solutions $x_3 = 3$ and $x_3 = -\frac{1}{3}$. Hence the system also has two solutions: $x_1 = 3$, $x_2 = 1$, $x_3 = 3$ and $x_1 = -\frac{1}{3}$, $x_2 = 13/3$, $x_3 = -\frac{1}{3}$. $\qquad\square$

(ii) $x_1(x_1 + x_2) = 9,$

$\quad x_2(x_1 + x_2) = 16.$

SOLUTION.   Adding the two equations, we obtain $(x_1 + x_2)^2 = 25$, and thus $x_1 + x_2 = \pm 5$. The system therefore has two solutions, which we obtain by substituting $x_1 + x_2$ into the original equations: $x_1 = \frac{9}{5}$, $x_2 = 16/5$ and $x_1 = -\frac{9}{5}$, $x_2 = -16/5$. $\qquad\square$

(iii) $x_1^2 + x_2^2 = \frac{7}{3},$

$\quad x_1^3 + x_2^3 = 3.$

SOLUTION.   We will try to eliminate the variable x_2. By the binomial theorem,

$$x_2^6 = (x_2^2)^3 = \left(\tfrac{7}{3} - x_1^2\right)^3 = \tfrac{343}{27} - \tfrac{49}{3}x_1^2 + 7x_1^4 - x_1^6,$$
$$x_2^6 = (x_2^3)^2 = (3 - x_1^3)^2 = 9 - 6x_1^3 + x_1^6,$$

and therefore
$$2x_1^6 - 7x_1^4 - 6x_1^3 + \tfrac{49}{3}x_1^2 - \tfrac{100}{27} = 0;$$

this, however, is an equation of degree 6 that we cannot solve. The only possibility, namely reducing its degree by using 3.23, does not work, because we have no root of this equation at all. Trying out all possibilities according to 3.26 (after multiplying by 27 so that we get a polynomial with integer coefficients), we find that this polynomial has no rational roots.

One method that leads to the solution of this problem will be described in 5.10, and example 5.8.(iii) will then be solved in 5.11.(ii).   □

5.9   Exercises

Solve the following systems of equations:

(i) $x_1^2 + x_2^2 + 6x_1 + 2x_2 = 0,$
$$x_1 + x_2 = -8.$$

(ii) $\frac{x_1}{x_2} + \frac{x_2}{x_1} = \frac{13}{6},$
$$x_1 + x_2 = 5.$$

(iii) $x_1^2 + 2x_2^2 = 9,$
$$(x_1 + 1)^2 + 2(x_2 + 1)^2 = 22.$$

(iv) $5x_1 + 5x_2 + 2x_1x_2 = -19,$
$$3x_1x_2 + x_1 + x_2 = -35.$$

(v) $x_2^2 - x_1 - 5 = 0,$
$$\frac{1}{x_2-1} - \frac{1}{x_2+1} = \frac{1}{x_1}.$$

(vi) $x_1^2 - x_1x_2 = 28,$
$$x_2^2 - x_1x_2 = -12.$$

*(vii) $x_1^2 - x_1x_2 + x_2^2 = 21,$
$$x_2^2 - 2x_1x_2 + 15 = 0.$$

*(viii) $2x_1^2 + 3x_1x_2 + x_2^2 = 12,$
$$2(x_1 + x_2)^2 - x_2^2 = 14.$$

(ix) $x_1^2 - x_1x_2 - x_2^2 + 3x_1 + 7x_2 + 3 = 0,$
$$2x_2^2 + x_1x_2 - x_1^2 = 0.$$

*(x) $3x_2^2 + x_1x_2 - x_1 + 4x_2 - 7 = 0,$
$$x_2^2 + 2x_1x_2 - 2x_1 - 2x_2 + 1 = 0.$$

*(xi) $x_1 + x_2 = 2,$
$$x_1x_3 + x_2x_4 = -8,$$
$$x_1x_3^2 + x_2x_4^2 = -4,$$
$$x_1x_3^3 + x_2x_4^3 = -128.$$

5.10   The Method of Symmetric Polynomials

The method of symmetric polynomials can be applied to the solution of systems of equations in which the unknowns occur as variables of symmetric polynomials. By the results of Section 4 we know that an arbitrary symmetric polynomial can be expressed as a polynomial in elementary symmetric polynomials. Therefore, the whole system can be rewritten as a system in which elementary symmetric polynomials of the original variables occur as new variables. The advantage of this method lies in the fact that as a rule, the new system is of lower degree than the original system. If we find the values of the elementary symmetric polynomials, then the values of the origial variables can be obtained as solutions of equations constructed with the help of Vieta's relations 3.18, which have degrees equal to the number of variables.

5.11 Examples

(i) Let us solve the system of equations

$$x_1^5 + x_2^5 = 464,$$
$$x_1 + x_2 = 4.$$

SOLUTION. It is true that by substituting the second equation into the first one we could obtain an equation in one variable, but we would be unable to solve it. Therefore, we use the method of symmetric polynomials, in the notation of (37). By the power sum relationship

$$s_5 = x_1^5 + x_2^5 = \sigma_1^5 - 5\sigma_1^3\sigma_2 + 5\sigma_1\sigma_2^2$$

from Section 4.4 we can rewrite our system as

$$\sigma_1^5 - 5\sigma_1^3\sigma_2 + 5\sigma_1\sigma_2^2 = 464,$$
$$\sigma_1 = 4,$$

which, after simplification, gives

$$\sigma_2^2 - 16\sigma_2 + 28 = 0,$$

and therefore $\sigma_2 = 2$ or $\sigma_2 = 14$. Thus we get two systems of equations in the original variables x_1, x_2:

$$x_1 + x_2 = 4, \qquad\qquad x_1 + x_2 = 4,$$
$$\text{resp.}$$
$$x_1 x_2 = 2, \qquad\qquad x_1 x_2 = 14,$$

from which by 3.18 it follows that x_1, x_2 are zeros of the polynomials $x^2 - 4x + 2$, resp. $x^2 - 4x + 14$. The first one of these has the real zeros $2 \pm \sqrt{2}$, while the second one has no real zeros. Our system has therefore the two soutions $x_1 = 2 + \sqrt{2}$, $x_2 = 2 - \sqrt{2}$ and $x_1 = 2 - \sqrt{2}$, $x_2 = 2 + \sqrt{2}$ in real numbers. (If we wanted to solve the system in the field of complex numbers, we would obtain the additional two solutions $x_1 = 2 + i\sqrt{10}$, $x_2 = 2 - i\sqrt{10}$, resp. $x_1 = 2 - i\sqrt{10}$, $x_2 = 2 + i\sqrt{10}$.) $\square$

(ii) Let us now solve 5.8.(iii) using the method of symmetric polynomials.

SOLUTION.   Since by Section 4.4 we have the identities $x_1^2 + x_2^2 = \sigma_1^2 - 2\sigma_2$ and $x_1^3 + x_2^3 = \sigma_1^3 - 3\sigma_1\sigma_2$, we can rewrite the original system as

$$\sigma_1^2 - 2\sigma_2 = \frac{7}{3},$$
$$\sigma_1^3 - 3\sigma_1\sigma_2 = 3.$$

By substituting σ_2 from the first equation into the second one, we obtain

$$\sigma_1^3 - 7\sigma_1 + 6 = 0,$$

which is an equation of degree three and has zeros 1, 2, and -3 (by 3.26). If $\sigma_1 = 1$, we get $\sigma_2 = -\frac{2}{3}$; if $\sigma_1 = 2$, then $\sigma_2 = \frac{5}{6}$, and for $\sigma_1 = -3$, we get $\sigma_2 = 10/3$. The original variables x_1, x_2 are then the zeros of the polynomial $x^2 - x - \frac{2}{3}$, resp. $x^2 - 2x + \frac{5}{6}$, resp. $x^2 + 3x + 10/3$. Thus we obtain the four real solutions $(x_1, x_2) \in \{((3 + \sqrt{33})/6; (3 - \sqrt{33})/6),$ $((3 - \sqrt{33})/6; (3 + \sqrt{33})/6), (1 + \sqrt{6}/6; 1 - \sqrt{6}/6), (1 - \sqrt{6}/6; 1 + \sqrt{6}/6)\}$. If we wanted to solve the system over the complex numbers, we would obtain the additional two solutions $x_1 = (-9 + i\sqrt{39})/6, x_2 = (-9 - i\sqrt{39})/6$ and $x_1 = (-9 - i\sqrt{39})/6, x_2 = (-9 + i\sqrt{39})/6$.   □

5.12   Exercises

Solve the following systems of equations; in (ix), $a \in \mathbb{R}$ is a parameter.

(i) $x_1^3 + x_2^3 = 35,$
$x_1 + x_2 = 5.$

(ii) $x_1 + x_2 = 7,$
$\dfrac{x_1}{x_2} + \dfrac{x_2}{x_1} = \dfrac{25}{12}.$

(iii) $x_1^2 + x_2^2 + x_1 + x_2 = 32,$
$12(x_1 + x_2) = 7x_1 x_2.$

(iv) $x_1^2 - x_1 x_2 + x_2^2 = 19,$
$x_1 - x_1 x_2 + x_2 = 7.$

(v) $\dfrac{x_1^2}{x_2} + \dfrac{x_2^2}{x_1} = 12,$
$\dfrac{1}{x_1} + \dfrac{1}{x_2} = \dfrac{1}{3}.$

*(vi) $\dfrac{x_1^5 + x_2^5}{x_1^3 + x_2^3} = \dfrac{31}{7},$
$x_1^2 + x_1 x_2 + x_2^2 = 3.$

*(iiv) $x_1^3 - x_2^3 = 19(x_1 - x_2),$
$x_1^3 + x_2^3 = 7(x_1 + x_2).$

*(viii) $x_1^2 + x_2 = 5,$
$x_1^6 + x_2^3 = 65.$

(ix) $x_1 + x_2 = a,$
$x_1^4 + x_2^4 = a^4.$

(x) $x_1 + x_2 = 3,$
$x_1^5 + x_2^5 = 33.$

*(xi) $x_1 + x_2 - x_3 = 7,$
$x_1^2 + x_2^2 - x_3^2 = 37,$
$x_1^3 + x_2^3 - x_3^3 = 1.$

5.13 Examples

Solve the following systems of equations:

(i) $x_1 + x_2 + x_3 = 2$,
$x_1^2 + x_2^2 + x_3^2 = 6$,
$x_1^3 + x_2^3 + x_3^3 = 8$.

SOLUTION.   We express the polynomials on the left sides in terms of the elementary symmetric polynomials (38), using Table 4.12, and we obtain the system

$$\sigma_1 = 2,$$
$$\sigma_1^2 - 2\sigma_2 = 6,$$
$$\sigma_1^3 - 3\sigma_1\sigma_2 + 3\sigma_3 = 8,$$

from which it follows that $\sigma_1 = 2$, $\sigma_2 = -1$, $\sigma_3 = -2$. By 3.18, x_1, x_2, x_3 are zeros of the polynomial $x^3 - 2x^2 - x + 2$. By 3.26, the rational zeros of this polynomial can be only $1, 2, -1, -2$. By substituting we find that the zeros are $-1, 1, 2$. The system has therefore six solutions, one of which is $x_1 = -1$, $x_2 = 1$, $x_3 = 2$, and the others are obtained by changing the order of the zeros.   □

(ii) $x_1 + x_2 + x_3 = 6$,
$x_1 x_2 + x_1 x_3 + x_2 x_3 = 11$,
$(x_1 - x_2)(x_2 - x_3)(x_3 - x_1) = 2$.

SOLUTION.   While the first and second equations have symmetric polynomials on the left-hand sides, the polynomial $L = (x_1 - x_2)(x_2 - x_3)(x_3 - x_1)$ is not symmetric, since upon interchanging x_1 and x_2 the sign will change. However, if we square it, it becomes symmetric, and we have

$$
\begin{aligned}
L^2 &= [(x_1 - x_2)(x_2 - x_3)(x_3 - x_1)]^2 \\
&= (x_1^2 - 2x_1 x_2 + x_2^2)(x_2^2 - 2x_2 x_3 + x_3^2)(x_3^2 - 2x_1 x_3 + x_1^2) \\
&= [s_2 - (2x_1 x_2 + x_3^2)][s_2 - (2x_2 x_3 + x_1^2)][s_2 - (2x_1 x_3 + x_2^2)],
\end{aligned}
$$

where $s_2 = x_1^2 + x_2^2 + x_3^2$, consistent with the notation in 4.12. We further set $a = 2x_1 x_2 + x_3^2$, $b = 2x_2 x_3 + x_1^2$, $c = 2x_1 x_3 + x_2^2$. Then

$$
\begin{aligned}
L^2 &= (s_2 - a)(s_2 - b)(s_2 - c) \\
&= s_2^3 - (a + b + c)s_2^2 + (ab + bc + ca)s_2 - abc.
\end{aligned}
$$

We express $a + b + c$, $ab + bc + ca$, abc according to 4.12:

$$a + b + c = 2x_1 x_2 + x_3^2 + 2x_2 x_3 + x_1^2 + 2x_1 x_3 + x_2^2 = s_2 + 2\sigma_2.$$

Since

$$ab = (2x_1x_2 + x_3^2)(2x_2x_3 + x_1^2) = x_1^2x_3^2 + 4x_1x_2^2x_3 + 2(x_1^3x_2 + x_2x_3^3),$$

we obtain by symmetry

$$ab + bc + ca = (x_1^2x_2^2 + x_1^2x_3^2 + x_2^2x_3^2) + 4x_1x_2x_3(x_1 + x_2 + x_3)$$
$$+ 2(x_1^3x_2 + x_1x_2^3 + x_1^3x_3 + x_1x_3^3 + x_2^3x_3 + x_2x_3^3)$$
$$= 2\sigma_1^2\sigma_2 - 3\sigma_2^2,$$

and finally

$$abc = (ab)c = [x_1^2x_3^2 + 4x_1x_2^2x_3 + 2(x_1^3x_2 + x_2x_3^3)](2x_1x_3 + x_2^2)$$
$$= 9x_1^2x_2^2x_3^2 + 2(x_1^3x_2^3 + x_1^3x_3^3 + x_2^3x_3^3) + 4x_1x_2x_3(x_1^3 + x_2^3 + x_3^3)$$
$$= 27\sigma_3^2 + 2\sigma_2^3 - 18\sigma_1\sigma_2\sigma_3 + 4\sigma_1^3\sigma_3.$$

Therefore, we have

$$L^2 = s_2^3 - (s_2 + 2\sigma_2)s_2^2 + (2\sigma_1^2\sigma_2 - 3\sigma_2^2)s_2$$
$$- (27\sigma_3^2 + 2\sigma_2^3 - 18\sigma_1\sigma_2\sigma_3 + 4\sigma_1^3\sigma_3)$$
$$= -2\sigma_2(\sigma_1^2 - 2\sigma_2)^2 + (2\sigma_1^2\sigma_2 - 3\sigma_2^2)(\sigma_1^2 - 2\sigma_2)$$
$$- 27\sigma_3^2 - 2\sigma_2^3 + 18\sigma_1\sigma_2\sigma_3 - 4\sigma_1^3\sigma_3$$
$$= \sigma_1^2\sigma_2^2 - 4\sigma_2^3 - 4\sigma_1^3\sigma_3 + 18\sigma_1\sigma_2\sigma_3 - 27\sigma_3^2.$$

If we substitute $\sigma_1 = 6$, $\sigma_2 = 11$ from the first two equations into the equation $L^2 = 4$, we obtain

$$\sigma_3^2 - 12\sigma_3 + 36 = 0,$$

and thus $\sigma_3 = 6$. The solutions of the original system are therefore the zeros of the polynomial

$$x^3 - 6x^2 + 11x - 6 = (x - 1)(x - 2)(x - 3).$$

Since in the course of this solution we carried out some irreversible steps (we took the square of an equation), not all six permutations of the numbers 1, 2, 3 may be solutions of our system. Indeed, by checking we find that the system has exactly three solutions: $x_1 = 1$, $x_2 = 2$, $x_3 = 3$ and $x_1 = 2$, $x_2 = 3$, $x_3 = 1$ and $x_1 = 3$, $x_2 = 1$, $x_3 = 2$. $\square$

5.14 Exercises

Solve the following systems of equations, where in (vi) a is a real parameter.

(i)
$$x_1 + x_2 + x_3 = 9,$$
$$\frac{1}{x_1} + \frac{1}{x_2} + \frac{1}{x_3} = 1,$$
$$x_1x_2 + x_1x_3 + x_2x_3 = 27.$$

(ii)
$$x_1 + x_2 + x_3 = \frac{13}{3},$$
$$\frac{1}{x_1} + \frac{1}{x_2} + \frac{1}{x_3} = \frac{13}{3},$$
$$x_1x_2x_3 = 1.$$

(iii) $(x_1+x_2)(x_2+x_3)+(x_2+x_3)(x_3+x_1)+(x_3+x_1)(x_1+x_2) = 1,$
$$x_1^2(x_2 + x_3) + x_2^2(x_3 + x_1) + x_3^2(x_1 + x_2) = -6,$$
$$x_1 + x_2 + x_3 = 2.$$

(iv) $x_1x_2(x_1 + x_2) + x_2x_3(x_2 + x_3) + x_1x_3(x_1 + x_3) = 12,$
$$x_1x_2 + x_1x_3 + x_2x_3 = 0,$$
$$x_1x_2(x_1^2 + x_2^2) + x_2x_3(x_2^2 + x_3^2) + x_1x_3(x_1^2 + x_3^2) = 12.$$

(v) $x_1 + x_2 + x_3 = 0,$ **(vi)** $x_1 + x_2 + x_3 = a,$
$\quad\ x_1^2 + x_2^2 + x_3^2 = x_1^3 + x_2^3 + x_3^3,$ $x_1^2 + x_2^2 + x_3^2 = a^2,$
$\quad\quad\ x_1x_2x_3 = 2.$ $x_1^3 + x_2^3 + x_3^3 = a^3.$

(vii) $x_1^2 + x_2^2 + x_3^2 = 9,$
$$\tfrac{1}{x_1} + \tfrac{1}{x_2} + \tfrac{1}{x_3} = x_1 + x_2 + x_3,$$
$$x_1x_2 + x_1x_3 + x_2x_3 = -4.$$

*(viii) $x_1^3 + 8x_2^3 - x_3^3 = 3,$
$$2x_1x_2 - x_1x_3 - 2x_2x_3 = x_1 + 2x_2 - x_3,$$
$$x_1x_2x_3 = -\tfrac{1}{2}.$$

*(ix) $x_1^5 + x_2^5 + x_3^5 - x_4^5 = 210,$
$$x_1^3 + x_2^3 + x_3^3 - x_4^3 = 18,$$
$$x_1^2 + x_2^2 + x_3^2 - x_4^2 = 6,$$
$$x_1 + x_2 + x_3 - x_4 = 0.$$

*(x) $3x_1x_2x_3 - x_1^3 - x_2^3 - x_3^3 = -9,$
$$x_1 + x_2 + x_3 = 3,$$
$$x_1^2 + x_2^2 - x_3^2 = 5.$$

6   Irrational Equations

In this section we introduce several methods that can be used to solve *irrational* equations, namely equations that contain a variable underneath a *radical sign*. We will solve these equations in $\mathbb{R}$; therefore, we first recall the definition of a root of a real number:

If n is an even natural number, we define the nth root of a nonnegative number a as the (unique) nonnegative number b for which $b^n = a$ (written $b = \sqrt[n]{a}$); for $a < 0$ the symbol $\sqrt[n]{a}$ is not defined.

If n is an odd natural number, we define the nth root of a real number a as the (unique) real number b for which $b^n = a$ (again written $b = \sqrt[n]{a}$). We point out that the numbers a and $\sqrt[n]{a}$ have the same sign.

6.1   The Implication Method

Irrational equations will be solved, as a rule, by using what we will call the *implication method*. We consider as part of this not only the usual reversible manipulations of equations (interchanging the sides of an equation, adding the same number or expression to both sides of an equation, multiplying both sides by the same nonzero number or expression), but also several *irreversible* manipulations, especially those that are derived from the following theorem.

Theorem. *Let $L(x)$, $R(x)$ be expressions in the variable x defined on the set $D \subseteq \mathbb{R}$. Then for each $x \in D$ we have*

(i) $L(x) = R(x) \Longrightarrow [L(x)]^n = [R(x)]^n$ *for each $n \in \mathbb{N}$*,
(ii) $L(x) = R(x) \Longrightarrow L(x) \cdot G(x) = R(x) \cdot G(x)$,

if the expression $G(x)$ is defined for each $x \in D$.

Remark. The implications (i) and (ii) can in general not be reversed.
   Let now

$$L_0(x) = R_0(x) \tag{46}$$

be some equation in the variable $x \in \mathbb{R}$, which we will solve by way of the implication method; that is, we construct the chain of implications

$$L_0(x) = R_0(x) \Longrightarrow L_1(x) = R_1(x) \Longrightarrow \cdots \Longrightarrow L_n(x) = R_n(x) \tag{47}$$

such that we can easily determine the roots of the equation $L_n(x) = R_n(x)$; in fact, $L_n(x)$ and $R_n(x)$ are most often polynomials. If for $i = 0, 1, \ldots, n$ we denote by $\mathcal{K}_i$ the set of roots of the equation $L_i(x) = R_i(x)$, then by (47) we obviously have $\mathcal{K}_0 \subseteq \mathcal{K}_1 \subseteq \cdots \subseteq \mathcal{K}_n$, and therefore $\mathcal{K}_0 \subseteq \mathcal{K}_n$. The set $\mathcal{K}_n$ certainly contains all the roots of the equation (46) (if they exist at all) and possibly also several other numbers that we drop when we carry out a check by substituting all numbers of the set $\mathcal{K}_n$ into (46).
   If we solve an irrational equation (46) by the implication method, then this check *must* be part of the solution. Therefore, this method is also called the method of *analyzing and checking* (by analyzing we mean the chain of implications (47)).
   Using this method for solving irrational equations therefore frees us from the necessity to keep track of whether for individual manipulations of an equation its domain of definition increases, or whether upon taking even powers the expressions on both sides of the equation have the same sign. The members of the set $\mathcal{K}_n \setminus \mathcal{K}_0$ (which are sometimes called "false roots") will be determined by checking.

6.2 Examples

We now solve several simple irrational equations.

(i) $\sqrt{1+3x} = 1 - x$.
If we square both sides, we obtain the quadratic equation $1 + 3x = 1 - 2x + x^2$, that is, $x^2 - 5x = 0$, which has the roots $x_1 = 0$, $x_2 = 5$. Substituting these into the original equation, we obtain $L(0) = 1 = R(0)$, $L(5) = 4 \neq -4 = R(5)$. Therefore, the equation $\sqrt{1+3x} = 1 - x$ has exactly the one solution $x = 0$.

(ii) $\sqrt{1+3x} = x + 1$.
Again by squaring we obtain the quadratic equation $1 + 3x = x^2 + 2x + 1$, that is, $x^2 - x = 0$, with roots $x_1 = 0$, $x_2 = 1$. Since $L(0) = 1 = R(0)$, $L(1) = \sqrt{4} = 2 = R(1)$ (having substituted into the original equation), the given equation has the two roots 0 and 1.

(iii) $\sqrt{2+x^2} = x - 1$.
By squaring we obtain $2 + x^2 = x^2 - 2x + 1$, thus $2x = -1$ and $x = -\frac{1}{2}$. By substituting we see that the original equation has no root at all, since $L(-\frac{1}{2}) = \sqrt{\frac{9}{4}} = \frac{3}{2} \neq -\frac{3}{2} = R(-\frac{1}{2})$.

6.3 Exercises

Solve the following equations:

(i) $\sqrt{x^2 + 7} = 2x + 2$.

(ii) $\sqrt{31 + x - x^2} = 5 - x$.

(iii) $2x - 6 = \sqrt{6x - x^2 - 5}$.

(iv) $\dfrac{\sqrt{5 - x^2}}{x + 1} = 1$.

6.4 Remark

Even though by using the implication method one need not worry about the equivalence of the manipulations, one can sometimes significantly accelerate the solution of a problem by determining the domains of definition of the equation or checking the signs of its sides. Thus, for example, in problems (i)–(iii) below one can easily decide that the equation has no solution in $\mathbb{R}$, and in problem (iv) it is easy to "guess" the solution.

(i) The domain of definition of the equation $\sqrt{4 - x} = \sqrt{x - 6}$ is the empty set, since the inequalities $x \leq 4$ and $x \geq 6$ cannot be satisfied at the same time.

(ii) The equation $\sqrt{x + 2} = -2$ has no real root, for while its left-hand side is nonnegative, on the right we have the negative number -2.

(iii) The left-hand side of the equation $\sqrt{2x+3} + \sqrt{x+3} = 0$ is a sum of a nonnegative and a positive number (indeed, for $x \geq -\frac{3}{2}$ we have $\sqrt{2x+3} \geq 0$ and $\sqrt{x+3} > 0$, while for $x < -\frac{3}{2}$ the equation makes no sense), and is thus a positive number. Therefore, the equation has no solution.

(iv) Let us now solve the equation

$$\sqrt{11x+3} - \sqrt{2-x} - \sqrt{9x+7} + \sqrt{x-2} = 0.$$

We easily convince ourselves that the domain of definition of this equation is $D = \{2\}$. If we substitute $x = 2$, we see that $L(2) = \sqrt{25} - 0 - \sqrt{25} + 0 = 0 = R(2)$; the given equation therefore has exactly the one solution $x = 2$.

6.5 Exercises

Show that the following equations have no roots in $\mathbb{R}$:

(i) $\sqrt{x-6} + \sqrt{3-x} = 4x - 3x^2 + 1$.

(ii) $\sqrt{4x+7} + \sqrt{3-4x+x^2} + 2 = 0$.

(iii) $\sqrt{x+9} + \sqrt{x} = 2$.

(iv) $2 \cdot \sqrt{x^3 - 4x^2 + 1} + \sqrt{x^2 \cdot (x+1)^3} = -5$.

6.6 Examples

We solve now two equations in a standard way.

(i)
$$\sqrt{2-x} + \frac{4}{\sqrt{2-x}+3} = 2.$$

SOLUTION. If we multiply the whole equation by the expression $\sqrt{2-x} + 3$, we obtain

$$2 - x + 3 \cdot \sqrt{2-x} + 4 = 2 \cdot \sqrt{2-x} + 6,$$

which implies $\sqrt{2-x} = x$, and thus $x^2 + x - 2 = 0$; this quadratic equation has roots $x_1 = 1$, $x_2 = -2$. By trial we find that the original equation has the unique root $x = 1$.   □

(ii)
$$\sqrt{x+5} - \sqrt{x} = 1. \tag{48}$$

SOLUTION. By squaring equation (48) we obtain

$$x + 5 - 2 \cdot \sqrt{x(x+5)} + x = 1,$$

and thus $x + 2 = \sqrt{x(x+5)}$. By squaring once more we get $x^2 + 4x + 4 = x^2 + 5x$, so $x = 4$. Since in (48) we have $L(4) = 1 = R(4)$, the equation has the root $x = 4$.   □

6.7 Remark

It is often advantageous to begin solving an irrational equation by transforming it appropriately. This can be done, for example, by *separating* one of the roots from the remaining terms and putting it on one side of the equation before taking powers. The necessary calculations are often simpler this way. Let us convince ourselves of this by solving the following three equations:

(i) $\sqrt{x+5} - \sqrt{x} = 1$.
We have already solved this equation in 6.6.(ii). Here we transform it into $\sqrt{x+5} = 1 + \sqrt{x}$, then $x + 5 = 1 + 2\sqrt{x} + x$, and thus $\sqrt{x} = 2$ and $x = 4$. This approach for solving (48) requires less work than the one used in 6.6.(ii).

(ii) $1 + \sqrt{1 - \sqrt{x^4 - x^2}} = x$.
We put the equation in the form $\sqrt{1 - \sqrt{x^4 - x^2}} = x - 1$, thus $1 - \sqrt{x^4 - x^2} = x^2 - 2x + 1$, and by separating the root from the remaining terms we obtain the equivalent equation $2x - x^2 = \sqrt{x^4 - x^2}$, which upon squaring gives $x^2(2 - x)^2 = x^2(x^2 - 1)$. Therefore, $x^2(5 - 4x) = 0$, which means $x_1 = 0$, $x_2 = \frac{5}{4}$. By checking we convince ourselves that while $L(0) = 2 \neq 0 = R(0)$, we have $L(\frac{5}{4}) = \frac{5}{4} = R(\frac{5}{4})$, and the original equation has a unique root.

(iii) $\sqrt{x+1} + \sqrt{4x + 13} = \sqrt{3x + 12}$.
Here we rewrite the equation as $\sqrt{3x + 12} - \sqrt{x+1} = \sqrt{4x + 13}$, so that upon squaring we obtain the easiest possible consequence

$$3x + 12 - 2\sqrt{(3x + 12)(x + 1)} + x + 1 = 4x + 13.$$

Thus $\sqrt{(3x + 12)(x + 1)} = 0$, which implies $x_1 = -1$, $x_2 = -4$. While for $x_2 = -4$ the expression $\sqrt{x_2 + 1}$ is not defined, $x = -1$ is a solution of the original equation.

6.8 Exercises

Solve the following equations:

(i) $\sqrt{x - 1} + \sqrt{2x + 6} = 6$.
(ii) $\sqrt{x^2 + 1} + \sqrt{x^2 - 8} = 3$.
(iii) $\sqrt{x + 2} - \sqrt{2x - 3} = \sqrt{x - 1}$.
(iv) $x^2 + \sqrt{x^2 + 20} = 22$.
(v) $\sqrt{5x + 1} - \sqrt{6x - 2} - \sqrt{x + 6} + \sqrt{2x + 3} = 0$.
(vi) $\sqrt{2x - 4} - \sqrt{3x - 11} = \sqrt{x - 3}$.
(vii) $\sqrt{6x + 1} + \sqrt{4x + 2} - \sqrt{8x} - \sqrt{2x + 3} = 0$.

6.9 Multiplication by a Conjugate Expression

For solving certain irrational equations of the form $\sqrt{f(x)} + \sqrt{g(x)} = \varphi(x)$, it may be useful to multiply by an expression of the form $\sqrt{f(x)} - \sqrt{g(x)}$, which is called *conjugate* to the expression $\sqrt{f(x)} + \sqrt{g(x)}$. The same is true for equations of the form $\sqrt{f(x)} - \sqrt{g(x)} = \varphi(x)$; in both cases use of the formula

$$\left(\sqrt{f(x)} + \sqrt{g(x)}\right)\left(\sqrt{f(x)} - \sqrt{g(x)}\right) = f(x) - g(x)$$

may simplify the equation, especially in cases where $f(x)$ and $g(x)$ are "similar." After this it is useful to consider the following pair of equations together:

$$\sqrt{f(x)} \pm \sqrt{g(x)} = \varphi(x),$$
$$f(x) - g(x) = \varphi(x) \cdot \left(\sqrt{f(x)} \mp \sqrt{g(x)}\right).$$

We illustrate the use of this method by solving the equation

$$\sqrt{2x^2 + 3x + 5} + \sqrt{2x^2 - 3x + 5} = 3x. \tag{49}$$

If we multiply (49) by the difference of the square roots, we obtain after simplification,

$$2x = x \cdot (\sqrt{2x^2 + 3x + 5} - \sqrt{2x^2 - 3x + 5}),$$

and since $x = 0$ is not a solution of (49), we proceed to solve the system

$$\sqrt{2x^2 + 3x + 5} + \sqrt{2x^2 - 3x + 5} = 3x,$$
$$\sqrt{2x^2 + 3x + 5} - \sqrt{2x^2 - 3x + 5} = 2.$$

By adding the two equations we get $2 \cdot \sqrt{2x^2 + 3x + 5} = 3x + 2$, and after squaring and simplification, $x^2 - 16 = 0$; thus $x_1 = -4$, $x_2 = 4$.

Since the left-hand side of (49) is nonnegative, the right-hand side must also be nonnegative, and the number $x_1 = -4$ cannot be a root of (49). By substituting into (49) we verify that $x = 4$ is the (unique) root of (49).

6.10 Exercises

Use the method of 6.9 to solve the following equations:

(i) $\sqrt{3x^2 + 5x + 8} - \sqrt{3x^2 + 5x + 1} = 1$.

(ii) $\sqrt{x^3 + x^2 - 1} + \sqrt{x^3 + x^2 + 2} = 3$.

(iii) $\sqrt{2x^2 + 5x - 2} - \sqrt{2x^2 + 5x - 9} = 1$.

6.11   Method of Substitution

A very efficient method for solving some irrational equations is the use of *substitution*. We illustrate this with several examples.

(i) In the equation

$$\sqrt{\frac{3-x}{2+x}} + 3 \cdot \sqrt{\frac{2+x}{3-x}} = 4$$

we set $\sqrt{(3-x)/(2+x)} = t \geq 0$; then the equation becomes $t + \frac{3}{t} = 4$, which gives a quadratic equation with roots $t_1 = 1$, $t_2 = 3$. If $t_1 = \sqrt{(3-x_1)/(2+x_1)} = 1$, then $x_1 = \frac{1}{2}$, and if $t_2 = \sqrt{(3-x_2)/(2+x_2)} = 3$, then $x_2 = -\frac{3}{2}$. By checking we verify that both numbers are indeed roots of the original equation.

(ii) $x^2 + \sqrt{x^2 + 2x + 8} = 12 - 2x$. We begin by writing this equation in the form $x^2 + 2x + 8 + \sqrt{x^2 + 2x + 8} = 20$. If we then set $\sqrt{x^2 + 2x + 8} = z \geq 0$, we obtain the quadratic equation $z^2 + z - 20 = 0$ with roots $z_1 = 4$, $z_2 = -5$. In view of the condition $z \geq 0$, only the root z_1 makes sense, and the equation $\sqrt{x^2 + 2x + 8} = 4$ gives the numbers $x_1 = 2$ and $x_2 = -4$, which are also roots of the given equation.

(iii) To solve the equation

$$\sqrt{x - \sqrt{x-2}} + \sqrt{x + \sqrt{x-2}} = 3,$$

in addition to the substitution $\sqrt{x-2} = t \geq 0$ we use also the method of Section 6.9. We have $\sqrt{t^2 - t + 2} + \sqrt{t^2 + t + 2} = 3$, and by multiplying by the conjugate of the left-hand side of this equation we obtain the following system:

$$\sqrt{t^2 - t + 2} + \sqrt{t^2 + t + 2} = 3,$$
$$3 \cdot \left(\sqrt{t^2 - t + 2} - \sqrt{t^2 + t + 2} \right) = -2t.$$

This implies $6 \cdot \sqrt{t^2 - t + 2} = 9 - 2t$, and thus $32t^2 - 9 = 0$, $t^2 = 9/32$. Since $x = t^2 + 2$, the root of the original equation is $x = 73/32$, of which we can convince ourselves by checking that $L(73/32)^2 = 9$.

(iv) If the equation

$$x + \sqrt{x} + \sqrt{x+2} + \sqrt{x^2 + 2x} = 3$$

is given, it is convenient to use the substitution $y = \sqrt{x} + \sqrt{x+2}$. We then have $y^2 = x + 2 \cdot \sqrt{x(x+2)} + x + 2 = 2(x + 1 + \sqrt{x^2 + 2x})$, and the original equation becomes $y^2/2 - 1 + y = 3$, which has the number 2 as its unique nonnegative root. As solution of the equation $\sqrt{x} + \sqrt{x+2} = 2$ we find $x = \frac{1}{4}$, which can be easily checked to be the root of the original equation.

6.12   Exercises

Use an appropriate substitution to solve the following equations.

(i) $\sqrt{\frac{2x+1}{x-1}} - 2 \cdot \sqrt{\frac{x-1}{2x+1}} = 1.$

(ii) $\sqrt{\frac{x+1}{x-1}} - \sqrt{\frac{x-1}{x+1}} = \frac{3}{2}.$

(iii) $x^2 - 4x + 6 = \sqrt{2x^2 - 8x + 12}.$

(iv) $2x^2 + 6 - 2 \cdot \sqrt{2x^2 - 3x + 2} = 3(x+4).$

(v) $\sqrt{x^3 + 8} + \sqrt[4]{x^3 + 8} = 6.$

(vi) $3x^2 + 15x + 2 \cdot \sqrt{x^2 + 5x + 1} = 2.$

(vii) $\dfrac{x \cdot \sqrt[3]{x} - 1}{\sqrt[3]{x^2} - 1} - \dfrac{\sqrt[3]{x^2} - 1}{\sqrt[3]{x} + 1} = 4.$

(viii) $\sqrt{x + 3 + 4 \cdot \sqrt{x-1}} + \sqrt{x + 8 - 6 \cdot \sqrt{x-1}} = 5.$

(ix) $\sqrt{x + 2\sqrt{x-1}} - \sqrt{x - 2\sqrt{x-1}} = 2.$

(x) $\sqrt{x^2 + x + 4} + \sqrt{x^2 + x + 1} = \sqrt{2x^2 + 2x + 9}.$

(xi) $\sqrt{x^2 + x + 7} + \sqrt{x^2 + x + 2} = \sqrt{3x^2 + 3x + 19}.$

(xii) $\sqrt{x + \sqrt{x + 11}} + \sqrt{x - \sqrt{x + 11}} = 4.$

(xiii) $\sqrt{x - 1} + \sqrt{x + 3} + 2 \cdot \sqrt{(x-1)(x+3)} = 4 - 2x.$

6.13   Examples

We solve now two equations that are more complicated; again it will be convenient to use substitutions.

(i) $$x^3 + 1 = 2 \cdot \sqrt[3]{2x - 1}. \tag{50}$$

SOLUTION.   If we set $y = \sqrt[3]{2x - 1}$, then $y^3 = 2x-1$. We obtain the system of equations

$$x^3 + 1 = 2y,$$
$$y^3 + 1 = 2x.$$

By subtracting the second equation from the first, we get $x^3 - y^3 = 2(y-x)$, thus $(x-y)(x^2 + xy + y^2 + 2) = 0$, and so either $x = y$ or $x^2 + xy + y^2 = -2$. If $x = y$, then $x^3 + 1 = 2x$, which has roots $x_1 = 1$, $x_2 = -(1+\sqrt{5})/2$, $x_3 = -(1 - \sqrt{5})/2$; all three numbers also satisfy equation (50), which is easy to verify. The equation $x^2 + xy + y^2 = -2$ has no real roots, since for each pair x, y of real numbers we have $x^2 + xy + y^2 = (x+y/2)^2 + 3y^2/4 \geq 0$. Equation (50) therefore has the three roots 1, $-(1 + \sqrt{5})/2$, $-(1 - \sqrt{5})/2$.   □

(ii) $$4(\sqrt{1+x} - 1)(\sqrt{1-x} + 1) = x. \tag{51}$$

SOLUTION.   First we note that $x \in [-1, 1]$, and we set $u = \sqrt{1+x}$, $v = \sqrt{1-x}$. Therefore, $u^2 = 1+x$, $v^2 = 1-x$, and thus $u^2 + v^2 = 2$ and $u^2 - v^2 = 2x$, where $0 \le u, v \le \sqrt{2}$. By substituting into (51) we obtain after simplification $8 \cdot [uv + (u-v) - 1] = u^2 - v^2$. In this equation we substitute the expression uv: Since $(u-v)^2 = u^2 - 2uv + v^2$, and using $u^2 + v^2 = 2$, we obtain $2uv = 2 - (u-v)^2$. Hence

$$8 - 4(u-v)^2 + 8(u-v) - 8 - (u-v)(u+v) = 0,$$
$$(u-v)(8 - 5u + 3v) = 0.$$

The condition $u - v = 0$ implies $x = 0$, which is also a solution of (51). The equation $5u - 3v = 8$ has no solution, since from the conditions $0 \le u, v \le \sqrt{2}$ it follows that $5u - 3v \le 5u \le 5\sqrt{2} < 8$.   □

6.14   Exercises

Solve the following equations:

(i) $\sqrt{\dfrac{2x}{x+1}} - \sqrt{\dfrac{2(x+1)}{x}} = 1$.

(ii) $\sqrt{y - 2 + \sqrt{2y - 5}} + \sqrt{y + 2 + 3\sqrt{2y - 5}} = 7\sqrt{2}$.

(iii) $\dfrac{\sqrt{x^2 + 8x}}{\sqrt{x+1}} + \sqrt{x+7} = \dfrac{7}{\sqrt{x+1}}$.

*(iv) $\dfrac{1}{x} + \dfrac{1}{\sqrt{1-x^2}} = \dfrac{35}{12}$.

6.15   Equations of the Form $\sqrt[n]{a + \varphi(x)} + \sqrt{b - \varphi(x)} = c$

The method of "paired" substitution, which we have used to solve example 6.13.(ii), can be used in a standard way to solve equations of the type $\sqrt[n]{a + \varphi(x)} + \sqrt[n]{b - \varphi(x)} = c$, where $n \in \mathbb{N}$, $a, b, c \in \mathbb{R}$, $\varphi(x)$ is an expression in the variable $x \in \mathbb{R}$, or equations of a similar type.

If we set $u = \sqrt[n]{a + \varphi(x)}$, $v = \sqrt[n]{b - \varphi(x)}$, then simultaneously

$$u + v = c,$$
$$u^n + v^n = a + b.$$

Systems like this can be solved, for instance, by using the method of symmetric polynomials from Section 4. Let us again consider two examples.

(i) The equation

$$\sqrt[5]{32 - x} + \sqrt[5]{1 + x} = 3,$$

with the substitutions $u = \sqrt[5]{32 - x}$, $v = \sqrt[5]{1 + x}$, leads to the system

$$u + v = 3,$$
$$u^5 + v^5 = 33,$$

which we have solved in Exercise 5.12.(x); hence, $u_1 = 1$, $v_1 = 2$, or $u_2 = 2$, $v_2 = 1$.

From the equation $\sqrt[5]{32 - x} = 1$ we obtain the first root $x_1 = 31$ of the original equation, and from $\sqrt[5]{1 + x} = 1$ we get the second root $x_2 = 0$, of which we can easily convince ourselves.

(ii) To solve the equations

$$\sqrt[3]{8 + x} + \sqrt[3]{8 - x} = 1,$$

we use the substitutions $u = \sqrt[3]{8 + x}$, $v = \sqrt[3]{8 - x}$ and solve the system

$$u + v = 1,$$
$$u^3 + v^3 = 16.$$

If we set $\sigma_1 = \sigma_1(u, v) = u + v$, $\sigma_2 = \sigma_2(u, v) = u \cdot v$, then

$$1 = \sigma_1,$$
$$16 = \sigma_1^3 - 3\sigma_1\sigma_2,$$

and thus $\sigma_1 = 1$ and $\sigma_2 = -5$. Therefore, u, v are roots of the polynomial $F(t) = t^2 - \sigma_1 t + \sigma_2 = t^2 - t - 5$, that is, $\{u, v\} = \{(1 + \sqrt{21})/2, (1 - \sqrt{21})/2\}$. Since $x = u^3 - 8$, we obtain $x_1 = 3 \cdot \sqrt{21}$, $x_2 = -3 \cdot \sqrt{21}$.

We remark that the solution of these problems could be accelerated: Indeed, if $\sigma_2(u, v) = u \cdot v = -5$, then $-5 = \sqrt[3]{8 + x} \cdot \sqrt[3]{8 - x}$. Therefore, $\sqrt[3]{64 - x^2} = -5$, which gives again $x_{1,2} = \pm 3\sqrt{21}$.

By substituting we verify that both numbers are roots of the original equation; to do this, we can use the result of Example 3.24.

6.16   Exercises

Use the method of 6.15 to solve the following equations:

(i) $\sqrt[4]{1 - x} + \sqrt[4]{15 + x} = 2.$

(ii) $\sqrt[4]{2x - 2} + \sqrt[4]{6 - 2x} = \sqrt{2}.$

(iii) $\sqrt[5]{1 + 2x} + \sqrt[5]{1 - 2x} = \sqrt[5]{2}.$

(iv) $\sqrt[3]{1 + \sqrt{x}} = 2 - \sqrt[3]{1 - \sqrt{x}}.$

(v) $2 \cdot \sqrt[4]{(6 - x)(x - 2)} + \sqrt{6 - x} + \sqrt{x - 2} = 2.$

6.17   Equations of the Type $\sqrt[3]{f(x)} + \sqrt[3]{g(x)} = h(x)$

By using the formula

$$(u + v)^3 = u^3 + v^3 + 3uv(u + v),$$

we can modify the method introduced in Section 6.15 for $n = 3$ to include the solution of more complicated equations. Indeed, if in the equation $\sqrt[3]{f(x)} + \sqrt[3]{g(x)} = h(x)$ we set $u = \sqrt[3]{f(x)}$, $v = \sqrt[3]{g(x)}$, then according to the formula above we have

$$\sqrt[3]{f(x)} + \sqrt[3]{g(x)} = h(x) \implies f(x) + g(x) + 3h(x) \cdot \sqrt[3]{f(x) \cdot g(x)} = h^3(x).$$
$$(52)$$

We illustrate this approach with the following examples.

(i) The equation

$$\sqrt[3]{8 + x} + \sqrt[3]{8 - x} = 1$$

was already solved in 6.15.(ii). Now, using (52), we rewrite it in the form $8 + x + 8 - x + 3 \cdot 1 \cdot \sqrt[3]{(8 + x)(8 - x)} = 1$, which implies $\sqrt[3]{64 - x^2} = -5$; hence again $x_{1,2} = \pm 3 \cdot \sqrt{21}$.

(ii) Let us now solve the equation

$$\sqrt[3]{x - 1} + \sqrt[3]{x + 1} = x \cdot \sqrt[3]{2}.$$

Rewriting it according to (52), we get $x - 1 + x + 1 + 3x \cdot \sqrt[3]{2(x^2 - 1)} = 2x^3$, and thus $x \cdot [3\sqrt[3]{2(x^2 - 1)} - 2 \cdot (x^2 - 1)] = 0$, so $x_1 = 0$ or $3 \cdot \sqrt[3]{2(x^2 - 1)} = 2(x^2 - 1)$; this last equation has roots

$$x_2 = 1, \quad x_3 = -1, \quad x_{4,5} = \pm\sqrt{(2 + 3\sqrt{3})/2}.$$

The roots of the original equation are therefore some of the numbers x_i, $1 \leq i \leq 5$. While it is easy to check each of the first three numbers (by simple substitutions we verify that each one is indeed a solution of the original equation), we carry out this check for the remaining numbers x_4, x_5 with the help of the following trick: We set $a = \sqrt[3]{x_4 - 1}$, $b = \sqrt[3]{x_4 + 1}$, $c = x_4 \cdot \sqrt[3]{2}$ and show that $a + b = c$. Indeed, an easy calculation shows that the numbers a, b, c satisfy the relation $a^3 + b^3 + 3abc = c^3$; the desired equation $a + b = c$ then follows from the result of Example 3.16.(iii). This works similarly for x_5.

6.18 Exercises

Solve the following equations:

(i) $\sqrt[3]{12 - x} + \sqrt[3]{14 + x} = 2.$

(ii) $\sqrt[3]{x + 7} + \sqrt[3]{28 - x} = 5.$

(iii) $\sqrt[3]{2x - 1} + \sqrt[3]{x - 1} = 1.$

(iv) $\sqrt[3]{x + 1} + \sqrt[3]{x + 2} + \sqrt[3]{x + 3} = 0.$

(v) $\sqrt[3]{x} + \sqrt[3]{2x - 3} = \sqrt[3]{12(x - 1)}.$

*(vi) $\sqrt[3]{x} + \sqrt[3]{x - 16} = \sqrt[3]{x - 8}.$

6.19 *Irrational Equations with a Parameter*

We now turn our attention to solving several types of irrational equations with *parameters*. We will use all the methods described above, and pay special attention to those problems that often cause considerable complications.

(i) $\sqrt{x^2 + a} = a - x$, where x is a variable and a is a real parameter. We first square the equation: $x^2 + a = a^2 - 2ax + x^2$, and thus $2ax = a(a-1)$. For $a = 0$ we get the equation $0 = 0$, whose roots are all numbers $x \in \mathbb{R}$; for $a \neq 0$ it is $x = (a-1)/2$. We now do the necessary checking.

(a) $a = 0$: By substituting into the original equation we get $\sqrt{x^2} = -x$, which is equivalent to $|x| = -x$, and holds if and only if $x \leq 0$. For $a = 0$ we thus have an infinite set $\mathcal{K}$ of roots, with $\mathcal{K} = \mathbb{R}_0^-$.

(b) For $a \neq 0$ we have

$$L\left(\frac{a-1}{2}\right) = \sqrt{\frac{(a-1)^2}{4} + a} = \frac{|a+1|}{2}; \quad R\left(\frac{a-1}{2}\right) = \frac{a+1}{2},$$

so that $L = R$ if and only if $|a+1| = a + 1$, that is, $a \geq -1$. Therefore, the given equation has the unique root $x = (a-1)/2$ for each $a \in [-1, 0) \cup (0, \infty)$, while it has no root for $a \in (-\infty, -1)$.

(ii) The equation

$$\frac{a \cdot \sqrt{x+1} + b}{a - b \cdot \sqrt{x+1}} = \frac{a+b}{a-b}$$

with real parameters a, b makes sense only when $a \neq b$. We substitute $y = \sqrt{x+1}$ and rewrite it as $(ay + b)(a - b) = (a + b)(a - by)$; then after some manipulations we obtain $(a^2 + b^2) \cdot y = a^2 + b^2$. Thus $y = 1$ and $x = 0$ for each pair of parameters a, b, where $a \neq b$. The checking in this case is easy.

6.20 *Example*

Solve the following equation with parameter $a \in \mathbb{R}$:

$$x^2 - \sqrt{a-x} = a. \tag{53}$$

(i) FIRST SOLUTION. By separating the square root we obtain $\sqrt{a-x} = x^2 - a$, where $a - x \geq 0$ and at the same time $x^2 - a \geq 0$, hence $x \leq a \leq x^2$. After these observations, by squaring we obtain

$$x^4 - 2ax^2 + a^2 = a - x, \tag{54}$$

which is an equation of degree four in the variable x, where none of its zeros are easy to "guess." We will therefore use the following trick: Let us for a while consider equation (54) as an equation with variable a and parameter

x; for then we have a quadratic equation $a^2 - (2x^2 + 1)a + (x^4 + x) = 0$ with discriminant $D = (2x - 1)^2$. Therefore, we have either

$$a = \frac{2x^2 + 1 + (2x - 1)}{2} = x^2 + x,$$

or

$$a = \frac{2x^2 + 1 - (2x - 1)}{2} = x^2 - x + 1.$$

Now we return to the original variable and parameter, and distinguish between two cases:

(a) $a = x^2 + x$. Since $a \leq x^2$, we have $x \leq 0$. Then the quadratic equation $x^2 + x - a = 0$ has discriminant $D = 4a + 1$ (which means $a \geq -\frac{1}{4}$) and roots $x_{1,2} = \left(-1 \pm \sqrt{4a+1}\right)/2$. The root $x_1 = \left(-1 + \sqrt{4a+1}\right)/2$ makes sense only for $a \in [-\frac{1}{4}, 0]$, since for $a > 0$ we have $x_1 > 0$. The root $x_2 = \left(-1 - \sqrt{4a+1}\right)/2$ is negative for all $a \geq -\frac{1}{4}$.

(b) $a = x^2 - x + 1$. Since $a \leq x^2$ again, we have $(-x+1) \leq 0$, that is, $x \geq 1$, from which we furthermore obtain the condition $a \geq 1$. We will now solve the equation $x^2 - x + (1-a) = 0$ only for parameter values $a \geq 1$, and we find $x_{3,4} = \left(1 \pm \sqrt{4a-3}\right)/2$. Since $x \geq 1$, the number $x_4 = \left(1 - \sqrt{4a-3}\right)/2$ is not a solution. On the other hand, $x_3 = \left(1 + \sqrt{4a-3}\right)/2$ works, under the conditions $a \geq 1$ and $x \geq 1$.

In summary, we have that for $a < -\frac{1}{4}$ the equation (53) has no solutions; for $a = -\frac{1}{4}$ it has the unique root $-\frac{1}{2}$; for $a \in (-\frac{1}{4}, 0]$ it has the two roots x_1, x_2; for $a \in (0, 1)$ it has the unique root x_2; and finally, for $a \geq 1$ it has the two roots x_2, x_3.   □

Note: The roots of equation (53) are "unpleasant," and checking them by substituting would be difficult. Therefore, in the course of solving the problem above we made sure that all manipulations carried out were equivalences; we carefully followed the domains of definition of expressions, squaring only nonnegative expressions. We will follow this cautious approach also in the second solution of (53).

(ii) SECOND SOLUTION. We use the substitution $y = \sqrt{a - x}$; then $y \geq 0$, $a \geq x$, and we solve the system

$$x^2 = a + y,$$
$$y^2 = a - x.$$

Subtracting the equations, we obtain $x^2 - y^2 = y + x$; hence $(x + y)(x - y - 1) = 0$. Now we distinguish again between two cases:

(a) $x + y = 0$; then $x = -\sqrt{a - x}$, that is, $x^2 + x - a = 0$ under the condition that $x \leq 0$. We then continue as in 6.20.(i), case (a).

(b) $x - y - 1 = 0$; then $x - 1 = \sqrt{a - x}$, and thus $x \geq 1$. For $a \geq \frac{3}{4}$ the equation $x^2 - x + (1 - a) = 0$ has the roots $x_{3,4} = \left(1 \pm \sqrt{4a-3}\right)/2$.

Since $x_4 = \left(1 - \sqrt{4a - 3}\right)/2 \leq \frac{1}{2}$, the number x_4 is not a root of (53) for any value of the parameter a; the number $x_3 = \left(1 + \sqrt{4a - 3}\right)/2$ is a root exactly when $x_3 \geq 1$, that is, when $a \geq 1$.

A summary of these results can be found at the end of the first solution.   □

6.21 Exercises

Solve the following equations, where x is a variable and a is a real parameter:

(i) $\sqrt{x^2 + a^2} = x + a$.

(ii) $1 + \sqrt{x} = \sqrt{x - a}$.

(iii) $\sqrt{x - a} = a \cdot \sqrt{x}$.

(iv) $\sqrt{(x + a)^2 + 4a} + a = x$.

*(v) $\sqrt{a + \sqrt{a + x}} = x$.

6.22 Examples

At the end of this section we will solve a few *systems of irrational equations*. We limit ourselves to systems of two equations in two variables, and we apply some of the tricks that were used earlier.

(i) $\sqrt{x} + 2\sqrt{y} = 9$,

$\quad x - 4y = 9$.

SOLUTION.   Since $x \geq 0$, $y \geq 0$, the second equation of the system can be rewritten in the form $(\sqrt{x} + 2\sqrt{y})(\sqrt{x} - 2\sqrt{y}) = 9$, from which by the first equation it follows that $\sqrt{x} - 2\sqrt{y} = 1$. The system

$$\sqrt{x} + 2\sqrt{y} = 9,$$
$$\sqrt{x} - 2\sqrt{y} = 1,$$

has the solution $\sqrt{x} = 5$, $\sqrt{y} = 2$; hence $x = 25$, $y = 4$. By substituting we convince ourselves that the ordered pair $(x, y) = (25, 4)$ satisfies both equations of the system and is therefore its unique solution.   □

(ii) $\sqrt[4]{x^3} + \sqrt[5]{y^3} = 35$,

$\quad \sqrt[4]{x} + \sqrt[5]{y} = 5$.

SOLUTION.   We use substitution; if we set $u = \sqrt[4]{x}$, $v = \sqrt[5]{y}$, i.e., $x = u^4$, $y = v^5$, we obtain

$$u^3 + v^3 = 35,$$
$$u + v = 5,$$

which we solve using the method of symmetric polynomials (see Section 4). If $\sigma_1 = \sigma_1(u, v) = u + v$, $\sigma_2 = \sigma_2(u, v) = u \cdot v$, then we have: $\sigma_1 = 5$ and $35 = \sigma_1^3 - 3\sigma_1\sigma_2$, and thus $\sigma_2 = 6$. The numbers u, v are zeros of $F(t) = t^2 - 5t + 6$, so either $u = 2$ and $v = 3$, or $u = 3$ and $v = 2$. In the first case we have $x_1 = 16$, $y_1 = 243$, and in the second one, $x_2 = 81$ and $y_2 = 32$. By substituting we convince ourselves that both pairs $(16, 243)$ and $(81, 32)$ are indeed solutions of the original system. □

(iii) $\sqrt{\dfrac{2x - 1}{y + 2}} + \sqrt{\dfrac{y + 2}{2x - 1}} = 2,$ (55)

$$x + y = 12.$$

SOLUTION.   If we use the substitution $\sqrt{\frac{2x-1}{y+2}} = t \geq 0$ for the first equation of the system, then this equation appears in the form $t + \frac{1}{t} = 2$; this is a quadratic equation with a unique (double) root $t = 1$. The system (55) then takes the form of the following system of two linear equations in two variables:

$$2x - 1 = y + 2,$$
$$x + y = 12,$$

with the unique solution $x = 5$, $y = 7$. The pair $(5, 7)$ is then also a solution of the system (55). □

6.23 Exercises

Solve the following systems of equations, where in (viii) the numbers a and b are real parameters:

(i) $\sqrt[3]{x} + \sqrt[3]{y} = 4,$
$$xy = 27.$$

(ii) $\sqrt{\dfrac{x}{y}} + \sqrt{\dfrac{y}{x}} = \dfrac{5}{2},$
$$x + y = 10.$$

(iii) $\sqrt{x^2 + 4xy - 3y^2} = x + 1,$
$$x - y = 1.$$

(iv) $x + xy + y = 12,$
$$\sqrt[3]{x} + \sqrt[3]{xy} + \sqrt[3]{y} = 0.$$

(v) $10 \cdot \sqrt{xy} + 3x - 3y = 58,$
$$x - y = 6.$$

(vi) $\sqrt{\dfrac{x}{y}} + \sqrt{xy} = \dfrac{80}{\sqrt{xy}},$
$$x + y = 20.$$

*(vii)

$$\dfrac{x + \sqrt{x^2 - y^2}}{x - \sqrt{x^2 - y^2}} + \dfrac{x - \sqrt{x^2 - y^2}}{x + \sqrt{x^2 - y^2}} = \dfrac{17}{4},$$
$$x(x + y) + \sqrt{x^2 + xy + 4} = 52.$$

*(viii) $\dfrac{x - y\sqrt{x^2 - y^2}}{\sqrt{1 - x^2 + y^2}} = a,$

$$\dfrac{y - x\sqrt{x^2 - y^2}}{\sqrt{1 - x^2 + y^2}} = b.$$

7   Some Applications of Complex Numbers

Now, at the end of the first chapter, we will concentrate on problems that thematically belong to the preceding sections but that can be conveniently solved by use of *complex* numbers, although in the statements of the problems only *real* numbers appear. We stress that the use of complex numbers for solving these problems is not imperative; however, it replaces a number of artificial tricks required in the corresponding "real" solutions, and thus makes for clearer solutions.

We assume that the reader is familiar with the definition of complex numbers, knows how to work with their *algebraic* $(a+bi)$ and *trigonometric* $(r(\cos\varphi + i\sin\varphi))$ forms and knows *de Moivre's theorem*

$$(\cos\varphi + i\sin\varphi)^t = \cos t\varphi + i\sin t\varphi,$$

which holds for arbitrary numbers $\varphi \in \mathbb{R}$ and $t \in \mathbb{Z}$.

7.1   Sums of Binomial Coefficients

Already in Section 1.3.(i) we saw that by appropriate application of the binomial theorem we can obtain the sums of certain binomial coefficients. If in 1.2 we now set $A = 1$ and $B = i$, we get

$$(1+i)^n = \binom{n}{0} + \binom{n}{1}i - \binom{n}{2} - \binom{n}{3}i + \cdots + \binom{n}{n-1}i^{n-1} + \binom{n}{n}i^n.$$

We evaluate the left-hand side by way of de Moivre's theorem:

$$(1+i)^n = \left[\sqrt{2}\left(\cos\frac{\pi}{4} + i\sin\frac{\pi}{4}\right)\right]^n = \left(\sqrt{2}\right)^n\left(\cos\frac{n\pi}{4} + i\sin\frac{n\pi}{4}\right).$$

Comparing the real and imaginary parts on both sides, we obtain

$$\binom{n}{0} - \binom{n}{2} + \binom{n}{4} - \cdots = \left(\sqrt{2}\right)^n\cos\frac{n\pi}{4}, \tag{56}$$

$$\binom{n}{1} - \binom{n}{3} + \binom{n}{5} - \cdots = \left(\sqrt{2}\right)^n\sin\frac{n\pi}{4}. \tag{57}$$

Here, as in the following equations, the dots denote the terms in an obvious pattern, continuing as long as the binomial coefficients involved are defined (we have defined $\binom{n}{k}$ only for $k = 0, 1, \ldots, n$). By adding and subtracting (56), resp. (57), to/from 1.4.(i), resp. 1.4.(ii), we obtain the following formulas:

$$\binom{n}{0} + \binom{n}{4} + \binom{n}{8} + \cdots = 2^{n-2} + \left(\sqrt{2}\right)^{n-2} \cos\frac{n\pi}{4},$$

$$\binom{n}{1} + \binom{n}{5} + \binom{n}{9} + \cdots = 2^{n-2} + \left(\sqrt{2}\right)^{n-2} \sin\frac{n\pi}{4},$$

$$\binom{n}{2} + \binom{n}{6} + \binom{n}{10} + \cdots = 2^{n-2} - \left(\sqrt{2}\right)^{n-2} \cos\frac{n\pi}{4},$$

$$\binom{n}{3} + \binom{n}{7} + \binom{n}{11} + \cdots = 2^{n-2} - \left(\sqrt{2}\right)^{n-2} \sin\frac{n\pi}{4}.$$

7.2 Example

For an arbitrary natural number n, determine the sum

$$S = \binom{n}{0} + \binom{n}{6} + \binom{n}{12} + \cdots.$$

SOLUTION.   We set $\varepsilon = \cos\frac{\pi}{3} + i\sin\frac{\pi}{3}$. In the binomial theorem 1.2 we set $A = 1$, $B = \varepsilon^k$, where $k \in \{0, 1, \ldots, 5\}$, and we get

$$(1 + \varepsilon^k)^n = \binom{n}{0} + \binom{n}{1}\varepsilon^k + \binom{n}{2}\varepsilon^{2k} + \cdots + \binom{n}{n}\varepsilon^{nk}. \tag{58}$$

If t is an arbitrary natural number not divisible by 6, then $\varepsilon^t \neq 1$, and therefore by (5), and using the fact that $\varepsilon^6 = 1$, we have

$$1 + \varepsilon^t + \varepsilon^{2t} + \varepsilon^{3t} + \varepsilon^{4t} + \varepsilon^{5t} = \frac{\varepsilon^{6t} - 1}{\varepsilon^t - 1} = \frac{0}{\varepsilon^t - 1} = 0. \tag{59}$$

If, on the other hand, $t \in \mathbb{N}_0$ is divisible by 6, then

$$1 + \varepsilon^t + \varepsilon^{2t} + \varepsilon^{3t} + \varepsilon^{4t} + \varepsilon^{5t} = 1 + 1 + 1 + 1 + 1 + 1 = 6. \tag{60}$$

Adding (58) for all $k \in \{0, 1, \ldots, 5\}$, we get

$$\sum_{k=0}^{5}(1 + \varepsilon^k)^n = 6\binom{n}{0} + (1 + \varepsilon + \cdots + \varepsilon^5)\binom{n}{1} + \cdots$$
$$+ (1 + \varepsilon^n + \varepsilon^{2n} + \cdots + \varepsilon^{5n})\binom{n}{n}.$$

By (59) and (60) the sum on the right-hand side is $6S$. Hence

$$S = \frac{1}{6}[(1+1)^n + (1+\varepsilon)^n + \cdots + (1 + \varepsilon^5)^n].$$

To evaluate the numbers on the right-hand side, we first determine the numbers ε^k by de Moivre's theorem:

$$\varepsilon = \frac{1}{2} + i\frac{\sqrt{3}}{2}, \quad \varepsilon^2 = -\frac{1}{2} + i\frac{\sqrt{3}}{2}, \quad \varepsilon^3 = -1,$$

$$\varepsilon^4 = -\frac{1}{2} - i\frac{\sqrt{3}}{2}, \quad \varepsilon^5 = \frac{1}{2} - i\frac{\sqrt{3}}{2}.$$

Hence

$$S = \tfrac{1}{6}\Big[2^n + \Big(\tfrac{3}{2} + i\tfrac{\sqrt{3}}{2}\Big)^n + \Big(\tfrac{1}{2} + i\tfrac{\sqrt{3}}{2}\Big)^n + 0 + \Big(\tfrac{1}{2} - i\tfrac{\sqrt{3}}{2}\Big)^n + \Big(\tfrac{3}{2} - i\tfrac{\sqrt{3}}{2}\Big)^n \Big].$$

Now we again use de Moivre's theorem,

$$\Big(\tfrac{1}{2} \pm i\tfrac{\sqrt{3}}{2}\Big)^n = \big[\cos(\pm\tfrac{\pi}{3}) + i\sin(\pm\tfrac{\pi}{3})\big]^n = \cos\tfrac{n\pi}{3} \pm i\sin\tfrac{n\pi}{3},$$

$$\Big(\tfrac{3}{2} \pm i\tfrac{\sqrt{3}}{2}\Big)^n = \Big[\sqrt{3}\Big(\tfrac{\sqrt{3}}{2} \pm i\tfrac{1}{2}\Big)\Big]^n = \big[\sqrt{3}(\cos(\pm\tfrac{\pi}{6}) + i\sin(\pm\tfrac{\pi}{6}))\big]^n$$

$$= \big(\sqrt{3}\big)^n (\cos\tfrac{n\pi}{6} \pm i\sin\tfrac{n\pi}{6}),$$

and obtain the desired sum

$$S = \frac{1}{3}\Big[2^{n-1} + \cos\frac{n\pi}{3} + \big(\sqrt{3}\big)^n \cdot \cos\frac{n\pi}{6}\Big].$$

We remark that the choice of the number ε that we have used above is not as unnatural as it may appear at first sight. By comparing the formula in the binomial theorem and the sum S we see that we need a complex number whose sixth power is 1 and whose first to fifth powers are not equal to 1 (that is, a primitive sixth root of unity). It follows from de Moivre's theorem that the number $\varepsilon = \cos\frac{2\pi}{6} + i\sin\frac{2\pi}{6}$ has this property (apart from this number, so does ε^5, but no others). For a similar reason we have worked with the number $i = \cos\frac{2\pi}{4} + i\sin\frac{2\pi}{4}$ in 7.1.   □

7.3 Exercises

For arbitrary natural numbers n, find the following sums:

(i) $\binom{n}{0} + \binom{n}{3} + \binom{n}{6} + \cdots$.

(ii) $\binom{n}{1} + \binom{n}{4} + \binom{n}{7} + \cdots$.

(iii) $\binom{n}{2} + \binom{n}{5} + \binom{n}{8} + \cdots$.

(iv) $\binom{n}{0} - 3\binom{n}{2} + 3^2\binom{n}{4} - 3^3\binom{n}{6} + \cdots$.

(v) $\binom{n}{1} - 3\binom{n}{3} + 3^2\binom{n}{5} - 3^3\binom{n}{7} + \cdots$.

7.4 Trigonometric Functions

With the help of calculations based on the trigonometric form of complex numbers, we can obtain several interesting identities in which trigonometric functions occur. We illustrate this in the solutions of the following four examples. In the first two we apply a technique similar to the one in Sections 7.1 and 7.2, while in the remaining ones we will use the decomposition of polynomials into linear factors and Vieta's relations between their coefficients and zeros.

(i) Express $\cos n\alpha$ and $\sin n\alpha$ for each $n \in \mathbb{N}$ in terms of $\cos \alpha$ and $\sin \alpha$.

SOLUTION. By de Moivre's theorem and the binomial theorem we have

$$
\begin{aligned}
(\cos n\alpha + i\sin n\alpha) &= (\cos \alpha + i\sin \alpha)^n \\
&= \cos^n \alpha + i\binom{n}{1}\cos^{n-1}\alpha \sin \alpha - \binom{n}{2}\cos^{n-2}\alpha \sin^2\alpha \\
&\quad -i\binom{n}{3}\cos^{n-3}\alpha \sin^3\alpha + \binom{n}{4}\cos^{n-4}\alpha \sin^4\alpha + i\binom{n}{5}\cos^{n-5}\alpha \sin^5 \alpha - \cdots,
\end{aligned}
$$

from which we obtain the following two identities by comparing real and imaginary parts:

$$
\begin{aligned}
\cos n\alpha &= \cos^n \alpha - \binom{n}{2}\cos^{n-2}\alpha \sin^2 \alpha + \binom{n}{4}\cos^{n-4}\alpha \sin^4 \alpha - \cdots, \\
\sin n\alpha &= \binom{n}{1}\cos^{n-1}\alpha \sin \alpha - \binom{n}{3}\cos^{n-3}\alpha \sin^3 \alpha \\
&\quad + \binom{n}{5}\cos^{n-5}\alpha \sin^5 \alpha - \cdots .
\end{aligned}
$$
$\square$

(ii) Find the values of the sums

$$
\begin{aligned}
S_1 &= \sin \varphi + \sin(\varphi + \alpha) + \sin(\varphi + 2\alpha) + \cdots + \sin(\varphi + n\alpha), \\
S_2 &= \cos \varphi + \cos(\varphi + \alpha) + \cos(\varphi + 2\alpha) + \cdots + \cos(\varphi + n\alpha),
\end{aligned}
$$

for arbitrary $n \in \mathbb{N}$ and $\varphi, \alpha \in \mathbb{R}$.

SOLUTION. The problem is trivial for $\alpha = 2k\pi$ ($k \in \mathbb{Z}$), in which case we have $S_1 = (n+1)\sin \varphi$, $S_2 = (n+1)\cos \varphi$. For all other values of α we consider the sum $S = S_2 + iS_1$:

$$
\begin{aligned}
S &= [\cos \varphi + i\sin \varphi] + [\cos(\varphi + \alpha) + i\sin(\varphi + \alpha)] \\
&\quad + [\cos(\varphi + 2\alpha) + i\sin(\varphi + 2\alpha)] + \cdots \\
&\quad + [\cos(\varphi + n\alpha) + i\sin(\varphi + n\alpha)].
\end{aligned}
\tag{61}
$$

To simplify notation, we set $a = \cos \varphi + i\sin \varphi$, $b = \cos(\frac{\alpha}{2}) + i\sin(\frac{\alpha}{2})$. By de Moivre's theorem we have

$$
S = a + ab^2 + ab^4 + \cdots + ab^{2n} = a(1 + b^2 + \cdots + b^{2n}).
$$

From the assumption $\alpha \neq 2k\pi$ $(k \in \mathbb{Z})$ it follows that $b^2 \neq 1$, so that by formula (5) we can evaluate $1 + b^2 + \cdots + b^{2n}$ and thus obtain the expression

$$S = \frac{a}{b^2 - 1} \cdot (b^{2n+2} - 1) = \frac{a}{b(b - b^{-1})} \cdot b^{n+1}(b^{n+1} - b^{-n-1})$$

$$= \frac{b^{n+1} - b^{-n-1}}{b - b^{-1}} ab^n. \tag{62}$$

By de Moivre's theorem it follows that

$$b - b^{-1} = (\cos \tfrac{\alpha}{2} + i \sin \tfrac{\alpha}{2}) - (\cos \tfrac{\alpha}{2} - i \sin \tfrac{\alpha}{2}) = 2i \sin \tfrac{\alpha}{2},$$

$$b^{n+1} - b^{-n-1} = \left(\cos \tfrac{(n+1)\alpha}{2} + i \sin \tfrac{(n+1)\alpha}{2}\right) - \left(\cos \tfrac{(n+1)\alpha}{2} - i \sin \tfrac{(n+1)\alpha}{2}\right)$$

$$= 2i \sin \tfrac{(n+1)\alpha}{2},$$

$$a \cdot b^n = (\cos \varphi + i \sin \varphi) \cdot (\cos \tfrac{n\alpha}{2} + i \sin \tfrac{n\alpha}{2})$$

$$= \cos(\varphi + \tfrac{n\alpha}{2}) + i \sin(\varphi + \tfrac{n\alpha}{2}).$$

Substituting these into (62), we obtain

$$S = \frac{\sin \tfrac{(n+1)\alpha}{2}}{\sin \tfrac{\alpha}{2}} \cdot [\cos(\varphi + \tfrac{n\alpha}{2}) + i \sin(\varphi + \tfrac{n\alpha}{2})],$$

from which, upon comparing real and imaginary parts, we get with (61),

$$S_1 = \sin \varphi + \sin(\varphi + \alpha) + \cdots + \sin(\varphi + n\alpha) = \frac{\sin \tfrac{(n+1)\alpha}{2} \cdot \sin(\varphi + \tfrac{n\alpha}{2})}{\sin \tfrac{\alpha}{2}},$$

$$S_2 = \cos \varphi + \cos(\varphi + \alpha) + \cdots + \cos(\varphi + n\alpha) = \frac{\sin \tfrac{(n+1)\alpha}{2} \cdot \cos(\varphi + \tfrac{n\alpha}{2})}{\sin \tfrac{\alpha}{2}}$$

for each $\alpha \neq 2k\pi$ $(k \in \mathbb{Z})$. $\square$

(iii) Show that for each $n \in \mathbb{N}$ greater than 1 we have

$$\sin \frac{\pi}{n} \sin \frac{2\pi}{n} \sin \frac{3\pi}{n} \cdots \sin \frac{(n-1)\pi}{n} = \frac{n}{2^{n-1}}. \tag{63}$$

SOLUTION.   The polynomial $x^n - 1$ has the n complex zeros $1, \varepsilon, \varepsilon^2, \ldots,$ ε^{n-1}, where ε is the complex unit $\cos(2\pi/n) + i \sin(2\pi/n)$. Its decomposition into linear factors is therefore

$$x^n - 1 = (x - 1)(x - \varepsilon)(x - \varepsilon^2) \cdots (x - \varepsilon^{n-1}).$$

If we divide the polynomials on both sides by $x - 1$, we obtain

$$x^{n-1} + x^{n-2} + \cdots + x + 1 = (x - \varepsilon)(x - \varepsilon^2) \cdots (x - \varepsilon^{n-1}),$$

which by substituting $x = 1$ gives the equation

$$n = (1 - \varepsilon)(1 - \varepsilon^2) \cdots (1 - \varepsilon^{n-1}).$$

This implies $n = |1 - \varepsilon| \cdot |1 - \varepsilon^2| \cdots |1 - \varepsilon^{n-1}|$, since the absolute value of a product of complex numbers is equal to the product of the absolute values. If we substitute for each $k = 1, 2, \ldots, n - 1$ the result of the calculation

$$|1 - \varepsilon^k| = \sqrt{\left(1 - \cos \frac{2k\pi}{n}\right)^2 + \sin^2 \frac{2k\pi}{n}} = \sqrt{2 - 2\cos \frac{2k\pi}{n}} = 2\sin \frac{k\pi}{n}$$

(where we have used the well-known identity $\cos 2\alpha = 1 - 2\sin^2 \alpha$), we obtain the desired equation after dividing by 2^{n-1}. This result has an interesting geometrical interpretation. It arises when in the complex plane one considers a regular n-gon whose vertices are the images of the complex numbers $1, \varepsilon, \varepsilon^2, \ldots, \varepsilon^{n-1}$, and if one realizes that the absolute values of the complex numbers $1 - \varepsilon^k$ are equal to the lengths of the segments $A_1 A_k$. The reader is invited to continue this argument. □

(iv) For each $n \in \mathbb{N}$ verify the equality

$$\cot^2 \frac{\pi}{2n+1} + \cot^2 \frac{2\pi}{2n+1} + \cot^2 \frac{3\pi}{2n+1} + \cdots + \cot^2 \frac{n\pi}{2n+1} = \frac{n(2n-1)}{3}.$$
$$\tag{64}$$

SOLUTION.   We will find an equation of degree n with the n distinct zeros

$$x_k = \cot^2 \frac{k\pi}{2n+1} \qquad (k = 1, 2, \ldots, n).$$

To this end we use the expression for $\sin(2n + 1)\alpha$ in terms of $\sin \alpha$ and $\cos \alpha$, which we have derived in the solution of example (i):

$$\sin(2n+1)\alpha = \sum_{j=0}^{n} (-1)^j \binom{2n+1}{2j+1} \cos^{2n-2j} \alpha \sin^{2j+1} \alpha.$$

If on the right-hand side we factor out $\sin^{2n+1} \alpha$, we obtain

$$\sin(2n+1)\alpha = \sin^{2n+1} \alpha \cdot \sum_{j=0}^{n} (-1)^j \binom{2n+1}{2j+1} \cot^{2n-2j} \alpha$$

(supposing $\sin \alpha \neq 0$). We will use this last formula for each $\alpha = \frac{k\pi}{2n+1}$, where $k = 1, 2, \ldots, n$. Since for each such α we have $\sin(2n+1)\alpha = 0$ and $\sin \alpha \neq 0$, we conclude that all the numbers x_k defined above are roots of the algebraic equation

$$\binom{2n+1}{1} x^n - \binom{2n+1}{3} x^{n-1} + \binom{2n+1}{5} x^{n-2} - \cdots + (-1)^n \binom{2n+1}{2n+1} = 0.$$

The sum of the roots of this equation is, according to Vieta's relations,

$$\sum_{k=1}^{n} x_k = \frac{\binom{2n+1}{3}}{\binom{2n+1}{1}} = \frac{n(2n-1)}{3}.$$

This proves equality (64).   □

For the sake of interest we will now show how this result can easily be used to derive the sum of the important infinite series

$$1 + \frac{1}{2^2} + \frac{1}{3^2} + \frac{1}{4^2} + \cdots + \frac{1}{n^2} + \cdots = \frac{\pi^2}{6}.$$

To achieve this, we begin with the fact that for each α, $0 < \alpha < \frac{\pi}{2}$, we have

$$0 < \sin \alpha < \alpha < \tan \alpha.$$

(Justify this inequality geometrically by comparing the areas of three shapes: an isosceles triangle with length 1 of the equal sides and angle α between them, a circular sector with radius 1 and angle α, and a right-angled triangle with sides 1 and $\tan \alpha$ making the right angle). From these inequalities we get

$$\cot^2 \alpha < \frac{1}{\alpha^2} < 1 + \cot^2 \alpha.$$

If we add over all $\alpha = \frac{k\pi}{2n+1}$ ($k = 1, 2, \ldots, n$), we obtain with (64),

$$\frac{n(2n-1)}{3} < \sum_{k=1}^{n} \left(\frac{2n+1}{k\pi} \right)^2 < n + \frac{n(2n-1)}{3}.$$

From this, after multiplying by $(\frac{\pi}{2n+1})^2$, we reach the conclusion that for the partial sums

$$S(n) = 1 + \frac{1}{2^2} + \frac{1}{3^2} + \cdots + \frac{1}{n^2},$$

for each $n \in \mathbb{N}$ we have the bounds

$$\left(1 - \frac{1}{2n+1} \right)\left(1 - \frac{2}{2n+1} \right) \frac{\pi^2}{6} < S(n) < \left(1 - \frac{1}{2n+1} \right)\left(1 + \frac{1}{2n+1} \right) \frac{\pi^2}{6},$$

which confirms that $S(n) \to \pi^2/6$ for $n \to \infty$.

(v) Prove the identity

$$\sqrt[3]{\cos \frac{2\pi}{7}} + \sqrt[3]{\cos \frac{4\pi}{7}} + \sqrt[3]{\cos \frac{6\pi}{7}} = \sqrt[3]{\frac{1}{2} \cdot (5 - 3\sqrt[3]{7})},$$

where (as in Section 6) the third root is considered to be a real function, that is, a bijection on the set $\mathbb{R}$.

SOLUTION.   The equation $x^6 + x^5 + \cdots + x + 1 = 0$ has the complex units $\varepsilon, \varepsilon^2, \ldots, \varepsilon^6$ as roots, where $\varepsilon = \cos \frac{2\pi}{7} + i \sin \frac{2\pi}{7}$. These six complex numbers, upon substituting $y = x + \frac{1}{x}$ (a standard method of solution for these so-called reciprocal equations), turn into the three real numbers

$$y_1 = 2 \cos \frac{2\pi}{7}, \quad y_2 = 2 \cos \frac{4\pi}{7}, \quad y_3 = 2 \cos \frac{6\pi}{7},$$

which are solutions of the equation $y^3 + y^2 - 2y - 1 = 0$. This last equation can be derived by first dividing the original equation of degree 6 by x^3 and then using the expressions

$$x + \frac{1}{x} = y, \quad x^2 + \frac{1}{x^2} = y^2 - 2, \quad x^3 + \frac{1}{x^3} = y^3 - 3y.$$

Hence we have to verify the equality

$$\sqrt[3]{y_1} + \sqrt[3]{y_2} + \sqrt[3]{y_3} = \sqrt[3]{5 - 3\sqrt[3]{7}},$$

where y_j are real roots of an equation whose coefficients we know. Let us therefore formulate the following more general problem: Given the cubic equation $y^3 - ay^2 + by - c = 0$ with real roots y_1, y_2, y_3 (in our case $a = -1$, $b = -2$, $c = 1$), construct an equation $z^3 - Az^2 + Bz - C = 0$ with real roots $\sqrt[3]{y_1}$, $\sqrt[3]{y_2}$ $\sqrt[3]{y_3}$. From Vieta's relations

$$
\begin{aligned}
y_1 + y_2 + y_3 &= a, & \sqrt[3]{y_1} + \sqrt[3]{y_2} + \sqrt[3]{y_3} &= A, \\
y_1 y_2 + y_2 y_3 + y_3 y_1 &= b, & \sqrt[3]{y_1 y_2} + \sqrt[3]{y_2 y_3} + \sqrt[3]{y_3 y_1} &= B, \\
y_1 y_2 y_3 &= c, & \sqrt[3]{y_1 y_2 y_3} &= C,
\end{aligned}
$$

it is immediately clear that $C = \sqrt[3]{c}$. Next we use the following identity from Table 4.12:

$$(m + p + q)^3 = m^3 + p^3 + q^3 + 3(m + p + q)(mp + pq + qm) - 3mpq.$$

If we substitute the triple $\sqrt[3]{y_1}$, $\sqrt[3]{y_2}$, $\sqrt[3]{y_3}$, we obtain $A^3 = a + 3AB - 3C$, and by substituting $\sqrt[3]{y_1 y_2}$, $\sqrt[3]{y_2 y_3}$, $\sqrt[3]{y_3 y_1}$, we get $B^3 = b + 3BCA - 3C^2$. In the case of the coefficients $a = -1$, $b = -2$, $c = 1$ we therefore conclude that $C = 1$ and that A and B are the solutions of the pair of equations $A^3 = 3AB - 4$ and $B^3 = 3AB - 5$. Multiplying these two equations we obtain, for the new variable $w = AB$, the equation $w^3 = (3w - 4)(3w - 5)$, or $(w - 3)^3 = -7$. Hence $w = 3 - \sqrt[3]{7}$, and so $A^3 = 3w - 4 = 5 - 3\sqrt[3]{7}$ and $B^3 = 3w - 5 = 4 - 3\sqrt[3]{7}$. Since $A = \sqrt[3]{y_1} + \sqrt[3]{y_2} + \sqrt[3]{y_3}$, the proof is now complete.   □

7.5 Exercises

(i) Express $\tan 6\alpha$ in terms of $\tan \alpha$ for any $\alpha \in \mathbb{R}$.

(ii) For each $n \in \mathbb{N}$ prove the identity

$$\cos \frac{2\pi}{2n+1} + \cos \frac{4\pi}{2n+1} + \cos \frac{6\pi}{2n+1} + \cdots + \cos \frac{2n\pi}{2n+1} = -\frac{1}{2}.$$

(iii) Show that for arbitrary $n \in \mathbb{N}$, $\alpha \in \mathbb{R}$ we have

$$\binom{n}{0}\cos n\alpha + \binom{n}{1}\cos(n-2)\alpha + \binom{n}{2}\cos(n-4)\alpha + \cdots$$
$$+ \binom{n}{n}\cos(n-2n)\alpha = 2^n \cdot \cos^n \alpha.$$

*(iv) For arbitrary $n \in \mathbb{N}$, $\alpha, \beta \in \mathbb{R}$ find the sums

$S_1 = \binom{n}{0} \cos \beta + \binom{n}{1} \cos(\beta+\alpha) + \binom{n}{2} \cos(\beta+2\alpha) + \cdots + \binom{n}{n} \cos(\beta+n\alpha)$,

$S_2 = \binom{n}{0} \sin \beta + \binom{n}{1} \sin(\beta+\alpha) + \binom{n}{2} \sin(\beta+2\alpha) + \cdots + \binom{n}{n} \sin(\beta+n\alpha)$.

*(v) Let $r, \alpha, \varphi \in \mathbb{R}$, $\sin \alpha \neq 0$ and $n \in \mathbb{N}$. Prove the identitites

$$\sin \varphi + r \cdot \sin(\varphi+\alpha) + r^2 \cdot \sin(\varphi+2\alpha) + \cdots + r^n \cdot \sin(\varphi+n\alpha)$$
$$= \frac{\sin \varphi - r \cdot \sin(\varphi - \alpha) - r^{n+1} \sin(\varphi+(n+1)\alpha) + r^{n+2} \sin(\varphi+n\alpha)}{1 - 2r \cos \alpha + r^2},$$

$$\cos \varphi + r \cdot \cos(\varphi+\alpha) + r^2 \cdot \cos(\varphi+2\alpha) + \cdots + r^n \cos(\varphi+n\alpha)$$
$$= \frac{\cos \varphi - r \cdot \cos(\varphi - \alpha) - r^{n+1} \cos(\varphi+(n+1)\alpha) + r^{n+2} \cos(\varphi+n\alpha)}{1 - 2r \cos \alpha + r^2}.$$

*(vi) For each $n \in \mathbb{N}$ verify the identity

$$\tan^2 \frac{\pi}{4n} + \tan^2 \frac{3\pi}{4n} + \tan^2 \frac{5\pi}{4n} + \cdots + \tan^2 \frac{(2n - 1)\pi}{4n} = n(2n - 1).$$

(vii) Let us return to the polynomial identity

$$x^{n-1} + x^{n-2} + \cdots + x + 1 = (x - \varepsilon)(x - \varepsilon^2) \cdots (x - \varepsilon^{n-1}) \quad (65)$$

($\varepsilon = \cos(2\pi/n) + i \sin(2\pi/n)$) from Example 7.4.(iii), where we have substituted $x = 1$. In the case of an odd $n = 2m + 1$ it is worth also substituting $x = -1$. Do this, and thus derive the identity (for $m \in \mathbb{N}$)

$$\cos \frac{\pi}{2m + 1} \cos \frac{2\pi}{2m + 1} \cos \frac{3\pi}{2m + 1} \cdots \cos \frac{m\pi}{2m + 1} = \frac{1}{2^m}. \quad (66)$$

Let us note that the corresponding formula for even $n = 2m$,

$$\cos \frac{\pi}{2m} \cos \frac{2\pi}{2m} \cos \frac{3\pi}{2m} \cdots \cos \frac{(m - 1)\pi}{2m} = \frac{\sqrt{m}}{2^{m-1}}$$

($m \in \mathbb{N}$, $m \geq 2$), follows directly from (63) with the help of the identities

$$\cos \frac{k\pi}{2m} = \sin \frac{(m - k)\pi}{2m} = \sin \frac{(m + k)\pi}{2m}$$

used for $k = 1, 2, \ldots, m - 1$.

*(viii) Implement another method for deriving formulas (63) and (66), namely one based on the multiple-angle formulas for sine obtained in 7.4.(i): First determine
(a) the polynomial $A_m(x)$ of degree m with zeros
$x_k = \sin^2 \frac{k\pi}{2m+1}$ ($1 \leq k \leq m$),
(b) the polynomial $B_m(y)$ of degree m with zeros
$y_k = \sin^2 \frac{k\pi}{2m+2}$ ($1 \leq k \leq m$).
With the help of Vieta's relations then obtain again formulas (63) and (66).

*(ix) Using an approach similar to 7.4.(v), prove the identity

$$\sqrt[3]{\cos\frac{2\pi}{9}} + \sqrt[3]{\cos\frac{4\pi}{9}} + \sqrt[3]{\cos\frac{8\pi}{9}} = \sqrt[3]{\frac{1}{2} \cdot (3\sqrt[3]{9} - 6)}.$$

(x) Prove the following equalities: $\cos\frac{\pi}{5} = \frac{1+\sqrt{5}}{4}$, $\cos\frac{2\pi}{5} = \frac{-1+\sqrt{5}}{4}$.

7.6 Polynomials

Complex numbers can often also be used in problems involving polynomials with real coefficients, for instance in testing for divisibility by way of the criterion introduced in 3.15.(ii). We will illustrate this with the following examples.

(i) Suppose that the integer $n > 1$ is divisible neither by 2 nor by 3. Show that the polynomial

$$F(x) = \binom{n}{1}x^{n-2} + \binom{n}{2}x^{n-3} + \cdots + \binom{n}{n-2}x + \binom{n}{n-1}$$

is divisible by $G(x) = x^3 + 2x^2 + 2x + 1$.

SOLUTION.   Let us first find the roots of the polynomial $G(x)$. We easily convince ourselves that $G(-1) = 0$, and thus $G(x)$ is divisible by $x+1$. By dividing we obtain

$$G(x) = (x + 1)(x^2 + x + 1),$$

where the zeros of $x^2 + x + 1$ are $(-1 \pm i\sqrt{3})/2$. By 3.15.(ii) we are done if we can show that $F(x)$ has -1, $(-1 + i\sqrt{3})/2$, and $(-1 - i\sqrt{3})/2$ as zeros. By the binomial theorem, we have for each $x \in \mathbb{R}$, $x \neq 0$,

$$F(x) = \frac{1}{x} \cdot [(x + 1)^n - x^n - 1],$$

from which by substituting we get $F(-1) = -[0^n - (-1)^n - 1] = 0$, since n is odd. The values of the polynomial $F(x)$ at the numbers $(-1 + i\sqrt{3})/2$ and $(-1 - i\sqrt{3})/2$ can be computed together:

$$F\left(\frac{-1 \pm i\sqrt{3}}{2}\right) = \frac{2}{-1 \pm i\sqrt{3}}\left[\left(\frac{1}{2} \pm i\frac{\sqrt{3}}{2}\right)^n - \left(-\frac{1}{2} \pm i\frac{\sqrt{3}}{2}\right)^n - 1\right].$$

By de Moivre's theorem we have

$$\left(\frac{1}{2} \pm i\frac{\sqrt{3}}{2}\right)^n - \left(-\frac{1}{2} \pm i\frac{\sqrt{3}}{2}\right)^n$$

$$= \left(\cos\left(\pm\frac{\pi}{3}\right) + i\sin\left(\pm\frac{\pi}{3}\right)\right)^n - \left(\cos\left(\pm\frac{2\pi}{3}\right) + i\sin\left(\pm\frac{2\pi}{3}\right)\right)^n$$

$$= \cos\frac{n\pi}{3} \pm i\sin\frac{n\pi}{3} - \cos\frac{2n\pi}{3} \mp i\sin\frac{2n\pi}{3}.$$

In view of the fact that n is divisible neither by 2 nor by 3, there exists a $k \in \mathbb{N}$ such that $n = 6k \pm 1$, and thus

$$
\begin{aligned}
\sin \frac{n\pi}{3} &= \sin \left(2k\pi \pm \frac{\pi}{3} \right) = \sin \left(\pm \frac{\pi}{3} \right) \\
&= \sin \left(\pm \frac{2\pi}{3} \right) = \sin \left(4k\pi \pm \frac{2\pi}{3} \right) = \sin \frac{2n\pi}{3}, \\
\cos \frac{n\pi}{3} &= \cos \left(2k\pi \pm \frac{\pi}{3} \right) = \cos \left(\pm \frac{\pi}{3} \right) = \frac{1}{2}, \\
\cos \frac{2n\pi}{3} &= \cos \left(4k\pi \pm \frac{2\pi}{3} \right) = \cos \left(\pm \frac{2\pi}{3} \right) = -\frac{1}{2}.
\end{aligned}
$$

By substituting these values we get

$$
F\left(\frac{-1 \pm i\sqrt{3}}{2} \right) = 2 \frac{\cos \frac{n\pi}{3} \pm i \sin \frac{n\pi}{3} - \cos \frac{2n\pi}{3} \pm i \sin \frac{2n\pi}{3} - 1}{-1 \pm i\sqrt{3}} = 0.
$$

This completes the proof of the statement. □

(ii) Show that for arbitrary numbers $n \in \mathbb{N}$ and $\alpha \in \mathbb{R}$ satisfying the conditions $n > 1$ and $\sin \alpha \neq 0$, the polynomial

$$
P(x) = x^n \sin \alpha - x \sin n\alpha + \sin(n-1)\alpha
$$

is divisible by the polynomial $Q(x) = x^2 - 2x \cos \alpha + 1$.

SOLUTION.   It is easy to convince ourselves that $Q(x)$ has the zeros $x_{1,2} = \cos \alpha \pm i \sin \alpha$. Thus,

$$
P(x_{1,2}) = (\cos \alpha \pm i \sin \alpha)^n \sin \alpha - (\cos \alpha \pm i \sin \alpha) \sin n\alpha + \sin(n-1)\alpha.
$$

If we use the expression $(\cos \alpha \pm i \sin \alpha)^n = \cos n\alpha \pm i \sin n\alpha$, we obtain upon simplification

$$
P(x_1) = P(x_2) = \cos n\alpha \sin \alpha - \cos \alpha \sin n\alpha + \sin(n-1)\alpha.
$$

the expression on the right vanishes thanks to the formula $\sin(n-1)\alpha = \cos \alpha \sin n\alpha - \cos n\alpha \sin \alpha$, which we can derive by applying de Moivre's theorem to both sides of the equality

$$
(\cos \alpha + i \sin \alpha)^{n-1} = (\cos \alpha + i \sin \alpha)^n \cdot (\cos \alpha + i \sin \alpha)^{-1}
$$

and then equating the imaginary parts. We have thus come to the conclusion that $P(x_{1,2}) = 0$. By 3.15.(ii) this means that $Q(x)$ divides $P(x)$. □

7.7 Exercises

(i) Show that for arbitrary $n \in \mathbb{N}_0$ the polynomial $(x+1)^{2n+1} + x^{n+2}$ is divisible by $x^2 + x + 1$.

(ii) Show that the polynomial $(x \sin \alpha + \cos \alpha)^n - x \sin n\alpha - \cos n\alpha$ is divisible by $x^2 + 1$ for any $\alpha \in \mathbb{R}$ and any $n \in \mathbb{N}$, $n > 1$.

(iii) Show that for any $a, b, c, d \in \mathbb{N}_0$ the polynomial $x^{4a} + x^{4b+1} + x^{4c+2} + x^{4d+3}$ is divisible by $x^3 + x^2 + x + 1$.

*(iv) Find all real numbers $a \in (-2, 2)$ satisfying the following condition: The polynomial $x^{154} - ax^{77} + 1$ is a multiple of the polynomial $x^{14} - ax^7 + 1$.

To conclude this section we deal with one more problem concerning a polynomial with real coefficients, where again we apply complex numbers.

7.8 Example

Let $x_1, x_2, \ldots, x_n$ be the zeros of the polynomial

$$x^n + p_1 x^{n-1} + p_2 x^{n-2} + \cdots + p_{n-1}x + p_n$$

with real coefficients $p_1, p_2, \ldots, p_n$. Show that

$$(x_1^2 + 1) \cdots (x_n^2 + 1) = (1 - p_2 + p_4 - \cdots)^2 + (p_1 - p_3 + p_5 - \cdots)^2.$$

SOLUTION.   By 3.14 we have

$$x^n + p_1 x^{n-1} + \cdots + p_{n-1}x + p_n = (x - x_1)(x - x_2) \cdots (x - x_n).$$

We substitute $x = i$, $x = -i$ into this and multiply the two expressions thus obtained. Since $(i - x_k)(-i - x_k) = x_k^2 + 1$, we have

$$(x_1^2 + 1) \cdots (x_n^2 + 1) = (i^n + p_1 i^{n-1} + \cdots + p_{n-1}i + p_n)$$
$$\times ((-i)^n + p_1(-i)^{n-1} + \cdots + p_{n-1}(-i) + p_n).$$

If we set $a = 1 - p_2 + p_4 - \cdots$, $b = p_1 - p_3 + p_5 - \cdots$, then

$$(x_1^2 + 1) \cdots (x_n^2 + 1) = (i^n a + i^{n-1}b)((-i)^n a + (-i)^{n-1}b) = a^2 + b^2,$$

which was to be shown. (Furthermore, it follows that the identity holds also in the case where the coefficients $p_1, \ldots, p_n$ are complex numbers.)   □

7.9 Exercises

(i) Suppose that the polynomial $F(x) = x^n + a_{n-1}x^{n-1} + \cdots + a_1 x + a_0$ has zeros $x_1, x_2, \ldots, x_n$. Show that

$$(x_1^2 + x_1 + 1) \cdot (x_2^2 + x_2 + 1) \cdots (x_n^2 + x_n + 1)$$
$$= \frac{1}{2} \cdot [(A - B)^2 + (A - C)^2 + (B - C)^2],$$

with $A = a_0 + a_3 + \cdots$, $B = a_1 + a_4 + \cdots$, $C = a_2 + a_5 + \cdots$, where the sums are taken over indices not exceeding n, and where we understand that $a_n = 1$.

*(ii) Let x_1, x_2, x_3, x_4 be the roots of the polynomial

$$G(x) = x^4 + px^3 + qx^2 + rx + 1.$$

Show that

$$(x_1^4 + 1)(x_2^4 + 1)(x_3^4 + 1)(x_4^4 + 1) = (p^2 + r^2)^2 + q^4 - 4pq^2r.$$

2
Algebraic Inequalities

Inequalities are essential tools in many areas of mathematics. Orderings of numbers, which are obtained by way of inequalities, find applications even outside of mathematics in various theoretical and practical fields. Secondary-school curricula have therefore traditionally devoted a certain amount of attention to linear and quadratic inequalities, and to inequalities involving absolute values. Already less emphasis is given to the "calculus" of operations with inequalities and to methods and approaches for deriving estimates for algebraic expressions, and for finding proofs. The present chapter is devoted to exactly these questions.

The necessary basic results are summarized in Section 1. First we describe a way of introducing inequalities between real numbers, and then we derive some basic rules of arithmetic with inequalities. Further theorems and important results, connected with various methods, will be considered throughout the appropriate sections. The proofs in two particular sections (1.10 and 8.6) require a more exact concept of irrational numbers. Although in the text itself we restrict ourselves to the case of rational numbers, the reader can find in Section 9 the necessary supplementary remarks on the general case.

In the solutions to some examples, the letters L and R denote the left, resp. the right, sides of the inequalities under consideration.

1 Definitions and Properties

Already in early childhood, when our first ideas of the numbers $1, 2, 3, \ldots$ are formed, we know about "bigger" and "smaller"; this obviously occurs before we learn how to add or subtract these first numbers. Later we become aware of the fact that the difference $a - b$ makes sense only when "a is greater than b." We overcome this limitation by enlarging the collection of the original (positive) numbers $1, 2, 3, \ldots$ by new numbers, namely zero and the negative numbers $-1, -2, -3, \ldots$ (as numbers "opposite" to the positive numbers). Now we generalize this process: We construct the required theory of inequalities on the basis of dividing the integers into positive and negative numbers. In the following paragraphs we express the known properties of positive and negative numbers by way of the "rules of sign."

1.1 Positive and Negative Numbers

Definition. Every real number $x \neq 0$ is either positive or negative. The set $\mathbb{R}$ of all real numbers is therefore divided into three parts: the set $\mathbb{R}^+$ of all positive numbers, the set $\mathbb{R}^-$ of all negative numbers, and the singleton $\{0\}$. How is this subdivision preserved under arithmetic operations? We give only those rules that will be required later:

(i) $a \in \mathbb{R}^+ \wedge b \in \mathbb{R}^+ \implies (a + b) \in \mathbb{R}^+$,
(ii) $a \in \mathbb{R}^+ \wedge b \in \mathbb{R}^+ \implies ab \in \mathbb{R}^+$,
(iii) $a \in \mathbb{R}^- \iff (-a) \in \mathbb{R}^+$,
(iv) $a \in \mathbb{R}^+ \wedge b \in \mathbb{R}^- \implies ab \in \mathbb{R}^-$,
(v) $a \in \mathbb{R}^+ \iff \frac{1}{a} \in \mathbb{R}^+$.

A real number $x \notin \mathbb{R}^-$ is called nonnegative; the set of all nonnegative numbers is denoted by $\mathbb{R}_0^+$, so that $\mathbb{R}_0^+ = \mathbb{R}^+ \cup \{0\}$. We will often require a generalization of (i) to the case of a sum of several numbers from $\mathbb{R}_0^+$. We therefore list this as an additional rule:

(vi) If $a_1, a_2, \ldots, a_n$ are nonnegative numbers, then $s = a_1 + a_2 + \cdots + a_n$ is also nonnegative; furthermore, $s = 0$ if and only if $a_1 = a_2 = \cdots = a_n = 0$.

It is easy to derive (vi) from (i): The value of the sum s does not change if we leave out all summands a_i that are equal to zero.

1.2 Introducing Numerical Inequalities

Definition. We say that the number a is greater (resp. less) than the number b (written $a > b$, resp. $a < b$) if and only if $a - b$ is positive (resp.

negative):

$$a > b \iff (a - b) \in \mathbb{R}^+,$$
$$a < b \iff (a - b) \in \mathbb{R}^-. \tag{1}$$

Since the numbers $(a - b)$ and $(b - a)$ are mutually "opposite," it follows from 1.1.(iii) that $a > b$ is the same as $b < a$. From the decomposition $\mathbb{R} = \mathbb{R}^+ \cup \{0\} \cup \mathbb{R}^-$ we immediately obtain the *trichotomy law*: For an arbitrary pair of numbers $a, b \in \mathbb{R}$, one and only one of the relations $a > b$, $a = b$, $a < b$ is true. In the case that $a > b$ (resp. $a < b$) does not hold, we write $a \le b$ (resp. $a \ge b$). Inequalities with $<$ and $>$ (resp. $\le$ and $\ge$) are called *strict* (resp. *weak*). Thus, the weak inequality $a \ge b$ means that $(a - b) \in \mathbb{R}_0^+$.

If we represent the real numbers in the usual way by points on the number line oriented from left to right, then the inequality $a > b$ acquires geometrical meaning: the image of the number a lies to the right of the image of the number b. Thus, for example, we have $-1 > -2$, even though one sometimes inaccurately says "-2 is a larger negative number than -1." It is better to avoid this last expression.

Finally, we note that according to (1) with $b = 0$ a number is positive (resp. negative) if and only if it is greater (resp. less) than zero. Thus the notations $x \in \mathbb{R}^+$, $y \in \mathbb{R}_0^+$, and $z \in \mathbb{R}^-$ for real x, y, z also stand for $x > 0$, $y \ge 0$, and $z < 0$.

Now we use definition (1) and the rules 1.1.(i)–(vi) to derive all the basic properties of inequalities between numbers.

1.3   Transitivity

Theorem. *If $a > b$ and $b > c$, then $a > c$. More generally: If $a_1 \ge a_2$, $a_2 \ge a_3, \dots, a_{n-1} \ge a_n$, then $a_1 \ge a_n$, where $a_1 = a_n$ if and only if $a_1 = a_2 = \cdots = a_n$.*

PROOF.   If $a > b$ and $b > c$, that is, $(a - b) \in \mathbb{R}^+$ and $(b - c) \in \mathbb{R}^+$, then by 1.1.(i) we have $(a - b) + (b - c) \in \mathbb{R}^+$, that is, $a > c$. The more general statement follows from 1.1.(vi) and the equality $a_1 - a_n = (a_1 - a_2) + (a_2 - a_3) + \cdots + (a_{n-1} - a_n)$.   □

We remark that we can write the pair of inequalities $a > b$, $b > c$ more concisely as $a > b > c$; similarly, we often write the system of inequalities $a_1 \ge a_2$, $a_2 \ge a_3, \dots, a_{n-1} \ge a_n$ as the chain $a_1 \ge a_2 \ge \cdots \ge a_n$. However, the notation $x < y > z$ is not customary.

1.4   Adding a Number

Theorem. *If $a > b$, then $a + c > b + c$ for each c.*

PROOF.   If $a > b$, then $(a + c) - (b + c) = (a - b) \in \mathbb{R}^+$, that is, $a + c > b + c$.   □

We note that upon replacing the number c by $-c$ we obtain the rule of "subtracting a number": If $a > b$, then $a - c > b - c$ for any c.

1.5   Adding Inequalities

Theorem. *If $a > b$ and $c > d$, then $a + c > b + d$. More generally: If $a_1 \geq b_1$, $a_2 \geq b_2, \ldots, a_n \geq b_n$, then*

$$a_1 + a_2 + \cdots + a_n \geq b_1 + b_2 + \cdots + b_n, \tag{2}$$

where we have equality in (2) if and only if $a_1 = b_1, a_2 = b_2, \ldots, a_n = b_n$.

PROOF.   If $a > b$ and $c > d$, then by 1.4 we have $a + c > b + c$ and $b + c > b + d$, which by 1.3 gives $a + c > b + d$. The more general assertion follows from 1.1.(vi) and the equation

$$(a_1 + a_2 + \cdots + a_n) - (b_1 + b_2 + \cdots + b_n)$$
$$= (a_1 - b_1) + (a_2 - b_2) + \cdots + (a_n - b_n). \qquad □$$

1.6   Multiplying by a Number

Theorem. *If $a > b$, then $ac > bc$ for any $c > 0$, and $ac < bc$ for any $c < 0$.*

PROOF.   If $a > b$, then $(a - b) \in \mathbb{R}^+$. By 1.1.(ii) we have $(a - b)c \in \mathbb{R}^+$ whenever $c \in \mathbb{R}^+$, and by 1.1.(iv) we have $(a - b)c \in \mathbb{R}^-$ whenever $c \in \mathbb{R}^-$. Since $(a - b)c = ac - bc$, in the first case we get $ac > bc$, while in the second case it is $ac < bc$.   □

We note that upon replacing the number c by $1/c$ we obtain, in view of 1.1.(v), the rule of "dividing by a number": If $a > b$, then $a/c > b/c$ for $c > 0$, and $a/c < b/c$ for $c < 0$.

1.7   Subtracting Inequalities

Theorem. *If $a > b$ and $c > d$, then $a - d > b - c$. More generally: If $a \geq b$ and $c \geq d$, then $a - d \geq b - c$, where equality $a - d = b - c$ occurs if and only if $a = b$ and $c = d$.*

PROOF.   By 1.6 the inequality $c > d$ holds if and only if $-c < -d$. Hence 1.7 follows from 1.5 for the pair of inequalities $a \geq b$ and $-d \geq -c$.   □

1.8 Multiplying Inequalities

Theorem. *If $a > b > 0$ and $c > d > 0$, then $ac > bd$. More generally: If $a_1 \geq b_1 > 0, a_2 \geq b_2 > 0, \ldots, a_n \geq b_n > 0$, then*

$$a_1 a_2 \cdots a_n \geq b_1 b_2 \cdots b_n, \tag{3}$$

with equality if and only if $a_1 = b_1, a_2 = b_2, \ldots, a_n = b_n$.

PROOF. From $a_1 \geq b_1 > 0$ and $a_2 \geq b_2 > 0$ it follows by 1.6 that $a_1 a_2 \geq b_1 a_2$ and $b_1 a_2 \geq b_1 b_2$. Hence 1.3 implies $a_1 a_2 \geq b_1 b_2$. Since $b_1 b_2 > 0$ by 1.1.(ii), we can repeat this argument for the case $n \geq 3$. Thus we obtain $a_1 a_2 a_3 \geq b_1 b_2 b_3$, etc., and finally (3), where the last step is the chain of inequalities

$$a_1 a_2 \cdots a_{n-1} a_n \geq a_1 a_2 \cdots a_{n-1} b_n \geq b_1 b_2 \cdots b_n. \tag{4}$$

If equality occurs in (3), then by (4) we have $a_1 a_2 \cdots a_{n-1} a_n = a_1 a_2 \cdots a_{n-1} b_n$. This implies $a_n = b_n$ (since $a_1 a_2 \cdots a_{n-1} \neq 0$) and by (4), $a_1 a_2 \cdots a_{n-1} = b_1 b_2 \cdots b_{n-1}$. Repeating this argument, we obtain $a_{n-1} = b_{n-1}$, etc., and finally $a_1 = b_1$.   □

1.9 Dividing Inequalities

Theorem. *If $a > b > 0$ and $c > d > 0$, then $\frac{a}{d} > \frac{b}{c}$. More generally: If $a \geq b > 0$ and $c \geq d > 0$, then $\frac{a}{d} \geq \frac{b}{c}$, where equality occurs if and only if $a = b$ and $c = d$. In particular, for $a = b = 1$ we get that if $c > d > 0$, then $\frac{1}{d} > \frac{1}{c}$.*

PROOF. Since $cd > 0$, by 1.6 we have $a/d > b/c$ if and only if $ac > bd$. Hence 1.9 follows from 1.8.   □

1.10 Exponentiating

Theorem. *If $a > b > 0$, then $a^m > b^m$ and $\sqrt[m]{a} > \sqrt[m]{b}$ for each integer $m \geq 2$. More generally: If $a > b > 0$, then $a^r > b^r$ for each $r > 0$, and $a^r < b^r$ for each $r < 0$.*

PROOF. Let $a > b > 0$. Using 1.8 with m identical inequalities $a > b$, we obtain $a^m > b^m$. Let us assume that $\sqrt[m]{a} \leq \sqrt[m]{b}$; then by the above, $(\sqrt[m]{a})^m \leq (\sqrt[m]{b})^m$, that is, $a \leq b$, which is a contradiction. If $r > 0$ and $r \in \mathbb{Q}$ (the case $r \in \mathbb{R}^+ \setminus \mathbb{Q}$ will be discussed in Section 9), then $r = m/n$ ($m, n \in \mathbb{N}$), and by the above we obtain $a^m > b^m$; hence $\sqrt[n]{a^m} > \sqrt[n]{b^m}$, that is, $a^r > b^r$. If finally $r < 0$, then by the above we have $a^{-r} > b^{-r}$ (since $-r > 0$), which by 1.9 means that $1/a^{-r} < 1/b^{-r}$, that is, $a^r < b^r$.   □

1.11 Exponentiating Inequalities

Theorem. If $a > b$ and $0 < c < 1 < d$, then $c^a < c^b$ and $d^a > d^b$.

PROOF. Let $0 < c < 1 < d$ and $r = a - b > 0$. Then by 1.10 we have $c^r < 1^r < d^r$, that is, $c^{a-b} < 1$ and $d^{a-b} > 1$. If we multiply the last two inequalities by c^b, resp. d^b, then by 1.6 we obtain $c^a < c^b$ and $d^a > d^b$.   □

To conclude this section, the properties we have thus far obtained are summarized in the following table.

$$a > b \wedge b > c \implies a > c$$
$$a > b \implies a + c > b + c \wedge a - c > b - c$$
$$a > b \wedge c > d \implies a + c > b + d$$
$$a > b \wedge c > 0 \implies ac > bc \wedge \frac{a}{c} > \frac{b}{c}$$
$$a > b \wedge c < 0 \implies ac < bc \wedge \frac{a}{c} < \frac{b}{c}$$
$$a > b \wedge c > d \implies a - d > b - c$$
$$a > b > 0 \wedge c > d > 0 \implies ac > bd \wedge \frac{a}{d} > \frac{b}{c}$$
$$a > b > 0 \implies \frac{1}{a} < \frac{1}{b}$$
$$a > b > 0 \wedge r > 0 \implies a^r > b^r$$
$$a > b > 0 \wedge r < 0 \implies a^r < b^r$$
$$a > b \wedge 0 < c < 1 \implies c^a < c^b$$
$$a > b \wedge d > 1 \implies d^a > d^b$$

2 Basic Methods

We will now show how the basic rules obtained in Section 1 can be used to solve some easy problems involving inequalities.

2.1 Equivalent Transformations

We begin by considering five problems that will be solved as follows: The inequality that is to be proved will be transformed in a succession of steps until we reach an inequality that is obviously true. In doing so it is important to make sure that each step is *equivalent*, that is, the validity or failure of the inequality is not changed by the operation. According to Section 1, such operations include, for example, adding the same expression to both sides of the inequality, multiplying the inequality by a positive expression, or raising an inequality between positive terms to the rth power

$(r > 0)$. Also, we must not forget algebraic transformations of both sides of an inequality. As a "template" for a proof we often choose the opposite direction: We transform the final, "obvious," inequality in a succession of steps until we arrive at the original inequality that was to be proved.

(i) Let us show that $\sqrt[8]{8!} < \sqrt[9]{9!}$. Since both numbers are positive, by 1.10 it suffices to show that $(\sqrt[8]{8!})^{72} < (\sqrt[9]{9!})^{72}$, that is, $(8!)^9 < (9!)^8$. Since $9! = 9 \cdot 8!$, we can transform this inequality to $(8!)^9 < 9^8 \cdot (8!)^8$. Upon division by $(8!)^8$ (using 1.6) we obtain $8! < 9^8$. This last inequality holds by 1.8, since it is the product of the eight inequalities $1 < 9, 2 < 9, \ldots, 8 < 9$.

(ii) Let us decide which one of the three numbers

$$x = (a+b)(c+d), \quad y = (a+c)(b+d), \quad z = (a+d)(b+c)$$

is largest, under the assumption that $a < b < c < d$. We evaluate the difference $y - x$:

$$y - x = (ab + cb + ad + cd) - (ac + bc + ad + bd)$$
$$= d(c - b) - a(c - b) = (d - a)(c - b) > 0,$$

since both numbers $(d - a)$ and $(c - b)$ are positive. Hence $y > x$. Similarly, we convince ourselves that $z > y$. Therefore, z is the largest of the numbers.

(iii) We now show that for each $a > 1$ we have the inequality

$$\frac{1}{\sqrt{a}} < \sqrt{a+1} - \sqrt{a-1}. \tag{5}$$

Since $a + 1 > a - 1 > 0$, by 1.10 we have $\sqrt{a+1} > \sqrt{a-1}$. Both sides of (5) are therefore positive; hence by 1.10 we can compare their squares:

$$\frac{1}{a} < (\sqrt{a+1} - \sqrt{a-1})^2.$$

After multiplying out and simplifying we get

$$2a - \frac{1}{a} > 2\sqrt{a^2 - 1}.$$

Both sides are again positive $(2a > 2 > 1 > 1/a)$; by squaring we obtain

$$4a^2 - 4 + \frac{1}{a^2} > 4(a^2 - 1),$$

and after subtracting the expression $4(a^2 - 1)$ we get the correct inequality $1/a^2 > 0$.

(iv) Let the numbers $a, b, r, s \in \mathbb{R}^+$ satisfy $a > b$ and $r > s$. We will prove the inequality

$$(a^s + b^s)(a^r - b^r) > (a^r + b^r)(a^s - b^s).$$

Multiplying out both sides and subtracting $(a^{s+r} - b^{s+r})$, we obtain $a^r b^s - a^s b^r > a^s b^r - a^r b^s$, that is, $2a^r b^s > 2a^s b^r$, and after dividing by $2a^s b^s$ we finally get $a^{r-s} > b^{r-s}$. But this holds by 1.10, since $a > b$ and $r - s > 0$.

(v) We will show that for arbitrary $a, b, r, s \in \mathbb{R}^+$ we have

$$a^{r+s} + b^{r+s} > a^r b^s + a^s b^r$$

if $a \neq b$. The difference of the left and the right sides is

$$L - R = a^r(a^s - b^s) - b^r(a^s - b^s) = (a^r - b^r)(a^s - b^s).$$

What can we say about the signs of the two numbers $A = a^r - b^r$ and $B = a^s - b^s$? By 1.10 we know that if $a > b$ (resp. $a < b$), then both A and B are positive (resp. negative). Therefore, in the case $a \neq b$ we always have $L - R > 0$, that is, $L > R$.

We remark that similar "discussions" of the values of parameters are often used in proofs of inequalities (see also 2.3.(iv) below).

2.2 Exercises

(i) Generalize the result of 2.1.(i).

(ii) Which of the numbers 2^{700} and 5^{300} is larger?

(iii) Which number is larger:

$$\frac{10^{1997} + 1}{10^{1998} + 1} \quad \text{or} \quad \frac{10^{1998} + 1}{10^{1999} + 1} \quad ?$$

Prove the inequalities (iv)–(viii):

(iv) $\sqrt{b} > a + \dfrac{b - a^2}{2a + 1}$ $(0 \leq a < \sqrt{b} < a + 1)$.

(v) $\dfrac{1}{\sqrt{a}} > 2(\sqrt{a+1} - \sqrt{a})$ $(a > 0)$.

(vi) $\dfrac{c + a}{\sqrt{c^2 + a^2}} > \dfrac{c + b}{\sqrt{c^2 + b^2}}$ $(0 < b < a, \ \sqrt{ab} < c)$.

(vii) $\sqrt[12]{a^7} + \sqrt[12]{b^7} \geq \sqrt[3]{a}\sqrt[4]{b} + \sqrt[4]{a}\sqrt[3]{b}$ $(a, b \in \mathbb{R}^+)$.

(viii) $(n!)^2 < k!(2n - k)!$ $(n, k \in \mathbb{N}, \ n > k)$.

2.3 Irreversible Transformations

We will now deal with problems in which the course of solution from an original "obvious" inequality to the desired inequality cannot be reversed. Thus, for example, the inequality $L \geq R$ can be proved by finding decompositions $L = L_1 + L_2 + \cdots + L_n$ and $R = R_1 + R_2 + \cdots + R_n$ such that

$L_k \geq R_k$ for each $k = 1, 2, \ldots, n$. (This may be a tough nut to crack even in the case $n = 2$). A similar approach is provided by the multiplication law 1.8. Let us now consider the following eight problems.

(i) Prove the inequality

$$\frac{a+b}{1+a+b} < \frac{a}{1+a} + \frac{b}{1+b} \qquad (a, b \in \mathbb{R}^+).$$

SOLUTION. The fraction on the left has the largest denominator. We can therefore write

$$L = \frac{a+b}{1+a+b} = \frac{a}{1+a+b} + \frac{b}{1+a+b} < \frac{a}{1+a} + \frac{b}{1+b} = R. \qquad \square$$

(ii) Let c be the hypotenuse of a right triangle, and a, b the other two sides. Show that for every integer $k > 2$ we have $a^k + b^k < c^k$.

SOLUTION. By the Pythagorean theorem we have $c^2 = a^2 + b^2$. Multiplying by c^{k-2}, we obtain $c^k = a^2 c^{k-2} + b^2 c^{k-2}$. Since $a < c$ and $k > 2$, by 1.10 we have $a^{k-2} < c^{k-2}$; hence $a^k = a^2 a^{k-2} < a^2 c^{k-2}$. Similarly, we get $b^k < b^2 c^{k-2}$, and by adding these two inequalities,

$$a^k + b^k < a^2 c^{k-2} + b^2 c^{k-2} = c^k. \qquad \square$$

(iii) Let $a > b > 0$ and $n \in \mathbb{N}$. Determine which of the numbers

$$A = \frac{1 + a + a^2 + \cdots + a^n}{1 + a + a^2 + \cdots + a^{n-1}} \quad \text{and} \quad B = \frac{1 + b + b^2 + \cdots + b^n}{1 + b + b^2 + \cdots + b^{n-1}}$$

is larger.

SOLUTION. We rewrite both numbers according to the identity

$$\frac{1 + x + x^2 + \cdots + x^n}{1 + x + x^2 + \cdots + x^{n-1}} = 1 + \frac{x^n}{1 + x + x^2 + \cdots + x^{n-1}}$$

$$= 1 + \frac{1}{\frac{1}{x^n} + \frac{1}{x^{n-1}} + \cdots + \frac{1}{x}}.$$

From $a > b > 0$ it follows by 1.10 that $a^{-k} < b^{-k} (1 \leq k \leq n)$. By adding we get

$$\frac{1}{a} + \frac{1}{a^2} + \cdots + \frac{1}{a^n} < \frac{1}{b} + \frac{1}{b^2} + \cdots + \frac{1}{b^n},$$

which leads to the conclusion that $A > B$. $\qquad \square$

(iv) Show that $x^6 + 2 \geq x^4 + 2x$ for all $x \in \mathbb{R}$.

SOLUTION. The given inequality is an equality for $x = 1$; hence the polynomial $L - R$ is a multiple of $(x - 1)$ (see Chapter 1, Section 3). By dividing we find that $(x^6 + 2) - (x^4 + 2x) = (x - 1)Q(x)$, where $Q(x) = x^5 + x^4 - 2$. Therefore, it remains to show that $Q(x) \geq 0$ for $x > 1$ and $Q(x) \leq 0$ for $x < 1$. Since $Q(1) = 0$, one could repeat the division of the polynomial by $(x - 1)$; however, in view of the form of $Q(x)$, the following discussion suggests itself:

$$x > 1 \Longrightarrow x^5 > 1 \wedge x^4 > 1 \Longrightarrow Q(x) > 1 + 1 - 2 = 0,$$
$$-1 < x < 1 \Longrightarrow x^5 < 1 \wedge x^4 < 1 \Longrightarrow Q(x) < 1 + 1 - 2 = 0,$$
$$x \leq -1 \Longrightarrow x^4 > 0 \wedge 1 + x \leq 0 \Longrightarrow Q(x) = x^4(1 + x) - 2 \leq -2.$$

This completes the proof. $\square$

We remark that the search for linear factors of a polynomial is a basic method for solving problems involving polynomial inequalities. We will use this approach also for solving the following problem.

(v) Let $a, b, c \in \mathbb{R}^+$ be such that $c > a + b$. Show that

$$a^3 + b^3 + c^3 + 3abc > 2(a + b)^2 c.$$

SOLUTION.   With the notation $F(c) = L - R = c^3 + 3abc - 2(a+b)^2 c + a^3 + b^3$ we will show that $F(x) > 0$ for all $x > a + b$. Since

$$F(a + b) = (a + b)^3 + 3ab(a + b) - 2(a + b)^3 + a^3 + b^3 = 0,$$

the polynomial $F(x)$ is divisible by $(x - a - b)$; by dividing we obtain

$$F(x) = (x - a - b)[x^2 + (a + b)x - a^2 + ab - b^2].$$

Hence it suffices to show that the term in square brackets is positive for $x > a + b$. But this is easy: For such x and positive a, b we have

$$x^2 + (a + b)x - a^2 + ab - b^2 \geq (a + b)^2 + (a + b)^2 - a^2 + ab - b^2$$
$$= a^2 + 5ab + b^2 > 0.   \square$$

(vi) Show that for each $n \in \mathbb{N}$ and each $a \in \mathbb{R}^+$, $a \neq 1$, we have the inequality

$$n(a^{2n+1} + 1) > a + a^2 + \cdots + a^{2n}. \tag{6}$$

SOLUTION.   Since both sides of (6) are sums of $2n$ numbers (on the left we have n-times the number a^{2n+1} and n-times the number 1), we first check whether (6) is a sum of $2n$ valid inequalities. For $a > 1$ we have the chain $a^{2n+1} > a^{2n} > a^{2n-1} > \cdots > a^2 > a > 1$ (for positive $a < 1$ the chain goes in the opposite direction); this, however, does not imply (6). Therefore, we try another possibility, namely the sum of n inequalities $a^{2n+1} + 1 \geq R_k$ $(1 \leq k \leq n)$, where $R_1, R_2, \ldots, R_n$ are appropriate sums of two terms each from the right-hand side R of the inequality (6), with $R = R_1 + R_2 + \cdots + R_n$. Obviously, it is best to combine "symmetric" terms: the smallest with the largest, etc., that is, to set $R_k = a^k + a^{2n+1-k}$ for $k = 1, 2, \ldots, n$. This attempt is successful, since

$$a^{2n+1} + 1 - R_k = a^{2n+1} + 1 - (a^k + a^{2n+1-k})$$
$$= (a^{2n-k+1} - 1)(a^k - 1) > 0. \qquad \square$$

(vii) Show that if $0 < a < 1$, then $(1+a)^{1-a} \cdot (1-a)^{1+a} < 1$.

SOLUTION.   If we set $b = (1-a)/(1+a)$, then

$$(1+a)^{1-a} \cdot (1-a)^{1+a} = (1+a)(1-a)b^a = (1-a^2)b^a < b^a,$$

since $0 < 1 - a^2 < 1$ if $0 < a < 1$. Next, since $0 < b < 1$, 1.11 implies $b^a < 1$. $\qquad \square$

(viii) Show that for arbitrary numbers $a, b, c \in \mathbb{R}^+$ we have

$$(a^a \cdot b^b \cdot c^c)^2 \geq a^{b+c} \cdot b^{c+a} \cdot c^{a+b}.$$

SOLUTION.   First we note that for any $a, b \in \mathbb{R}^+$ we have $a^a \cdot b^b \geq a^b \cdot b^a$. Indeed, this inequality can be written in the equivalent form $a^{a-b} \geq b^{a-b}$; now it suffices to distinguish between the cases $a > b$, $a = b$, $a < b$ and to use rule 1.10. By multiplying the three valid inequalities

$$a^a \cdot b^b \geq a^b \cdot b^a, \quad b^b \cdot c^c \geq b^c \cdot c^b, \quad a^a \cdot c^c \geq a^c \cdot c^a,$$

we then obtain the desired inequality. $\qquad \square$

2.4   Exercises

Prove the inequalities (i)–(iii):

(i) $\dfrac{1}{a+b} + \dfrac{1}{b+c} + \dfrac{1}{c+a} > \dfrac{3}{a+b+c}$ $\qquad (a, b, c \in \mathbb{R}^+)$.

(ii) $\dfrac{x_1 + x_2 + \cdots + x_6}{6} \leq \dfrac{x_1 + x_2 + \cdots + x_{10}}{10}$ $\qquad (x_1 \leq \cdots \leq x_{10})$.

(iii) $\dfrac{(1+a)(1+b)}{2+a+b} < \dfrac{1+a+b}{2}$ $\qquad (a, b \in \mathbb{R}^+)$.

(iv) Show that if the numbers a, b, c are the side lengths of some triangle, then the same is true for the triples $\sqrt{a}$, $\sqrt{b}$, $\sqrt{c}$ and $(a+b)^{-1}$, $(b+c)^{-1}$, $(c+a)^{-1}$.

Prove the inequalities (v)–(ix):

(v) $x^8 - x^5 + x^2 - x + 1 > 0$ $\qquad (x \in \mathbb{R})$.

(vi) $\dfrac{1+a+a^2+\cdots+a^n}{a+a^2+a^3+\cdots+a^{n-1}} \geq \dfrac{n+1}{n-1}$ $(a \in \mathbb{R}^+,\ n \in \mathbb{N},\ n \geq 2)$.

*(vii) $\left(m+\frac{3}{2}\right)\left(m+\frac{7}{2}\right)\cdots\left(m+\frac{4k-1}{2}\right) > \sqrt{\frac{(m+2k)!}{m!}}$ $(k, m \in \mathbb{N})$.

*(viii) $a^3+b^3+c^3+3abc \geq ab(a+b)+bc(b+c)+ac(a+c)$ $(a, b, c \in \mathbb{R}^+)$.

(ix) $(1+na)^{n+1} > [1+(n+1)a]^n$ $(a \in \mathbb{R}^+,\ n \in \mathbb{N})$.
(Hint: Use the identity

$$1+(n+1)a = \dfrac{1+(n+1)a}{1+na} \cdot \dfrac{1+na}{1+(n-1)a} \cdots \dfrac{1+2a}{1+a} \cdot \dfrac{1+a}{1},$$

and change the product of the $n+1$ terms on the right to a power of the smallest one of these terms.)

(x) Which one of the numbers 31^{11} and 17^{14} is larger?

*(xi) Which one of the numbers $2^{3^{100}}$ and $3^{2^{150}}$ is larger?

(xii) Using a similar trick as in example 2.3.(iii), show that the equation

$$x^n - a_1 x^{n-1} - a_2 x^{n-2} - \cdots - a_{n-1}x - a_n = 0$$

with nonnegative coefficients a_j cannot have two distinct positive roots.

(xiii) Use the method of Example 2.3.(v) to show that if $a, b, c \in \mathbb{R}^+$ and $c < a+b$, then

$$a^3 + b^3 + c^3 + 3abc > 2(a+b)c^2.$$

*(xiv) Let $a, b, c, d \in \mathbb{R}^+$. Show that of the three inequalities

$$a+b < c+d, \quad (a+b)(c+d) < ab+cd, \quad (a+b)cd < ab(c+d),$$

at least one is false.

*(xv) Let $A, B, C, a, b, c \in \mathbb{R}^+$ and $s \geq a+b+c$. Show that the numbers

$$\dfrac{A+B+a+b}{A+B+s}, \quad \dfrac{B+C+b+c}{B+C+s}, \quad \dfrac{A+C+a+c}{A+C+s}$$

are the side lengths of some triangle. (Hint: Use the fact that if $0 < x < y$, then $x/y < (x+p)/(y+p)$ for any $p > 0$.)

*(xvi) For fixed positive numbers $a_1, a_2, \ldots, a_n$ set

$$A(r) = a_1^r + a_2^r + \cdots + a_n^r \quad (r \in \mathbb{R}).$$

Show that if $r < s \leq t < u$ and $r+u = s+t$, then $A(r)A(u) > A(s)A(t)$, with the exceptional case $a_1 = a_2 = \cdots = a_n$.

2.5 The Estimation Method

In mathematics it is often required to estimate the value of some "complicated" expression Q. This can be done, for example, by replacing it with a simpler expression (with an error as small as possible). The numbers or, more generally, the expressions L and U satisfying the inequality $L \leq Q \leq U$ are called the *lower* and *upper bound* of the expression Q; the lower bound L_1 is called *sharper* than the bound L if $L < L_1 \leq Q$. Similarly, we define a sharper upper bound. The following seven examples will illustrate some simple approaches to obtaining bounds (especially of finite sums).

(i) Let us find bounds for the expression

$$R = \frac{a}{a+b+d} + \frac{b}{a+b+c} + \frac{c}{b+c+d} + \frac{d}{a+c+d} \qquad (a,b,c,d \in \mathbb{R}^+).$$

It is clear that $0 < R < 4$; by numerical experimentation we are unable to find concrete values a, b, c, d such that $R \leq 1$ or $R \geq 2$. In fact, such values do not exist, for we derive the bounds $R > 1$ and $R < 2$:

$$R > \frac{a}{a+b+c+d} + \frac{b}{a+b+c+d} + \frac{c}{a+b+c+d} + \frac{d}{a+b+c+d} = 1,$$

$$R < \frac{a}{a+b} + \frac{b}{a+b} + \frac{c}{c+d} + \frac{d}{c+d} = 2.$$

By appropriate choices of numbers a, b, c, d one can show that there are no sharper bounds (see [7], p. 165).

(ii) The easiest approach to obtaining bounds for the sum $a_1 + a_2 + \cdots + a_n$ is as follows: We find the smallest and the largest among the values $a_1, a_2, \ldots, a_n$; these, multiplied by n, are then bounds for the given sum. Thus, for the sum

$$S(n) = 1 + \frac{1}{\sqrt{2}} + \frac{1}{\sqrt{3}} + \cdots + \frac{1}{\sqrt{n}} \qquad (n > 1)$$

we obtain $S(n) > n \cdot (1/\sqrt{n}) = \sqrt{n}$ and $S(n) < n \cdot 1 = n$. Sharper bounds for $S(n)$ will be obtained in 2.5.(iv) and 2.6.(iv) below.

(iii) The method of the previous example can sometimes be improved by first subdividing the terms a_j of the sum $a_1 + a_2 + \cdots + a_n$ into smaller groups (for instance pairs); then we find bounds for the sums of these groupings, and finally we add these bounds. We use this method to prove the inequality

$$\frac{1}{n+1} + \frac{1}{n+2} + \frac{1}{n+3} + \cdots + \frac{1}{2n} < \frac{3}{4}.$$

For $n > 1$ we pair the summands as follows:

$$\left(\frac{1}{n+1} + \frac{1}{2n-1}\right) + \left(\frac{1}{n+2} + \frac{1}{2n-2}\right) + \left(\frac{1}{n+3} + \frac{1}{2n-3}\right) + \cdots .$$

To avoid having to distinguish between even and odd n, we multiply the original inequality by two and write it in the form

$$\sum_{k=1}^{n-1}\left(\frac{1}{n+k} + \frac{1}{2n-k}\right) < \frac{3}{2} - \frac{1}{n}. \tag{7}$$

For $1 \le k \le n-1$ we have

$$\frac{1}{n+k} + \frac{1}{2n-k} = \frac{3n}{2n^2 + k(n-k)} < \frac{3n}{2n^2 + 0} = \frac{3}{2n}.$$

Hence the sum on the left-hand side of (7) is less than

$$(n-1)\frac{3}{2n} = \frac{3n-3}{2n} < \frac{3n-2}{2n} = \frac{3}{2} - \frac{1}{n}.$$

(iv) If we sum the inequalities

$$\frac{1}{\sqrt{k}} < \sqrt{k+1} - \sqrt{k-1} \qquad (2 \le k \le n)$$

(see 2.1.(iii)), we obtain a new upper bound for the sum $S(n)$ in 2.5.(ii) that is considerably sharper: $S(n) < \sqrt{n+1} + \sqrt{n} - \sqrt{2}$. It follows from 2.6.(iv) below that the error of this estimate for $n = 10^6$ is less than 1.

(v) We show that for arbitrary $n, k \in \mathbb{N}$ we have

$$\frac{1}{(n+1)!} + \frac{1}{(n+2)!} + \cdots + \frac{1}{(n+k)!} \le \frac{1}{n}\left[\frac{1}{n!} - \frac{1}{(n+k)!}\right]. \tag{8}$$

Indeed, if $1 \le j \le k$, then

$$\frac{n}{(n+j)!} \le \frac{n+j-1}{(n+j)!} = \frac{1}{(n+j-1)!} - \frac{1}{(n+j)!}.$$

Adding these estimates and dividing by n, we obtain (8).

(vi) A number of estimates for finite sums can be derived from results of Chapter 1, Section 2. Thus, for example, we obtain the bound

$$T(n) = \frac{1}{1^2} + \frac{1}{2^2} + \frac{1}{3^2} + \cdots + \frac{1}{n^2} < \frac{7}{4} - \frac{1}{n} \qquad (n > 2)$$

as follows: We keep the first two terms in $T(n)$ without change and estimate the remaining summands using the inequality $1/k^2 < 1/(k^2-k)$, $3 \le k \le n$; finally, we apply the result of Exercise 2.40.(i) in Chapter 1 for $k = 2$.

(vii) Finally we mention an example of an estimate for a finite product. For the expression

$$Q(n) = \frac{1}{2} \cdot \frac{3}{4} \cdot \frac{5}{6} \cdots \frac{2n-1}{2n} \qquad (n \ge 2)$$

we will prove the bounds

$$\frac{1}{2\sqrt{n}} < Q(n) < \frac{1}{\sqrt{2n+1}}. \tag{9}$$

Since $\frac{1}{2} < \frac{2}{3}$ and more generally $(2k-1)/2k < 2k/(2k+1)$ for each $k \ge 1$, we have

$$Q(n) = \frac{1}{2} \cdot \frac{3}{4} \cdot \frac{5}{6} \cdots \frac{2n-1}{2n} < \frac{2}{3} \cdot \frac{4}{5} \cdots \frac{2n}{2n+1} = \frac{1}{(2n+1)Q(n)}.$$

This implies $(2n+1)[Q(n)]^2 < 1$, which gives the right-hand side of (9). If we use the inequalities $\frac{3}{4} > \frac{2}{3}$ and more generally $(2k-1)/2k > (2k-2)/(2k-1)$ for each $k \ge 2$, we get

$$2Q(n) = \frac{3}{4} \cdot \frac{5}{6} \cdots \frac{2n-1}{2n} > \frac{2}{3} \cdot \frac{4}{5} \cdots \frac{2n-2}{2n-1} = \frac{1}{2nQ(n)};$$

hence $4n[Q(n)]^2 > 1$, which is the left-hand part of (9).

2.6 Exercises

Prove the inequalities (i)–(iii) for integers $n > 2$:

(i) $(n-1)\sqrt{2} < \sqrt{2} + \sqrt{3} + \cdots + \sqrt{n} < (n-1)\sqrt{n}$.

(ii) $(n!)^2 > n^n$.

(iii) $\displaystyle\sum_{k=n+1}^{n^2} \frac{1}{k} > \sum_{k=2}^{n} \frac{1}{k}$ and $\displaystyle\sum_{k=n}^{n^2-1} \frac{1}{k} < \sum_{k=1}^{n-1} \frac{1}{k}$.

(iv) Using the result of 2.2.(v), find a lower bound for the sum $S(n)$ in Example 2.5.(ii). Then determine the integer N for which the estimate $N < S(10^6) < N+1$ holds.

(v) With the help of the estimate in 2.5.(vi) prove, for each integer $n \ge 2$, the inequality

$$\frac{1}{2 \cdot 5} + \frac{1}{3 \cdot 7} + \frac{1}{4 \cdot 9} + \cdots + \frac{1}{n(2n+1)} < \frac{3}{8} - \frac{1}{2n}.$$

(vi) Using the result of Exercise 2.40.(v) in Chapter 1, prove the following estimate for each integer $n \ge 2$:

$$\frac{1}{2^3} + \frac{1}{3^3} + \cdots + \frac{1}{n^3} < \frac{1}{4} - \frac{1}{2(n^2+n)}.$$

(vii) Using the result of Exercise 2.40.(vi) in Chapter 1, prove the following estimate for each integer $n \geq 2$:

$$\frac{1}{2 \cdot 3} + \frac{1}{3 \cdot 5} + \frac{1}{4 \cdot 7} + \cdots + \frac{1}{n(2n-1)} < \frac{1}{2} - \frac{4n-1}{2(4n^2-1)}.$$

(viii) Show that for each integer $n \geq 1$ we have the bounds

$$n(2n+1) \leq \frac{1 \cdot 3}{1} + \frac{3 \cdot 5}{2} + \frac{5 \cdot 7}{3} + \cdots + \frac{(2n-1)(2n+1)}{n} < 2n(n+1).$$

(ix) For each integer $n \geq 1$, prove the inequality

$$\frac{3}{5} \cdot \frac{7}{9} \cdot \frac{11}{13} \cdots \frac{4n-1}{4n+1} < \sqrt{\frac{3}{4n+3}}.$$

(x) Let $p \in \mathbb{N}$. Show that for the expression $Q(n)$ from 2.5.(vii) we have

$$Q(p)\sqrt{\frac{p}{n}} \leq Q(n) \leq Q(p)\sqrt{\frac{2p+1}{2n+1}} \quad \text{for each } n \geq p.$$

2.7 Symmetry and Homogeneity

We will now mention two important properties of inequalities, which will be used in the following sections.

The first two of the inequalities

$$a+b+c > abc, \qquad \frac{a}{b} + \frac{b}{c} + \frac{c}{a} > 1, \qquad a+2b+c > \frac{ac}{b}$$

are (as opposed to the third one) *symmetric* in the variables a, b, c, that is, they do not change if the order of the triple a, b, c is changed. (Note that the third inequality is symmetric in the variables a, c.) Symmetries in the variables $x_1, x_2, \ldots, x_n$ are sometimes used as follows: We may assume that the variables $x_1, x_2, \ldots, x_n$ are appropriately *ordered*, for instance $x_1 \geq x_2 \geq \cdots \geq x_n$. Such an assumption can considerably shorten the arguments in the course of a proof. We illustrate this by the example of the inequality

$$(a+b+c)^{100} < 3^{100}\left(a^{100} + b^{100} + c^{100}\right), \tag{10}$$

which, as we will see, holds for arbitrary $a, b, c \in \mathbb{R}^+$. If $a \geq b \geq c$, then $a+b+c \leq a+a+a = 3a$, and thus

$$(a+b+c)^{100} \leq (3a)^{100} < 3^{100}\left(a^{100} + b^{100} + c^{100}\right).$$

This completes the proof. It may be interesting to note that, excepting the case $a = b = c$, a variant of (10), still with a strict inequality, holds even when the number 3^{100} on the right is replaced by 3^{99} (see 7.8.(v) below).

We explain the term *homogeneous inequality* by way of the example

$$a^8 + b^8 + c^8 \geq \left(\frac{1}{a} + \frac{1}{b} + \frac{1}{c} \right) a^3 b^3 c^3.$$

We find out how both sides of this inequality change if we replace a, b, c by ta, tb, tc for an arbitrary $t \in \mathbb{R}^+$:

$$L' = (ta)^8 + (tb)^8 + (tc)^8 = t^8 L,$$

$$R' = \left(\frac{1}{ta} + \frac{1}{tb} + \frac{1}{tc} \right) (ta)^3 (tb)^3 (tc)^3 = t^8 R.$$

This leads to an equivalent transformation: Both sides change by a factor of t^8, thus $L \geq R$ holds if and only if $L' \geq R'$. In general, the inequality $Q(x_1, x_2, \ldots, x_n) \geq 0$ is called *homogeneous* (in the variables $x_1, x_2, \ldots, x_n$) if it is equivalent to $Q(tx_1, tx_2, \ldots, tx_n) \geq 0$ for each $t \in \mathbb{R}^+$, especially if

$$Q(tx_1, tx_2, \ldots, tx_n) = t^r Q(x_1, x_2, \ldots, x_n)$$

for some $r \in \mathbb{R}$. (The inequality $L \geq R$ can be written as $Q = L - R \geq 0$.) Through the choice of the number $t \in \mathbb{R}^+$ we can appropriately "normalize" the variables $x_1, x_2, \ldots, x_n$ in a homogeneous inequality such that, for example, we get $x_n = 1$ (and thus reduce the number of unknowns by 1), or $x_1 + x_2 + \cdots + x_n = 1$, etc. As an illustration we prove the following assertion: If $0 < r < s$, then for arbitrary numbers $a, b \in \mathbb{R}^+$ we have

$$(a^s + b^s)^{1/s} < (a^r + b^r)^{1/r}. \tag{11}$$

This inequality is homogeneous in the variables a, b; through an appropriate choice of $t \in \mathbb{R}^+$ (the reader should determine this value) we may assume that $a^r + b^r = 1$. Then $a^r < 1$, which by 1.11 gives $a^s = (a^r)^{s/r} < a^r$, since $s/r > 1$. Similarly, we have $b^s < b^r$, and thus $a^s + b^s < a^r + b^r = 1$, that is, $(a^s + b^s)^{1/s} < 1$, which is (11) under the assumption that $a^r + b^r = 1$. This completes the proof.

A similar homogenization approach leads to the so-called *Jensen inequality*

$$\left(\sum_{k=1}^{n} a_k^s \right)^{1/s} < \left(\sum_{k=1}^{n} a_k^r \right)^{1/r} \qquad (0 < r < s)$$

for positive numbers $a_1, a_2, \ldots, a_n$ ($n \geq 2$). From this we derive the following interesting property: If $n \geq 2$ and $p, a_1, a_2, \ldots, a_n \in \mathbb{R}^+$, $p \neq 1$, then

$$(a_1 + a_2 + \cdots + a_n)^p > a_1^p + a_2^p + \cdots + a_n^p, \quad \text{if } p > 1, \tag{12}$$

respectively

$$(a_1 + a_2 + \cdots + a_n)^p < a_1^p + a_2^p + \cdots + a_n^p, \quad \text{if } 0 < p < 1. \tag{13}$$

In closing we remark that symmetric and homogeneous inequalities will be very prominent throughout this chapter.

2.8 *Exercises*

For arbitrary numbers $a, b, c \in \mathbb{R}^+$ prove the symmetric inequalities

(i) $a(a-b)(a-c) + b(b-c)(b-a) + c(c-a)(c-b) \geq 0$,

(ii) $(a-b)^2(a+b-c) + (a-c)^2(a+c-b) + (b-c)^2(b+c-a) \geq 0$.

Show that the inequality (ii) can be transformed into

$$a^4 + b^4 + c^4 + abc(a+b+c) \geq 2(a^2b^2 + a^2c^2 + b^2c^2)$$

by multiplication by $a+b+c$ and further equivalent transformations. It is very difficult to prove this last interesting inequality "directly" (that is, without recognizing the above connection); the authors know of no other approach than the use of (ii).

(iii) Prove the homogeneous inequality

$$\frac{a+b}{2} - \sqrt{ab} \geq \frac{(a-b)^2(a+3b)(b+3a)}{8(a+b)(a^2+6ab+b^2)} \qquad (a, b \in \mathbb{R}^+),$$

which provides a nonnegative lower bound for the difference between the arithmetic and the geometric means of the positive numbers a and b. (Hint: Setting $b = 1$ and $a = t^2$, where $t > 0$, you get a polynomial inequality that can be proved by way of factorization.)

3   The Use of Algebraic Formulas

We now add formulas for $A^n - B^n$ and $(A+B)^n$, which we derived in Chapter 1, to our collection of tools for proving inequalities. For the entire section we assume that $n, k \in \mathbb{N}$.

3.1 *The Decomposition of $A^n - B^n$*

In the following three examples we illustrate the use of the formula

$$A^n - B^n = (A-B)(A^{n-1} + A^{n-2}B + A^{n-3}B^2 + \cdots + AB^{n-2} + B^{n-1}) \tag{14}$$

for algebraic transformations of expressions that occur in the given inequalities.

(i) Show that if $a, b \in \mathbb{R}^+$, $a \neq b$, then $a^{n+1} + nb^{n+1} > (n+1)ab^n$.

SOLUTION.   By (14) we get

$$
\begin{aligned}
L - R &= a(a^n - b^n) - nb^n(a - b) \\
&= a(a-b)(a^{n-1} + a^{n-2}b + \cdots + ab^{n-2} + b^{n-1}) - nb^n(a-b) \\
&= (a-b)(a^n + a^{n-1}b + \cdots + a^2b^{n-2} + ab^{n-1} - nb^n).
\end{aligned}
$$

Upon adding $a^k b^{n-k} > b^n$ (if $a > b$), resp. $a^k b^{n-k} < b^n$ (if $a < b$), for $k = 1, 2, \ldots, n$ we obtain in both cases $L - R > 0$, that is, $L > R$.   □

(ii) Suppose the numbers $a, b, c \in \mathbb{R}^+$ have the following property: For each $n \geq 1$ there exists a triangle T_n with sides a^n, b^n, and c^n. Show that all triangles T_n are equilateral.

SOLUTION.   Let $a \geq b \geq c > 0$. The triangle inequality for T_n means that $c^n > a^n - b^n$, that is,

$$c^n > (a - b)(a^{n-1} + a^{n-2}b + \cdots + ab^{n-2} + b^{n-1}).$$

The last term in parentheses is at least nc^{n-1}, since $a \geq c$ and $b \geq c$ implies $a^{n-1-k}b^k \geq c^{n-1-k}c^k = c^{n-1}$, $0 \leq k \leq n - 1$. We thus obtain $c^n > (a-b)nc^{n-1}$, that is, $c > (a-b)n$. Since this last inequality holds for all $n \geq 1$ and furthermore we have $a \geq b$, it follows that $a = b$. (In the case $a > b$ we would obtain a contradiction when $n > \frac{a-b}{c}$.)   □

(iii) Show that if $a > b > 0$, then

$$\frac{n+1}{n}a > \frac{a^{n+1} - b^{n+1}}{a^n - b^n} > \frac{n+1}{n}b. \qquad (15)$$

SOLUTION.   By the identity (14) we have

$$\begin{aligned}
\frac{a^{n+1} - b^{n+1}}{a^n - b^n} &= \frac{(a - b)(a^n + a^{n-1}b + \cdots + ab^{n-1} + b^n)}{(a - b)(a^{n-1} + a^{n-2}b + \cdots + ab^{n-2} + b^{n-1})} \\
&= \frac{a^n + a^{n-1}b + \cdots + ab^{n-1} + b^n}{a^{n-1} + a^{n-2}b + \cdots + ab^{n-2} + b^{n-1}}.
\end{aligned}$$

We note that if we multiply the denominator D of the last term by the number a (resp. b), we get the first (resp. last) n terms of its numerator. Thus,

$$\frac{a^{n+1} - b^{n+1}}{a^n - b^n} = a + \frac{b^n}{D} = b + \frac{a^n}{D},$$

where

$$D = a^{n-1} + a^{n-2}b + \cdots + ab^{n-2} + b^{n-1}.$$

Since the "outer" expressions in (15) are $a + a/n$, resp. $b + b/n$, it suffices to show that $a/n > b^n/D$ and $b/n < a^n/D$, that is, $aD > nb^n$ and $bD < na^n$. But this is easy: $a > b$ implies

$$aD = a^n + a^{n-1}b + \cdots + a^2b^{n-2} + ab^{n-1} > b^n + b^n + \cdots + b^n = nb^n,$$

and similarly we obtain $bD < na^n$.   □

3.2   Exercises

Prove the assertions (i)–(iv):

(i) $0 < a < 1 \wedge n \geq 1 \implies \dfrac{1 - a^n}{n} > \dfrac{1 - a^{n+1}}{n+1}$.

(ii) $a > b > 0 \wedge c > 0 \wedge n \geq 2 \implies \sqrt[n]{a^n + c} - \sqrt[n]{b^n + c} < a - b$.

(iii) $a > 1 \implies \dfrac{1}{a^1 - a^{-1}} > \dfrac{2}{a^2 - a^{-2}} > \dfrac{3}{a^3 - a^{-3}} > \cdots$.

*(iv) $a > \sqrt{2} \wedge b > \sqrt{2} \implies a^4 - a^3 b + a^2 b^2 - ab^3 + b^4 > a^2 + b^2$.

(v) Suppose that $a, b \in \mathbb{R}^+$ satisfy $a^3 + b^3 = a - b$. Show that $a^2 + b^2 < 1$.

*(vi) Show that for arbitrary $a, b, c \in \mathbb{R}^+$,

$$\frac{a^3}{a^2 + ab + b^2} + \frac{b^3}{b^2 + bc + c^2} + \frac{c^3}{c^2 + ca + a^2} \geq \frac{a + b + c}{3}.$$

3.3   Estimates for $A^n - B^n$

The proofs of a number of inequalities rely on the estimates

$$n(A - B)B^{n-1} < A^n - B^n < n(A - B)A^{n-1}, \tag{16}$$

$(A > B \geq 0,\ n \geq 2)$, which are easy to obtain from formula (14): Just as in 3.1.(i) it is clear that

$$nB^{n-1} < A^{n-1} + A^{n-2}B + \cdots + AB^{n-2} + B^{n-1} < nA^{n-1}.$$

We give three examples.

(i) Show that if $0 < a < 1$, then $n + (1 + a)^n < na + 2^n$ for $n \geq 2$.

SOLUTION.   If $1 < 1 + a < 2$, then by the left part of (16) with $A = 2$ and $B = 1 + a$ we have

$$2^n - (1 + a)^n > n(1 - a)(1 + a)^{n-1} > n(1 - a) \cdot 1^{n-1},$$

and after an easy simplification we get the statement.   □

(ii) Show that the sums $S_k(n)$ (studied in Chapter 1, Section 2) satisfy the lower bound

$$S_k(n) = 1^k + 2^k + \cdots + n^k > \frac{1}{k+1} \cdot n^{k+1}.$$

SOLUTION.   By the right part of (16) with $A = j$, $B = j - 1$, and $n = k + 1$ we have for each $j \geq 1$ the inequality $j^{k+1} - (j-1)^{k+1} < (k+1)j^k$. Summing over $j = 1, 2, \ldots, n$ we obtain

$$n^{k+1} = (1^{k+1} - 0^{k+1}) + (2^{k+1} - 1^{k+1}) + \cdots + [n^{k+1} - (n-1)^{k+1}]$$
$$< (k + 1)(1^k + 2^k + \cdots + n^k),$$

from which upon division by $k + 1$ we obtain the desired bound.   □

(iii) Prove the so-called *Bernoulli inequality*

$$(1 + x)^n \geq 1 + nx \qquad (x \geq -1,\ n \geq 2), \tag{17}$$

with equality if and only if $x = 0$.

SOLUTION. If $x > 0$, then (17) follows from the left inequality of (16) with $A = 1 + x$ and $B = 1$. If $-1 \leq x < 0$, then (17) follows from the right part of (16) with $A = 1$ and $B = 1 + x$. In both cases the inequality (17) is strict, and for $x = 0$ we get equality in (17).   □

3.4   Exercises

Show that the inequalities (i)–(iii) hold for $n \geq 2$:

(i) $2(a + 1)^n + n2^n > (na + 2)2^n$ $(a > 1)$.

(ii) $n + (1 + a)^n < na^2 + 2^n$ $(0 < a < 1)$.

(iii) $(1 - a)^n + (2n - 1)a^n \geq na^{n-1}$ $(0 < a < 1)$.

(iv) The arithmetic means $A_1, A_2, \ldots, A_N$ of the real numbers a_i are defined as follows:

$$A_n = \frac{a_1 + a_2 + \cdots + a_n}{n} \qquad (1 \leq n \leq N).$$

Show that if $0 \leq a_1 \leq a_2 \leq \cdots \leq a_N$, then for each $n = 2, 3, \ldots, N$ we have the inequality $(A_n)^n > a_n(A_{n-1})^{n-1}$, unless $a_1 = a_2 = \cdots = a_n$. (Hint: Use (16) with $A = A_n$ and $B = A_{n-1}$.)

Using (17), show that for each $k \geq 2$ the numbers $c_k = (1 + \frac{1}{k})^k$ satisfy (v) and (vi):

(v) $c_k > 2$.

*(vi) $1 < \dfrac{c_{k+1}}{c_k} < 1 + \dfrac{1}{k^2 + 2k}$.

(Further properties of the numbers c_k will be derived in 3.8.(iv).)

(vii) Using (17), show that for $a, b \in \mathbb{R}^+$ and $n \geq 2$ we have

$$a^n + (n - 1)b \geq na \cdot \sqrt[n]{b^{n-1}}. \tag{18}$$

(Hint: Set $x = ab^r - 1$ in (17) for an appropriate r.) When is (18) strict?

Remark. The inequality (18) follows easily from the so-called AM–GM inequality, which will be proved later, in Section 8. Indeed, we can apply the inequality (84) to the n-tuple $a^n, b, b, \ldots, b$.

(viii) Show that *(vi) follows from (18) with $a = 1$ and appropriate b, n.

*****(ix)** Let the largest one of the nonnegative numbers $d_1, d_2, \ldots, d_k$ be equal to D. Show that for each $n \geq 2$ we have

$$\frac{d_1^n + d_2^n + \cdots + d_k^n}{k} \leq \left(\frac{d_1 + d_2 + \cdots + d_k}{k}\right)^n + \alpha_n \cdot D^n,$$

where $\alpha_n = (n-1)n^{\frac{n}{1-n}}$.

3.5 Estimates Using Geometric Series

From the formula (14) with $A = 1$ and $B = q$ it follows that

$$1 + q + q^2 + \cdots + q^m = \frac{1 - q^{m+1}}{1 - q} < \frac{1}{1 - q} \tag{19}$$

for each positive $q < 1$ and each $m = 1, 2, 3, \ldots$ (the number $\frac{1}{1-q}$ is the sum of the *infinite* geometric series $1 + q + q^2 + \cdots$). The inequality (19) can be used to obtain an upper bound for the sum $a_1 + a_2 + \cdots + a_k$ in the case that we can find numbers $b_j \geq a_j$ such that $b_{j+1} \leq q b_j$ for each j, where $0 < q < 1$. Using this approach, we will prove the bound

$$\frac{1}{(n+1)!} + \frac{1}{(n+2)!} + \cdots + \frac{1}{(n+k)!} < \frac{n+2}{n+1} \cdot \frac{1}{(n+1)!}. \tag{20}$$

Indeed, from the inequality $(n+j)! > (n+2)!(n+2)^{j-2}$ for $j \geq 3$ and from (19) with $q = 1/(n+2)$ it follows that

$$\frac{1}{(n+1)!} + \frac{1}{(n+2)!} + \cdots + \frac{1}{(n+k)!}$$

$$\leq \frac{1}{(n+1)!}\left[1 + \frac{1}{(n+2)} + \frac{1}{(n+2)^2} + \cdots + \frac{1}{(n+2)^{k-1}}\right]$$

$$< \frac{1}{(n+1)!} \cdot \frac{1}{1 - \frac{1}{n+2}} = \frac{n+2}{n+1} \cdot \frac{1}{(n+1)!}.$$

The reader should verify that (20) is sharper than the bound

$$\frac{1}{(n+1)!} + \frac{1}{(n+2)!} + \cdots + \frac{1}{(n+k)!} < \frac{1}{n \cdot n!},$$

which is the result of Example 2.5.(v).

3.6 Exercises

Prove the inequalities (i)–(iii), where $a_1, a_2, \ldots, a_n \in \mathbb{R}^+$:

(i) $\dfrac{1}{(n+2)!} + \dfrac{1}{(n+4)!} + \cdots + \dfrac{1}{(n+2k)!} < \dfrac{n^2 + 6n + 9}{n^2 + 6n + 8} \cdot \dfrac{1}{(n+2)!}$.

(ii) $\dfrac{1}{3 \cdot 1!} + \dfrac{1 \cdot 3}{3^2 \cdot 2!} + \dfrac{1 \cdot 3 \cdot 5}{3^3 \cdot 3!} + \cdots + \dfrac{1 \cdot 3 \cdot 5 \cdots \cdots (2n-1)}{3^n \cdot n!} < 1$.

*(iii) $(a_1 + a_2 + \cdots + a_n)(\sqrt[k]{2} - 1) < \sqrt[k]{2a_1^k + 2^2 a_2^k + \cdots + 2^n a_n^k}$.

3.7   Using the Expansion of $(A+B)^n$

The binomial expansion

$$(A+B)^n = A^n + \binom{n}{1}A^{n-1}B + \binom{n}{2}A^{n-2}B^2 + \cdots + \binom{n}{n-1}AB^{n-1} + B^n$$

$$(21)$$

can be used in the proofs of various inequalities that contain powers of integer degree $n \geq 2$. If we leave out several summands on the right-hand side of (21), we obtain a lower bound for the power $(A+B)^n$, for positive A, B. Thus, for instance, we get

$$(A+B)^n > A^n + B^n \qquad (A, B \in \mathbb{R}^+, n \geq 2) \qquad (22)$$

(compare with (12)). We give three examples.

(i) Suppose that the numbers $x, y, z \in \mathbb{R}$ satisfy $xyz > 0$ and $x + y + z > 0$. Show that the inequality $x^n + y^n + z^n > 0$ holds for all $n \geq 2$.

SOLUTION.   Since $xyz > 0$, either all three numbers x, y, z are positive (then the assertion is obvious), or one is positive and the remaining two are negative. For example, let $x = a$, $y = -b$ and $z = -c$, where $a, b, c \in \mathbb{R}^+$. From $x + y + z = a - b - c > 0$ it follows that $a > b + c$, which by (22) implies $a^n > (b+c)^n > b^n + c^n$; for odd n this can be written as $x^n + y^n + z^n = a^n - b^n - c^n > 0$. For even n the inequality $x^n + y^n + z^n > 0$ is obvious.   □

(ii) Show that $(1 + na)^{n+1} > [1 + (n+1)a]^n$ for all $a \in \mathbb{R}^+$.

SOLUTION.   We use (21) to expand both sides of the inequality, and then we compare the coefficients of equal powers a^k ($0 \leq k \leq n$) on both sides. It suffices to show that

$$\binom{n+1}{k}n^k \geq \binom{n}{k}(n+1)^k$$

holds for such k. (On the left side of the original inequality we have the additional positive term $n^{n+1}a^{n+1}$.) It is easy to verify by substituting that this inequality holds for $k = 0, 1$; for $k > 1$ it can be rewritten as

$(n+1)^k - n^k \leq k(n+1)^{k-1}$. But this is true by the right side of (16), and the proof is complete.

Another possible approach is based on the direct use of the right part of (16) with $A = 1 + (n+1)a$, $B = 1 + na$:

$$[1 + (n+1)a]^n - (1+na)^n < na[1 + (n+1)a]^{n-1}$$
$$= [1 + (n+1)a]^n - (1+a)[1 + (n+1)a]^{n-1},$$

which implies $(1+na)^n > (1+a)[1 + (n+1)a]^{n-1}$. If we multiply this last inequality by $1 + na$, we get

$$(1+na)^{n+1} > [1 + (n+1)a + na^2][1 + (n+1)a]^{n-1} > [1 + (n+1)a]^n.$$

We note that the original inequality was derived earlier, in Exercise 2.4.(ix), with a somewhat artificial trick. Also, it is worth mentioning two consequences: With $a = 1/(n^2 + n)$ we obtain the left part of 3.4.(vi), and with $a = 1$ we get

$$2 > \sqrt[2]{3} > \sqrt[3]{4} > \cdots > \sqrt[n]{n+1} > \cdots. \qquad \square$$

(iii) Show that for each $n \geq 3$ we have $(2n+1)^n > (2n)^n + (2n-1)^n$.

SOLUTION.   Using (21), we expand the powers $(2n+1)^n$ and $(2n-1)^n$:

$$(2n+1)^n - (2n-1)^n$$
$$= \sum_{k=0}^{n} \binom{n}{k}(2n)^{n-k} - \sum_{k=0}^{n} \binom{n}{k}(-1)^k(2n)^{n-k}$$
$$= 2\left[\binom{n}{1}(2n)^{n-1} + \binom{n}{3}(2n)^{n-3} + \binom{n}{5}(2n)^{n-5} + \cdots\right]$$
$$> 2\binom{n}{1}(2n)^{n-1} = (2n)^n. \qquad \square$$

3.8   Exercises

Prove the assertions (i)–(iii):

(i) If the positive numbers a, b, c satisfy $a^n + b^n = c^n$ for some $n \geq 2$, then a, b, c are the side lengths of some triangle.

(ii) If $0 < a < 1$ and $n \geq 3$, then $(1+a)^n + 2a^{n-1} > 1 + a^{n-1}(2^n + a)$.

(iii) If $2n \geq a > 0$, then $(a-1)^n + a^n \leq (a+1)^n$; if $(1+\sqrt{2})n < a$, then the opposite inequality holds. (This covers the result of 3.7.(iii).)

(iv) Given the numbers $c_k = (1 + \frac{1}{k})^k$, $k \geq 2$, derive the expression

$$c_k = 2 + \frac{1}{2!}\left(1 - \frac{1}{k}\right) + \frac{1}{3!}\left(1 - \frac{1}{k}\right)\left(1 - \frac{2}{k}\right) + \cdots$$

$$+ \frac{1}{k!}\left(1 - \frac{1}{k}\right)\left(1 - \frac{2}{k}\right)\cdots\left(1 - \frac{k-1}{k}\right). \tag{23}$$

Show that (23) again implies the inequality $c_{k+1} > c_k$ (see 3.4.(vi)) and prove the bound

$$c_k < 2 + \frac{1}{2!} + \frac{1}{3!} + \cdots + \frac{1}{k!} < 3 \quad \text{for each } k \geq 2. \tag{24}$$

The smallest constant e with the property $c_k < e$ for all $k \in \mathbb{N}$ is called the *base of the natural logarithm*. This irrational number with the approximate value $e \approx 2.71828$ is one of the most important mathematical constants.

4 The Method of Squares

The inequality $x^2 > 0$ $(x \in \mathbb{R}, x \neq 0)$ stands out among all other inequalities because of its simplicity and, at the same time, its wide applicability in many parts of mathematics. This inequality provides us with another method of proof of the inequality $L \geq R$: We try to find expressions $Q_1, Q_2, \ldots, Q_N$ such that

$$L - R = Q_1^2 + Q_2^2 + \cdots + Q_N^2,$$

or

$$L - R = S_1 Q_1^2 + S_2 Q_2^2 + \cdots + S_N Q_N^2,$$

where $S_1, S_2, \ldots, S_N$ are nonnegative expressions. In 4.1 we describe approaches for finding such expressions; they are mainly based on transforming "incomplete" squares $X^2 + 2XY$ using the formula

$$X^2 + 2XY = (X + Y)^2 - Y^2,$$

which is called *completing the square*. Consequences of the basic inequality (25) in 4.1 are of such importance that we devote the separate sections 4.3, 4.5, and 4.7 to them.

4.1 Examples

We now give ten examples of the use of the method of squares.

(i) The inequality

$$X^2 + Y^2 > 2XY \qquad (X, Y \in \mathbb{R},\ X \neq Y) \qquad (25)$$

follows easily from $(X - Y)^2 > 0$. Let us ask a more general question: Which numbers $p \in \mathbb{R}$ have the property that the inequality $x^2 + y^2 \geq pxy$ (which we denote by $L \geq R$) holds for all numbers $x, y \in \mathbb{R}$? The trinomial $L - R$ contains the binomial $x^2 - pxy$, which we complete as a square. Thus we get

$$L - R = x^2 + y^2 - pxy = \left(x - \frac{py}{2}\right)^2 + \frac{1}{4}(4 - p^2)y^2. \qquad (26)$$

Choosing $x = py/2$, $y \neq 0$, we obtain the necessary condition $4 - p^2 \geq 0$, which, in view of (26), is also sufficient. Solving the inequality $4 - p^2 \geq 0$, we obtain the answer: $-2 \leq p \leq 2$.

(ii) Let us now prove the inequality

$$\frac{a}{b^2} + \frac{b}{a^2} \geq \frac{1}{a} + \frac{1}{b} \qquad (a, b \in \mathbb{R}^+).$$

Upon multiplying by $a^2 b^2$ we obtain $a^3 + b^3 \geq ab(a + b)$; dividing by $a + b$ we get $a^2 - ab + b^2 \geq ab$, and after subtracting ab, finally $(a - b)^2 \geq 0$.

(iii) Determine the integers $n \geq 2$ for which the inequality

$$a_1^2 + a_2^2 + \cdots + a_n^2 \geq (a_1 + a_2 + \cdots + a_{n-1})a_n$$

holds for arbitrary $a_1, a_2, \ldots, a_n \in \mathbb{R}_0^+$. Completing squares, we obtain

$$L - R = \left(a_1 - \frac{a_n}{2}\right)^2 + \left(a_2 - \frac{a_n}{2}\right)^2 + \cdots + \left(a_{n-1} - \frac{a_n}{2}\right)^2 + \left(1 - \frac{n-1}{4}\right)a_n^2.$$

The inequality therefore holds when $1 - (n - 1)/4 \geq 0$, that is, $n \leq 5$; for $n \geq 6$ the inequality is generally not true: It suffices to consider the n-tuple $a_j = 1$ $(1 \leq j \leq n - 1)$ and $a_n = 2$.

(iv) We prove that for each $x, y \in \mathbb{R}$ we have the inequality

$$x^2 - 2xy + 6y^2 - 12x + 2y + 41 \geq 0.$$

The terms containing x will be completed as squares:

$$x^2 - 2xy - 12x = (x - y - 6)^2 - (y + 6)^2.$$

We thus obtain

$$L = (x - y - 6)^2 + 5y^2 - 10y + 5 = (x - y - 6)^2 + 5(y - 1)^2 \geq 0.$$

(v) We prove that whenever $p > q > 0$, the following inequalities hold for all $x \in \mathbb{R}$:

$$\frac{p+q}{p-q} \geq \frac{x^2 - 2qx + p^2}{x^2 + 2qx + p^2} \geq \frac{p-q}{p+q}.$$

The trinomial in the denominator satisfies

$$x^2 + 2qx + p^2 = (x+q)^2 + (p^2 - q^2) > 0.$$

If we multiply each of the inequalities by the (positive) product of both relevant denominators, we obtain after simplification the two inequalities $2q(x+p)^2 \geq 0$ and $2q(x-p)^2 \geq 0$.

(vi) We prove the following assertion: If $x^3 + y^3 = 2$ for some $x, y \in \mathbb{R}$, then $x + y \leq 2$. We note that for $x = y = 1$ we have $x^3 + y^3 = x + y = 2$; we will not find other such pairs x, y. In such a situation it is convenient to set $x = 1 + u$, $y = 1 + v$, and to derive $u + v \leq 0$ from the assumption $(1+u)^3 + (1+v)^3 = 2$. Indeed, the equation

$$2 = (1+u)^3 + (1+v)^3 = 2 + 3(u+v) + 3(u^2 + v^2) + u^3 + v^3$$

implies

$$(u+v)(u^2 - uv + v^2 + 3) = -3(u^2 + v^2) \leq 0.$$

From this we easily obtain $u + v \leq 0$, since by the result of part (i) we have $u^2 - uv + v^2 \geq 0$, and thus $u^2 - uv + v^2 + 3 > 0$.

(vii) To prove the inequality $a^3 + 3a^2 - 13a + 10 > 0$ $(a \in \mathbb{R}^+)$, we try to write its left-hand side in the form $L = a(a+p)^2 + b(a+q)^2 + c$ with appropriate constants $b, c \in \mathbb{R}_0^+$ and $p, q \in \mathbb{R}$. By comparing the coefficients of equal powers a^j we obtain the conditions $2p + b = 3$, $p^2 + 2bq = -13$, and $bq^2 + c = 10$. It is not difficult to find *some* solution of this system of three equations: We choose an arbitrary p and then successively compute b, q, c. But we require solutions with the property $b, c \in \mathbb{R}_0^+$! The first equation implies $p = (3-b)/2 \leq \frac{3}{2}$. Let us try such integers p: While the values $p = 1$, $p = 0$ do not work (we get $c = -39$ resp. $c \doteq -4.08$), we are successful with $p = -1$: We get $c = \frac{1}{5} > 0$ ($b = 5$ and $q = -\frac{7}{5}$). Thus we obtain

$$L = a(a-1)^2 + 5\left(a - \frac{7}{5}\right)^2 + \frac{1}{5} > 0.$$

(viii) The inequality

$$X^2 + Y^2 + Z^2 \geq XY + YZ + ZX \qquad (X, Y, Z \in \mathbb{R}) \qquad (27)$$

has numerous applications in proofs of other inequalities. It can be obtained by rewriting the inequality $(X-Y)^2 + (Y-Z)^2 + (Z-X)^2 \geq 0$. Thus, for example, we verify

$$x^4 + y^4 + z^4 \geq xyz(x+y+z) \qquad (28)$$

for arbitrary $x, y, z \in \mathbb{R}$ as follows: Using (27) with $X = x^2$, $Y = y^2$, and $Z = z^2$, we obtain $x^4 + y^4 + z^4 \geq x^2 y^2 + y^2 z^2 + z^2 x^2$; using (27) again, this time with $X = xy$, $Y = yz$, and $Z = zx$, we get

$$x^2 y^2 + y^2 z^2 + y^2 x^2 \geq xy^2 z + yz^2 x + zx^2 y = xyz(x + y + z).$$

Now (28) follows from these two consequences of (27).

The reader should verify that the inequality

$$a + b + c \leq \frac{bc}{a} + \frac{ca}{b} + \frac{ab}{c} \qquad (a, b, c \in \mathbb{R}^+)$$

also follows from (27), with

$$X = \sqrt{\frac{bc}{a}}, \quad Y = \sqrt{\frac{ca}{b}}, \quad \text{and} \quad Z = \sqrt{\frac{ab}{c}}.$$

(ix) We show that if $x + y + z \geq 0$, then $x^3 + y^3 + z^3 \geq 3xyz$. For this purpose we use the elementary symmetric polynomials $\sigma_1 = x + y + z$, $\sigma_2 = xy + xz + yz$, and $\sigma_3 = xyz$ (see Chapter 1, Section 4). Since

$$x^3 + y^3 + z^3 - 3xyz = (\sigma_1^3 - 3\sigma_1 \sigma_2 + 3\sigma_3) - 3\sigma_3 = \sigma_1(\sigma_1^2 - 3\sigma_2)$$

and $\sigma_1 \geq 0$, it suffices to show that $\sigma_1^2 - 3\sigma_2 \geq 0$, that is, $\sigma_1^2 - 2\sigma_2 \geq \sigma_2$. But this is just (27), since $x^2 + y^2 + z^2 = \sigma_1^2 - 2\sigma_2$.

(x) Let us find the smallest possible value of the expression

$$V = \frac{ab}{c} + \frac{bc}{a} + \frac{ca}{b},$$

if the positive numbers a, b, c satisfy the condition $a^2 + b^2 + c^2 = 1$. First we determine the square of V:

$$\begin{aligned}
V^2 &= \frac{a^2 b^2}{c^2} + \frac{b^2 c^2}{a^2} + \frac{c^2 a^2}{b^2} + 2\left(\frac{ab}{c} \cdot \frac{bc}{a} + \frac{ab}{c} \cdot \frac{ca}{b} + \frac{bc}{a} \cdot \frac{ca}{b}\right) \\
&= \frac{a^2 b^2}{c^2} + \frac{b^2 c^2}{a^2} + \frac{c^2 a^2}{b^2} + 2(a^2 + b^2 + c^2) \\
&= \frac{a^2 b^2}{c^2} + \frac{b^2 c^2}{a^2} + \frac{c^2 a^2}{b^2} + 2.
\end{aligned}$$

Since (27) implies

$$\left(\frac{ab}{c}\right)^2 + \left(\frac{bc}{a}\right)^2 + \left(\frac{ca}{b}\right)^2 \geq \frac{ab}{c} \cdot \frac{bc}{a} + \frac{ab}{c} \cdot \frac{ca}{b} + \frac{bc}{a} \cdot \frac{ca}{b}$$
$$= a^2 + b^2 + c^2 = 1,$$

we obtain the bound $V^2 \geq 3$, that is, $V \geq \sqrt{3}$. Equality $V = \sqrt{3}$ occurs in the case $a = b = c = \sqrt{3}/3$; hence we have found $\sqrt{3}$ as the smallest value of V.

4.2 Exercises

Prove (i)–(xxviii) for arbitrary $a, b, c \in \mathbb{R}^+$ and $x, y, z \in \mathbb{R}$:

(i) $2xyz \le x^2 + y^2 z^2$.

(ii) $(x^2 - y^2)^2 \ge 4xy(x - y)^2$.

(iii) $x^4 + y^4 \ge x^3 y + xy^3$.

(iv) $\dfrac{a}{\sqrt{b}} + \dfrac{b}{\sqrt{a}} \ge \sqrt{a} + \sqrt{b}$.

(v) $\dfrac{x^2}{1 + x^4} \le \dfrac{1}{2}$.

(vi) $x^2(1 + y^4) + y^2(1 + x^4) \le (1 + x^4)(1 + y^4)$.

(vii) $x^2 + 4y^2 + 3z^2 + 14 > 2x + 12y + 6z$.

(viii) $x^2 + 2y^2 + 2xy + y + 1 > 0$.

(ix) $2 + x^2(1 + y^2) \ge 2x(1 + y)$.

(x) $x^4 + y^4 + z^2 + 1 \ge 2x(xy^2 - x + z + 1)$.

*(xi) $\sqrt{a^2 + b^2} \ge a + b - (2 - \sqrt{2})\sqrt{ab}$.

(xii) $2x^2 + 4y^2 + z^2 \ge 4xy + 2xz$.

(xiii) $2(x^4 + y^4) - 12xy + 10 > 0$.

(xiv) $2(\sqrt[4]{a} + \sqrt[4]{b} + \sqrt[4]{c}) \le 3 + \sqrt{a} + \sqrt{b} + \sqrt{c}$.

(xv) $a(1 + b) + b(1 + c) + c(1 + a) \ge 6\sqrt{abc}$.

(xvi) $2(a^2 + b^2) + a + b \ge 2(ab + a\sqrt{b} + b\sqrt{a})$.

(xvii) $a^4 + 2a^3 b + 2ab^3 + b^4 \ge 6a^2 b^2$.

(xviii) $4x^2 y^2 + (z + x + y)(z + x - y)(z - x + y)(z - x - y) \ge 0$.

(xix) $x^4 - x^2 - 3x + 4 > 0$.

(xx) $a^4 + a^3 - 8a^2 + 4a + 4 > 0$.

(xxi) $\dfrac{x^2}{4} + y^2 + z^2 \ge xy - xz + 2yz$.

(xxii) $a + b + c \ge 2(\sqrt{ac} + \sqrt{bc} - \sqrt{ab})$.

*(xxiii) $\dfrac{a^2}{3} + b^2 + c^2 > ab + bc + ac$,   if $abc = 1$ and $a^3 > 36$.

(xxiv) $(1 + x + y)^2 \ge 3(x + y + xy)$.

(xxv) $ab + bc + ca \ge \sqrt{3abc(a + b + c)}$.

*(xxvi) $\dfrac{a^8 + b^8 + c^8}{a^3 b^3 c^3} \ge \dfrac{1}{a} + \dfrac{1}{b} + \dfrac{1}{c}$.

(xxvii) $xy + yz + zx \ge -\dfrac{1}{2}$,   if $x^2 + y^2 + z^2 = 1$.

*(xxviii) $4x(x + y)(x + z)(x + y + z) + y^2 z^2 \ge 0$.

(xxix) Assuming that $p, q, r, x, y, z \in \mathbb{R}$ satisfy $pz - 2qy + rx = 0$ and $pr - q^2 > 0$, prove the inequality $xz - y^2 \le 0$.

(xxx) Show that the polynomial $F(x) = x^4 + ax^3 + bx^2 + cx + d$ has no real zeros, where $a, b, c, d \in \mathbb{R}$, $d > 0$, and $c^2 + a^2 d < 4bd$.

*(xxxi) Show that for arbitrary positive numbers a, b, c we have

$$a^{p+2} + b^{p+2} + c^{p+2} \geq a^p bc + b^p ac + c^p ab \qquad (p \in \mathbb{R}_0^+).$$

This is a generalization of (27), which we obtain by setting $p = 0$.

*(xxxii) For arbitrary $a, b, c \in \mathbb{R}^+$ decide which of the two numbers

$$A = 3 + (a + b + c) + \left(\frac{1}{a} + \frac{1}{b} + \frac{1}{c}\right) + \left(\frac{a}{b} + \frac{b}{c} + \frac{c}{a}\right),$$

$$B = \frac{3(a+1)(b+1)(c+1)}{abc + 1}$$

is larger.

*(xxxiii) Find the largest $p \in \mathbb{R}$ such that the inequality

$$x^4 + y^4 + z^4 + xyz(x + y + z) \geq p(xy + xz + yz)^2$$

holds for all $x, y, z \in \mathbb{R}$.

*(xxxiv) For each $p \in \left\{1, \frac{4}{3}, \frac{6}{5}\right\}$, determine the $n \geq 2$ for which the inequality

$$x_1^2 + x_2^2 + \cdots + x_n^2 \geq p(x_1 x_2 + x_2 x_3 + \cdots + x_{n-1} x_n)$$

holds for all $x_1, x_2, \ldots, x_n \in \mathbb{R}$.

4.3 A Lower Bound for $A + B$

Using (25) with $X = \sqrt{A}$ and $Y = \sqrt{B}$, we obtain

$$A + B \geq 2\sqrt{AB} \qquad (A, B \in \mathbb{R}_0^+), \tag{29}$$

with equality only when $A = B$. We use this lower bound for the sum $A + B$ to solve the following seven problems.

(i) Show that for any $a, b, c \in \mathbb{R}^+$ the inequality $ab + \frac{c}{b} \geq 2\sqrt{ac}$ holds.

SOLUTION.   The result follows from (29) with $A = ab$ and $B = c/b$.   □

(ii) Show that if $a, b \in \mathbb{R}^+$, then $ab(12 - 2a - 5b) < 2a + 5b$.

SOLUTION.   We note that the term $2a + 5b$ occurs on both sides of the inequality; we therefore rewrite it as $12ab < (2a + 5b)(1 + ab)$. Using (29), both terms on the right of this last inequality can be estimated from below with the help of $\sqrt{ab}$: $(2a + 5b) \geq 2\sqrt{10ab}$ and $1 + ab \geq 2\sqrt{ab}$. Multiplying these together, we obtain $(2a + 5b)(1 + ab) \geq 4\sqrt{10}ab > 12ab$, since $\sqrt{10} > 3$.   □

(iii) Show that if $a, b, c, d \in \mathbb{R}_0^+$, then $a + b + c + d \geq 4 \cdot \sqrt[4]{abcd}$.

SOLUTION. Since by (29) we have $a + b \geq 2\sqrt{ab}$ and $c + d \geq 2\sqrt{cd}$, it suffices to show that $\sqrt{ab} + \sqrt{cd} \geq 2 \cdot \sqrt[4]{abcd}$. But this is just (29) with $A = \sqrt{ab}$ and $B = \sqrt{cd}$. $\square$

(iv) Under the assumption that the numbers $a_1, a_2, b_1, b_2, c_1, c_2 \in \mathbb{R}_0^+$ satisfy $a_1 c_1 \geq b_1^2$ and $a_2 c_2 \geq b_2^2$, prove the inequality $(a_1 + a_2)(c_1 + c_2) \geq (b_1 + b_2)^2$.

SOLUTION. We rewrite the desired inequality as

$$(a_1 c_1 - b_1^2) + (a_2 c_2 - b_2^2) + (a_1 c_2 + a_2 c_1 - 2 b_1 b_2) \geq 0.$$

Thus it remains to show that $a_1 c_2 + a_2 c_1 - 2 b_1 b_2 \geq 0$, and indeed, by (29) with $A = a_1 c_2$ and $B = a_2 c_1$ we have

$$a_1 c_2 + a_2 c_1 \geq 2\sqrt{a_1 c_2 a_2 c_1} = 2\sqrt{a_1 c_1} \cdot \sqrt{a_2 c_2} \geq 2 b_1 b_2. \qquad \square$$

(v) Show that if $a, b \in \mathbb{R}^+$ and $ab \leq 1$, then $(1 + \frac{1}{a})(1 + \frac{1}{b}) \geq 4$.

SOLUTION. By (29) we have $1 + \frac{1}{a} \geq \frac{2}{\sqrt{a}}$ and $1 + \frac{1}{b} \geq \frac{2}{\sqrt{b}}$, and thus

$$\left(1 + \frac{1}{a}\right)\left(1 + \frac{1}{b}\right) \geq \frac{2}{\sqrt{a}} \cdot \frac{2}{\sqrt{b}} = \frac{4}{\sqrt{ab}} \geq 4, \quad \text{since } ab \leq 1. \qquad \square$$

(vi) Show that if $a, b, c \in \mathbb{R}^+$, then $(a + b)(b + c)(a + c) \geq 8abc$.

SOLUTION. By (29) we have $a + b \geq 2\sqrt{ab}$, $b + c \geq 2\sqrt{bc}$, and $a + c \geq 2\sqrt{ac}$; multiplying these three inequalities together, we obtain

$$(a + b)(b + c)(a + c) \geq 8\sqrt{a^2 b^2 c^2} = 8abc. \qquad \square$$

(vii) Suppose that the real numbers $x_1, x_2, \ldots, x_n$ satisfy the condition $x_1 x_2 \cdots x_k \geq 1$ for each index $k = 1, 2, \ldots, n$. Prove the inequality

$$\frac{1}{1 + x_1} + \frac{2}{(1 + x_1)(1 + x_2)} + \cdots + \frac{n}{(1 + x_1)(1 + x_2) \cdots (1 + x_n)} < 2. \quad (30)$$

SOLUTION. Obviously, $x_k > 0$ for each $k = 1, 2, \ldots, n$. Hence by (29) with $A = 1$ and $B = x_k$ we have $1 + x_k \geq 2\sqrt{x_k}$, and thus

$$\frac{k}{(1 + x_1)(1 + x_2) \cdots (1 + x_k)} \leq \frac{k}{2^k \sqrt{x_1 x_2 \cdots x_k}} \leq \frac{k}{2^k}$$

for each $k = 1, 2, \ldots, n$. Therefore, the left-hand side of (30) does not exceed

$$\frac{1}{2} + \frac{2}{2^2} + \cdots + \frac{n}{2^n} = 2 - \frac{n + 2}{2^n} < 2$$

(see Chapter 1, Example 2.37). $\square$

4.4   Exercises

Show that the inequalities (i)–(viii) hold for all $a, b, c, d \in \mathbb{R}^+$:

(i) $a^2 b + \dfrac{1}{b} \geq 2a$.

(ii) $6a + (c+2)(2c+3)b > 6(c+2)\sqrt{ab}$.

(iii) $a^4 + a^3 b - 4a^2 b + ab + b^2 \geq 0$.

(iv) $(a+1)(b+1)(a+c)(b+c) \geq 16abc$.

(v) $\sqrt{(a+b)(c+d)} + \sqrt{(a+c)(b+d)} + \sqrt{(a+d)(b+c)} \geq 6 \cdot \sqrt[4]{abcd}$.

(vi) $\dfrac{1}{ab} + \dfrac{1}{cd} \geq \dfrac{8}{(a+b)(c+d)}$.

(vii) $(\sqrt{a} + \sqrt{b})^2 \geq 2\sqrt{2(a+b)\sqrt{ab}}$.

(viii) $2(a+b)^2 + (a+b) \geq 4(a\sqrt{b} + b\sqrt{a})$.

*(ix) Show that for arbitrary $a_1, a_2, \ldots, a_n \in \mathbb{R}^+$ we have

$$\binom{n}{2} \sum_{i=1}^{n-1} \sum_{j=i+1}^{n} \frac{1}{a_i a_j} \geq 4 \left(\sum_{i=1}^{n-1} \sum_{j=i+1}^{n} \frac{1}{a_i + a_j} \right)^2 .$$

4.5   Upper Bound for $A \cdot B$

Rewriting (29), we obtain

$$AB \leq \left(\frac{A+B}{2} \right)^2 \qquad (A, B \in \mathbb{R}_0^+), \tag{31}$$

with equality occurring only if $A = B$. Using the bound (31), we now solve the following five problems.

(i) Show that if $a, b \in \mathbb{R}^+$ and $a + b = 1$, then $(1 + \frac{1}{a})(1 + \frac{1}{b}) \geq 9$.

SOLUTION.   By (31) with $a + b = 1$ we get $ab \leq \frac{1}{4}$. Hence

$$\left(1 + \frac{1}{a} \right) \left(1 + \frac{1}{b} \right) = 1 + \frac{a+b+1}{ab} = 1 + \frac{2}{ab} \geq 1 + \frac{2}{\frac{1}{4}} = 9. \qquad \square$$

(ii) Let $b_1, b_2, \ldots, b_n$ be an arbitrary ordering of the numbers $a_1, a_2, \ldots, a_n$, with $0 < a_k < 1$ for $k = 1, 2, \ldots, n$. Show that at least one of the numbers

$$a_1(1 - b_1), \quad a_2(1 - b_2), \quad \ldots, \quad a_n(1 - b_n)$$

does not exceed $\frac{1}{4}$.

SOLUTION.   Suppose to the contrary that $a_k(1 - b_k) > \frac{1}{4}$ for *all* indices $k = 1, 2, \ldots, n$. Then

$$\left(\frac{1}{4}\right)^n < a_1(1 - b_1)a_2(1 - b_2)\cdots a_n(1 - b_n)$$
$$= a_1(1 - a_1)a_2(1 - a_2)\cdots a_n(1 - a_n),$$

which means that $a_k(1 - a_k) > \frac{1}{4}$ for *some* $k = 1, 2, \ldots, n$. But this is a contradiction, since by (31) we have $a(1-a) \leq \frac{1}{4}$ whenever $0 < a < 1$.   □

(iii) Show that if $0 < b < a$, then $a + 1/(ab - b^2) \geq 3$.

SOLUTION.   By (31) with $A = a - b$ and $B = b$ we have the inequality $(a-b)b \leq a^2/4$, which implies $a+1/(ab-b^2) \geq a+4/a^2$. Therefore, it suffices to show that $a+4/a^2 \geq 3$. But this is easy to rewrite as $(a+1)(a-2)^2 \geq 0$. Furthermore, we see that the original inequality becomes an equality only if $a - b = b$ and $a - 2 = 0$, that is, $a = 2$ and $b = 1$.   □

(iv) Show that if $a \geq 0$, then $4\sqrt{a}(\sqrt{a} - 2)(1 + 2\sqrt{a}) \leq (a + 1)^2$.

SOLUTION.   It suffices to examine the case $\sqrt{a} > 2$. From (31) with $A = \sqrt{a}(\sqrt{a} - 2) = a - 2\sqrt{a}$ and $B = 1 + 2\sqrt{a}$ we get

$$\sqrt{a}(\sqrt{a} - 2)(1 + 2\sqrt{a}) \leq \left(\frac{a - 2\sqrt{a} + 1 + 2\sqrt{a}}{2}\right)^2 = \frac{(a + 1)^2}{4},$$

which gives the desired inequality. It is left to the reader to show that equality occurs only in the case $a = 9 + 4\sqrt{5}$.   □

(v) Show that for arbitrary $a, b, c, d \in \mathbb{R}^+$ we have

$$\frac{a}{b + c} + \frac{b}{c + d} + \frac{c}{d + a} + \frac{d}{a + b} \geq 2.$$

SOLUTION.   On the left-hand side of this inequality we add the first and the third terms, and the second and the fourth terms. Then, using (31), we estimate the common denominators $(b + c)(d + a)$ and $(c + d)(a + b)$ from above by $\frac{1}{4}(a + b + c + d)^2$:

$$L = \frac{a^2 + c^2 + ad + bc}{(b + c)(d + a)} + \frac{b^2 + d^2 + ab + cd}{(c + d)(a + b)}$$
$$\geq \frac{4(a^2 + b^2 + c^2 + d^2 + ad + bc + ab + cd)}{(a + b + c + d)^2}$$
$$= 2 \cdot \frac{(a + b + c + d)^2 + (a - c)^2 + (b - d)^2}{(a + b + c + d)^2} \geq 2.$$ □

4.6 *Exercises*

Prove the inequalities (i)–(v) for arbitrary $a, b, c, d \in \mathbb{R}^+$:

(i) $(1 + a + b)^2 \geq (1 + 2a)(1 + 2b)$.

(ii) $\sqrt{b} \leq a + \dfrac{b - a^2}{2a + 1} + \dfrac{1}{4(2a + 1)}$, if $a < \sqrt{b} < a + 1$.

(iii) $2\sqrt{a^3 + b^3} + 1 \leq a(a + 1) + b(b + 1)$, if $ab = 1$.

(iv) $\dfrac{ab}{a + b} + \dfrac{bc}{b + c} + \dfrac{ca}{c + a} \leq \dfrac{a + b + c}{2}$.

*(v) $\dfrac{a + b + c + d}{4} \geq \sqrt[3]{\dfrac{abc + abd + acd + bcd}{4}}$.

*(vi) For each integer $n \geq 2$ find the largest number $p_n \in \mathbb{R}_0^+$ such that for all n-tuples $x_1, x_2, \ldots, x_n \in \mathbb{R}_0^+$ we have

$$\left(\sum_{k=1}^{n} x_k\right)^2 \cdot \left(\sum_{k=1}^{n} x_k x_{k+1}\right) \geq p_n \cdot \sum_{k=1}^{n} x_k^2 x_{k+1}^2, \quad \text{where} \quad x_{n+1} = x_1.$$

*(vii) Let $n \geq 4$, and suppose that the sum of the nonnegative real numbers $a_1, a_2, \ldots, a_n$ is equal to 1. Prove the inequality

$$a_1 a_2 + a_2 a_3 + \cdots + a_{n-1} a_n + a_n a_1 \leq \frac{1}{4}.$$

Remark. The results of 2.2.(iv) and 4.6.(ii) present bounds for the value of $\sqrt{b}$ if we know an interval of length 1 in which this value lies. Thus the fact that $\sqrt{2}$ lies between 1 and 2 implies $\frac{4}{3} < \sqrt{2} < \frac{17}{12}$.

4.7 *Lower Bound for* $A + A^{-1}$

If we set $B = 1/A$ in (29), we get the inequality

$$A + \frac{1}{A} \geq 2 \quad (A \in \mathbb{R}^+), \tag{32}$$

with equality only if $A = 1$. This inequality is useful for solving a number of problems; from among them we will choose four examples. We remark that as a rule it is necessary to divide the desired inequality by an appropriate expression before (32) can be used.

(i) Show that for arbitrary $a, b, c \in \mathbb{R}^+$ we have

$$ab(a + b) + bc(b + c) + ac(a + c) \geq 6abc.$$

SOLUTION. After dividing by abc we rewrite the inequality as

$$\left(\frac{a}{b} + \frac{b}{a}\right) + \left(\frac{a}{c} + \frac{c}{a}\right) + \left(\frac{b}{c} + \frac{c}{b}\right) \geq 6,$$

which is the sum of the three inequalities obtained when (32) is applied to $A = \frac{a}{b}, \frac{a}{c}$, and $\frac{b}{c}$. $\qquad\qquad\qquad\square$

(ii) Show that for arbitrary $a, b, c, d \in \mathbb{R}^+$ we have

$$(a^2 + a + 1)(b^2 + b + 1)(c^2 + c + 1)(d^2 + d + 1) \geq 81abcd.$$

SOLUTION.   After dividing by $abcd$ we rewrite the inequality as

$$\left(a + 1 + \frac{1}{a}\right)\left(b + 1 + \frac{1}{b}\right)\left(c + 1 + \frac{1}{c}\right)\left(d + 1 + \frac{1}{d}\right) \geq 81.$$

By (32), none of the four expressions in parentheses is less than 3; their product is therefore at least $3^4 = 81$. $\qquad\qquad\qquad\square$

(iii) Prove the inequality $\sqrt{a}(a + 1) + a(a - 4) + 1 \geq 0$ for each $a \in \mathbb{R}^+$.

SOLUTION.   After dividing by a we rewrite the inequality as

$$\left(\sqrt{a} + \frac{1}{\sqrt{a}}\right) + \left(a + \frac{1}{a}\right) \geq 4.$$

By (32), none of the expressions in parentheses on the left is less than 2. $\qquad\qquad\qquad\square$

(iv) Show that for arbitrary $a, b, c \in \mathbb{R}^+$ we have

$$\frac{a}{b + c} + \frac{b}{c + a} + \frac{c}{a + b} \geq \frac{3}{2}.$$

SOLUTION.   If we set $u = b + c$, $v = c + a$, and $w = a + b$, we obtain $a = (v + w - u)/2$, $b = (u + w - v)/2$, $c = (u + v - w)/2$. Upon simplification we obtain

$$\frac{a}{b + c} + \frac{b}{c + a} + \frac{c}{a + b} = \frac{1}{2}\left[\left(\frac{u}{v} + \frac{v}{u}\right) + \left(\frac{v}{w} + \frac{w}{v}\right) + \left(\frac{w}{u} + \frac{u}{w}\right) - 3\right],$$

and it suffices to note that by (32) each of the three terms in parentheses is at least 2. $\qquad\qquad\qquad\square$

4.8   Exercises

Show that for $a, b, c, d \in \mathbb{R}^+$ the following inequalities hold:

(i) $\dfrac{a^2 + 3}{\sqrt{a^2 + 2}} > 2.$

(ii) $\dfrac{2a^2 + 1}{\sqrt{4a^2 + 1}} > 1.$

(iii) $(ab + cd)\left(\dfrac{1}{ac} + \dfrac{1}{bd}\right) \geq 4.$

(iv) $4ab(3 - a) - 4a(1 + b^2) \leq b.$

(v) $a^2 + b + \sqrt{a} + \sqrt{ab}\left(a\sqrt{b} - 4\sqrt{a}\right) \geq 0.$

5 The Discriminant and Cauchy's Inequality

Let us consider the quadratic polynomial

$$F(x) = ax^2 + bx + c \qquad (a, b, c \in \mathbb{R},\ a \neq 0). \tag{33}$$

Completing the square (see the introduction to Section 4), we obtain the identity

$$F(x) = a\left(x + \frac{b}{2a}\right)^2 - \frac{D}{4a}, \quad \text{where } D = b^2 - 4ac. \tag{34}$$

The number D is called the *discriminant* of the polynomial $F(x)$. The expression (34) implies not only the known formula for the roots $x_{1,2}$ of the equation $F(x) = 0$, but also the following result.

5.1 Values of a Quadratic Polynomial

Theorem. *Let $F(x)$ be the polynomial (33) with positive leading coefficient a and discriminant D as in (34). Put $x_0 = \frac{-b}{2a}$. Then*

(i) $F(x) > F(x_0)$ *for any $x \in \mathbb{R}$, $x \neq x_0$,*

(ii) $F(x) > 0$ *for all $x \in \mathbb{R}$ if and only if $D < 0$,*

(iii) *if $D = 0$, then $F(x) \geq 0$ for all $x \in \mathbb{R}$; in this case $F(x) = 0$ if and only if $x = x_0$,*

(iv) *if $D > 0$, the equation $F(x) = 0$ has two different roots, say $x_1 > x_2$; in this case $F(x) > 0$ when $x < x_2$ or $x > x_1$, and $F(x) < 0$ when $x_2 < x < x_1$.*

(The reader should think about how (i)–(iv) change in the case $a < 0$.)

PROOF. If $a > 0$, then by (34) we have

$$F(x) \geq -\frac{D}{4a} \qquad (x \in \mathbb{R}),$$

where equality occurs only for $x = x_0$. This implies the assertions (i)–(iii); the roots x_1, x_2 in part (iv) are given by the well-known quadratic formula, and the last assertion about the sign of $F(x)$ follows easily from the factorization $F(x) = a(x - x_1)(x - x_2)$ (see 3.14 in Chapter 1). $\square$

5.2 Examples

In the following seven examples we will practice the use of the properties of the discriminant stated and proved in 5.1.

(i) Show that the positive numbers a, b, c are side lengths of some triangle if and only if they satisfy the inequality

$$(a^2 + b^2 + c^2)^2 > 2(a^4 + b^4 + c^4). \tag{35}$$

SOLUTION.   The above inequality is equivalent to

$$a^4 - 2(b^2 + c^2)a^2 + (b^2 - c^2)^2 < 0.$$

The equation $x^2 - 2(b^2 + c^2)x + (b^2 - c^2)^2 = 0$ has the two roots $x_1 = (b+c)^2$ and $x_2 = (b - c)^2$. By 5.1.(iv), the inequality (35) holds if and only if $(b - c)^2 < a^2 < (b + c)^2$, that is, if and only if $|b - c| < a < b + c$, which is the known criterion for side lengths of a triangle.   □

(ii) Let $x_1, x_2, \ldots, x_n$ be given real numbers. Determine the $x \in \mathbb{R}$ for which the sum $S = (x - x_1)^2 + (x - x_2)^2 + \cdots + (x - x_n)^2$ is minimal.

SOLUTION.   We rewrite $S = nx^2 + bx + c$, where $b = -2(x_1 + x_2 + \cdots + x_n)$. Then by part (i) of Theorem 5.1, S attains its smallest value for

$$x = -\frac{b}{2n} = \frac{x_1 + x_2 + \cdots + x_n}{n}. \qquad □$$

(iii) Let $x, y, z \in \mathbb{R}$. Show that the smallest of the numbers $(x-y)^2$, $(y-z)^2$, and $(z - x)^2$ is greater than or equal to $(x^2 + y^2 + z^2)/2$.

SOLUTION.   In view of the symmetry we may assume without loss of generality that $x \le y \le z$, and we set $a = y - x \ge 0$, $b = z - y \ge 0$. Then $x^2 + y^2 + z^2 = (y - a)^2 + y^2 + (y + b)^2$. By 5.2.(ii) this sum is (for fixed values of a, b) smallest when $y = (a + 0 - b)/3 = (a - b)/3$. Hence

$$\frac{x^2 + y^2 + z^2}{2} \ge \frac{\left(\frac{a-b}{3} - a\right)^2 + \left(\frac{a-b}{3}\right)^2 + \left(\frac{a-b}{3} + b\right)^2}{2} = \frac{a^2 + ab + b^2}{3}.$$

Clearly, $(a^2 + ab + b^2)/3 \ge (c^2 + c^2 + c^2)/3 = c^2$, where c is the smallest of the two numbers a, b. Therefore, $(x^2 + y^2 + z^2)/2 \ge c^2$, where c^2 is the smallest of the two numbers $(y - x)^2$ and $(z - y)^2$.   □

(iv) Suppose that the numbers $a, b, c \in \mathbb{R}^+$ satisfy $a + b + c = abc$ and $a^2 = bc$. Show that $a \ge \sqrt{3}$.

SOLUTION.   By Vieta's relations (Chapter 1, Section 3) the numbers b and c are roots of the quadratic equation

$$x^2 - (a^3 - a)x + a^2 = 0,$$

since $bc = a^2$ and $b+c = abc-a = a \cdot a^2 - a = a^3 - a$. It therefore has a non-negative discriminant: $D = (a^3-a)^2 - 4a^2 \geq 0$. Since $D = a^2(a^2-3)(a^2+1)$, we obtain $a^2 > 3$, which implies the desired inequality $a \geq \sqrt{3}$.   □

(v) Suppose that x_1 is the smallest and x_n the largest among the n real numbers $x_1, x_2, \ldots, x_n$. Show that if $x_1 + x_2 + \cdots + x_n = 0$, then the sum $x_1^2 + x_2^2 + \cdots + x_n^2 + n x_1 x_n$ is not positive.

SOLUTION.   The polynomial $f(x) = x^2 - (x_1 + x_n)x + x_1 x_n$ has x_1 and x_n as its zeros. Hence Theorem 5.1 implies that if $x_1 \leq x \leq x_n$, then $0 \geq f(x)$. Adding the inequalities $0 \geq f(x_j)$ for $j = 1, 2, \ldots, n$, we obtain (in view of the condition $x_1 + x_2 + \cdots + x_n = 0$)

$$0 \geq \sum_{j=1}^{n} f(x_j) = \sum_{j=1}^{n} x_j^2 - (x_1 + x_n) \sum_{j=1}^{n} x_j + n x_1 x_n = \sum_{j=1}^{n} x_j^2 + n x_1 x_n. \quad □$$

(vi) The evaluation of the discriminant replaces several artificial tricks that were necessary for solving problems by way of the method of squares in Section 4. We illustrate this with the following example. Adding the inequalities $a + b + c \geq \sqrt{ab} + \sqrt{ac} + \sqrt{bc}$, $b + c \geq 2\sqrt{bc}$, $2(a + b) \geq 4\sqrt{ab}$, and $3(a+c) \geq 6\sqrt{ac}$ (see (27) and (29)) we obtain for arbitrary $a, b, c \in \mathbb{R}^+$,

$$6a + 4b + 5c \geq 5\sqrt{ab} + 7\sqrt{ac} + 3\sqrt{bc}. \tag{36}$$

"Discovering" this proof of (36) is certainly not easy; the reader should try to prove this inequality with the method of completing the square. However, the inequality (36) is quadratic in the variables $x = \sqrt{a}$, $y = \sqrt{b}$, and $z = \sqrt{c}$; with the help of the discriminant we will show that

$$6x^2 + 4y^2 + 5z^2 \geq 5xy + 7xz + 3yz \tag{37}$$

for arbitrary $x, y, z \in \mathbb{R}$, that is,

$$6x^2 - (5y + 7z)x + 4y^2 + 5z^2 - 3yz \geq 0.$$

(Note. The authors know that this will work. However, sometimes such a generalization can be misleading: It could happen that the inequality (37) holds only for $x, y, z \in \mathbb{R}^+$, since $x = \sqrt{a}$, etc. Then the discriminant of the last polynomial could be positive.) It therefore suffices to show that the discriminant $D = (5y + 7z)^2 - 24(4y^2 + 5z^2 - 3yz)$ is not positive for any $y, z \in \mathbb{R}$. By an easy calculation we find that $D = -71(y - z)^2$. This completes the proof of (37), and therefore also of (36).

(vii) In contrast to the previous example we now deal with a situation where the use of the discriminant is simply unsuitable. We show that the equation

$$F(x) = (x - a)(x - c) + 2(x - b)(x - d) = 0, \tag{38}$$

where $a < b < c < d$, has two distinct real roots. The following approach suggests itself: Since $F(x) = 3x^2 - (a+2b+c+2d)x + ac + 2bd$, the assertion holds if and only if $D = (a + 2b + c + 2d)^2 - 12(ac + 2bd) > 0$. The reader should attempt to derive this last inequality from the assumption $a < b < c < d$; this is not at all easy. Compare this with the following alternative solution: Substituting $x = b$, we obtain $F(b) = (b - a)(b - c) < 0$, since $a < b < c$. Therefore, the polynomial $F(x)$, with the positive coefficient 3 of x^2, attains a negative value at some point (namely when $x = b$). By Theorem 5.1 this means that $F(x)$ has two distinct real zeros. (Note that we required only the inequalities $a < b < c$ in this proof.) We finally remark that the zeros $x_1 > x_2$ of $F(x)$ satisfy the inequalities $a < x_2 < b$ and $c < x_1 < d$, since $F(a) > 0$, $F(b) < 0$, $F(c) < 0$, and $F(d) > 0$.

5.3   Exercises

(i) Suppose that the numbers $p, q, r, s \in \mathbb{R}$ are such that $x^2 + px + q \geq 0$ and $x^2 + rx + s \geq 0$ hold for all $x \in \mathbb{R}$. Show that $2x^2 + prx + 2qs \geq 0$ for all $x \in \mathbb{R}$.

(ii) Let $a, b, c \in \mathbb{R}^+$. Show that a, b, c are the side lengths of some triangle if and only if all $p, q \in \mathbb{R}$ with $p + q = 1$ satisfy the inequality

$$pa^2 + qb^2 > pqc^2.$$

(iii) Show that if a, b, c are the side lengths of some triangle, then the inequality $a^2 + b^2 + c^2 < 2(ab + bc + ca)$ holds.

(iv) Show that if $x < y < z < u$, then $(x + y + z + u)^2 > 8(xz + yu)$.

*(v) Determine the values $p \in \mathbb{R}$ for which

$$-2 < \frac{x^2 + 2px - 2}{x^2 - 2x + 2} < 2$$

holds for all $x \in \mathbb{R}$.

5.4   Cauchy's Inequality

Theorem. *The two arbitrary n-tuples of real numbers $u_1, u_2, \ldots, u_n$ and $v_1, v_2, \ldots, v_n$ satisfy the inequality*

$$(u_1v_1 + u_2v_2 + \cdots + u_nv_n)^2 \leq (u_1^2 + u_2^2 + \cdots + u_n^2)(v_1^2 + v_2^2 + \cdots + v_n^2). \tag{39}$$

Equality occurs in (39) only if $u_k = 0$ $(1 \leq k \leq n)$, or if there is a $t \in \mathbb{R}$ such that $v_k = tu_k$ $(1 \leq k \leq n)$.

PROOF 1.   If $u_k = 0$ $(1 \leq k \leq n)$, we have equality in (39). Let therefore $u_k \neq 0$ for some $k = 1, 2, \ldots, n$. We set

$$F(x) = (u_1x - v_1)^2 + (u_2x - v_2)^2 + \cdots + (u_nx - v_n)^2. \tag{40}$$

It is easy to see that $F(x)$ is a quadratic polynomial $ax^2 - 2bx + c$, where

$$a = \sum_{k=1}^{n} u_k^2 > 0, \quad b = \sum_{k=1}^{n} u_k v_k, \quad \text{and} \quad c = \sum_{k=1}^{n} v_k^2,$$

with discriminant $D = 4(b^2 - ac)$. By (40) we have $F(x) \geq 0$ for all $x \in \mathbb{R}$, and thus Theorem 5.1 implies $D \leq 0$, that is, $b^2 \leq ac$, but this is (39). Equality in (39) occurs if and only if $D = 0$, that is, if and only if $F(t) = 0$ for some $t \in \mathbb{R}$. By (40), this last condition means that $u_k t - v_k = 0$ for all $k = 1, 2, \ldots, n$. $\quad\square$

PROOF 2.   It suffices to consider the case where neither $u_1 = u_2 = \cdots = u_n = 0$ nor $v_1 = v_2 = \cdots = v_n = 0$. The inequality (39) is homogeneous (see 2.7) both in the variables $u_1, u_2, \ldots, u_n$ and in the variables $v_1, v_2, \ldots, v_n$. We may therefore normalize both n-tuples in such a way that

$$u_1^2 + u_2^2 + \cdots + u_n^2 = v_1^2 + v_2^2 + \cdots + v_n^2 = 1. \tag{41}$$

As in the first proof we set $b = u_1 v_1 + u_2 v_2 + \cdots + u_n v_n$. Adding the inequalities $2u_k v_k \leq u_k^2 + v_k^2$ (see (25)) for $k = 1, 2, \ldots, n$ we get, in view of (41), $2b \leq 2$, that is, $b \leq 1$. Similarly, the inequality $2(-u_k)v_k \leq (-u_k)^2 + v_k^2$ implies $b \geq -1$. Altogether we have therefore $-1 \leq b \leq 1$, that is, $b^2 \leq 1$, and this is (39) in the case (41). Equality in (39) occurs if and only if either $b = 1$ or $b = -1$, that is, if either $u_k = v_k$ $(1 \leq k \leq n)$, or $u_k = -v_k$ $(1 \leq k \leq n)$. If we return to the general n-tuple, we obtain the condition $v_k = t u_k$ $(1 \leq k \leq n)$ for appropriate $t \neq 0$. $\quad\square$

Remark 1. Cauchy's inequality (39) implies the weaker inequality

$$u_1 v_1 + u_2 v_2 + \cdots + u_n v_n \leq \sqrt{(u_1^2 + u_2^2 + \cdots + u_n^2)(v_1^2 + v_2^2 + \cdots + v_n^2)}. \tag{42}$$

(This inequality is trivial when the left-hand side is not positive; in the case of nonnegative numbers u_k, v_k $(1 \leq k \leq n)$, both forms (39) and (42) of Cauchy's inequality are equivalent.) From Proof 2 above it follows that equality in (42) occurs if and only if $u_k = 0$ $(1 \leq k \leq n)$ or $v_k = t u_k$ $(1 \leq k \leq n)$ for some $t \geq 0$.

Remark 2. Cauchy's inequality (39) can be written with the help of the vector $\vec{u}$ (with components u_i) and the vector $\vec{v}$ (with components v_i) in the form $(\vec{u}, \vec{v})^2 \leq (\vec{u}, \vec{u}) \cdot (\vec{v}, \vec{v})$, where $(\cdot, \cdot)$ stands for the *scalar product* in the n-dimensional Euclidean space. Equality then occurs if and only if the vectors $\vec{u}$ and $\vec{v}$ are *linearly dependent*. Thanks to this inequality one can introduce into these spaces the concepts of distance and of size of an angle, with similar properties as in elementary plane geometry (see, e.g., [10], pp. 217 f.).

5.5 Examples

We will now convince ourselves that Cauchy's inequality (39) can be used to prove a number of inequalities.

(i) Show that if $2x + 4y = 1$ for some $x, y \in \mathbb{R}$, then $x^2 + y^2 \geq \frac{1}{20}$.

SOLUTION. By (39) with $n = 2$ we have $(2x + 4y)^2 \leq (2^2 + 4^2)(x^2 + y^2)$. If $2x + 4y = 1$, we get $1 \leq 20(x^2 + y^2)$, that is, $x^2 + y^2 \geq \frac{1}{20}$. □

(ii) Show that for arbitrary $x, y, z \in \mathbb{R}$ we have

$$\left(\frac{x}{2} + \frac{y}{3} + \frac{z}{6}\right)^2 \leq \frac{x^2}{2} + \frac{y^2}{3} + \frac{z^2}{6}. \tag{43}$$

Furthermore, determine when equality occurs in (43).

SOLUTION. In (39) with $n = 3$ we set $u_1 = 1/\sqrt{2}$, $u_2 = 1/\sqrt{3}$, $u_3 = 1/\sqrt{6}$, $v_1 = x/\sqrt{2}$, $v_2 = y/\sqrt{3}$, and $v_3 = z/\sqrt{6}$. We thus obtain

$$\left(\frac{1}{\sqrt{2}} \cdot \frac{x}{\sqrt{2}} + \frac{1}{\sqrt{3}} \cdot \frac{y}{\sqrt{3}} + \frac{1}{\sqrt{6}} \cdot \frac{z}{\sqrt{6}}\right)^2 \leq \left(\frac{1}{2} + \frac{1}{3} + \frac{1}{6}\right)\left(\frac{x^2}{2} + \frac{y^2}{3} + \frac{z^2}{6}\right),$$

that is, (43). Since $v_1 = x u_1$, $v_2 = y u_2$, $v_3 = z u_3$, and $u_k \neq 0$ $(1 \leq k \leq 3)$, equality in (43) occurs if and only if $x = y = z$ (see Theorem 5.4). □

(iii) Show that for arbitrary $a_1, a_2, \ldots, a_n \in \mathbb{R}^+$ we have

$$\frac{a_1^2}{a_2} + \frac{a_2^2}{a_3} + \cdots + \frac{a_{n-1}^2}{a_n} + \frac{a_n^2}{a_1} \geq a_1 + a_2 + \cdots + a_n. \tag{44}$$

SOLUTION. By (39) with $u_k = a_k/\sqrt{a_{k+1}}$ and $v_k = \sqrt{a_{k+1}}$ $(1 \leq k \leq n$, $a_{n+1} = a_1)$ we have

$$\left(\frac{a_1}{\sqrt{a_2}} \cdot \sqrt{a_2} + \frac{a_2}{\sqrt{a_3}} \cdot \sqrt{a_3} + \cdots + \frac{a_n}{\sqrt{a_1}} \cdot \sqrt{a_1}\right)^2$$
$$\leq \left(\frac{a_1^2}{a_2} + \frac{a_2^2}{a_3} + \cdots + \frac{a_n^2}{a_1}\right)(a_2 + a_3 + \cdots + a_1),$$

and upon division by $a_1 + a_2 + \cdots + a_n$ we obtain (44). It is left to the reader to verify that equality in (44) occurs if and only if $a_1 = a_2 = \cdots = a_n$. □

(iv) Prove the so-called *triangle inequality*

$$\sqrt{\sum_{k=1}^{n}(u_k + v_k)^2} \leq \sqrt{\sum_{k=1}^{n} u_k^2} + \sqrt{\sum_{k=1}^{n} v_k^2} \tag{45}$$

for two arbitrary n-tuples of real numbers $u_1, \ldots, u_n$ and $v_1, \ldots, v_n$. (This inequality can be rewritten in vector form as $|\vec{u} + \vec{v}| \leq |\vec{u}| + |\vec{v}|$, where $|\cdot|$ denotes the *length* of the vector; the meaning of the attribute "triangle" is now clear from the geometrical construction of the vector sum $\vec{u} + \vec{v}$.)

SOLUTION. One can prove (45) as follows: Upon squaring, then subtracting $u_1^2 + \cdots + u_n^2 + v_1^2 + \cdots + v_n^2$, and finally dividing by two, we obtain the inequality (42). (From this we also obtain a criterion for equality in (45); see Remark 1 in 5.4 above.)

However, we give another proof with a trick that can also be used to prove the more general so-called *Minkowski inequality* (see Exercise 8.8.(xii)). If we add the two inequalities

$$\sum_{k=1}^{n}(u_k + v_k)u_k \leq \sqrt{\sum_{k=1}^{n}(u_k + v_k)^2 \sum_{k=1}^{n} u_k^2},$$

$$\sum_{k=1}^{n}(u_k + v_k)v_k \leq \sqrt{\sum_{k=1}^{n}(u_k + v_k)^2 \sum_{k=1}^{n} v_k^2},$$

of type (42), we obtain

$$\sum_{k=1}^{n}(u_k + v_k)^2 \leq \sqrt{\sum_{k=1}^{n}(u_k + v_k)^2} \cdot \left(\sqrt{\sum_{k=1}^{n} u_k^2} + \sqrt{\sum_{k=1}^{n} v_k^2} \right).$$

We have thus derived the inequality $L^2 \leq L \cdot R$, where $L \leq R$ is (45). Hence $L(L - R) \leq 0$. If $L = 0$, then (45) is trivial; if $L \neq 0$, then $L > 0$, and therefore the inequality $L(L - R) \leq 0$ implies $L - R \leq 0$, which proves (45).   □

(v) Show that for arbitrary $a_1, a_2, \ldots, a_n \in \mathbb{R}^+$ we have

$$(a_1 + a_2 + \cdots + a_n)\left(\frac{1}{a_1} + \frac{1}{a_2} + \cdots + \frac{1}{a_n} \right) \geq n^2, \qquad (46)$$

where equality occurs if and only if $a_1 = a_2 = \cdots = a_n$.

SOLUTION. (46) follows from (39) with $u_k = \sqrt{a_k}$ and $v_k = 1/\sqrt{a_k}$ for $k = 1, 2, \ldots, n$; since u_k and v_k are positive, equality in (46) occurs if and only if $v_k = tu_k$, that is, $1/\sqrt{a_k} = t\sqrt{a_k}$ $(1 \leq k \leq n)$ for some $t \in \mathbb{R}^+$. But this means that $a_1 = a_2 = \cdots = a_n \,(= 1/t)$, and the proof is complete.

We add that the inequality (46), which combines mathematical "beauty" with usefulness, is a special case of the inequality in Exercise 2.4.(xvi) with $r = -1$, $s = t = 0$, $u = 1$. It is also worth mentioning another generalization of (46):

$$(a_1 b_1 + a_2 b_2 + \cdots + a_n b_n)\left(\frac{b_1}{a_1} + \frac{b_2}{a_2} + \cdots + \frac{b_n}{a_n} \right) \geq (b_1 + b_2 + \cdots + b_n)^2,$$

with arbitrary $b_1, b_2, \ldots, b_n \in \mathbb{R}_0^+$. This follows from Cauchy's inequality (39) with the choice $u_k = \sqrt{a_k b_k}$ and $v_k = \sqrt{b_k/a_k}$, $k = 1, 2, \ldots, n$.   □

(vi) Show that for arbitrary $a, b, c \in \mathbb{R}^+$ we have

$$\frac{1}{1+a} + \frac{1}{1+b} + \frac{1}{1+c} \geq \frac{9}{3+a+b+c}.$$

SOLUTION.   We obtain this inequality from (46) with $n = 3$, $a_1 = 1 + a$, $a_2 = 1 + b$, and $a_3 = 1 + c$ upon dividing by $a_1 + a_2 + a_3 = 3 + a + b + c$.   □

(vii) Show that for two arbitrary n-tuples of positive numbers $b_1, b_2, \ldots, b_n$ and $c_1, c_2, \ldots, c_n$ we have

$$\sum_{k=1}^{n}(b_k + c_k)^2 \cdot \sum_{k=1}^{n}\frac{1}{b_k c_k} \geq 4n^2. \tag{47}$$

Furthermore, determine when equality occurs in (47).

SOLUTION.   By 4.3 we have $b_k + c_k \geq 2\sqrt{b_k c_k}$, that is, $(b_k + c_k)^2 \geq 4b_k c_k$, with equality if and only if $b_k = c_k$. This implies

$$\sum_{k=1}^{n}(b_k + c_k)^2 \sum_{k=1}^{n}\frac{1}{b_k c_k} \geq 4\sum_{k=1}^{n}b_k c_k \sum_{k=1}^{n}\frac{1}{b_k c_k}$$

(with equality if and only if $b_k = c_k$, $1 \leq k \leq n$). This inequality, together with (46) for $a_k = b_k c_k$ $(1 \leq k \leq n)$, gives (47). Equality in (47) occurs if and only if $b_k = c_k$ $(1 \leq k \leq n)$ and at the same time $b_1 c_1 = b_2 c_2 = \cdots = b_n c_n$ (see the criterion for equality in (46)), that is, if and only if $b_1 = b_2 = \cdots = b_n = c_1 = c_2 = \cdots = c_n$.   □

(viii) We obtain an important special case of Cauchy's inequality (39) by setting $v_1 = v_2 = \cdots = v_n = 1$:

$$(u_1 + u_2 + \cdots + u_n)^2 \leq n(u_1^2 + u_2^2 + \cdots + u_n^2), \tag{48}$$

with equality if and only if $u_1 = u_2 = \cdots = u_n$. (In the literature this inequality (48) is often referred to as Cauchy's inequality.)

The relation (48) between the sum of several numbers and the sum of their squares is useful in a number of situations, some of which we will mention now.

(ix) Show that if all the n zeros of the polynomial

$$F(x) = x^n - c_1 x^{n-1} + c_2 x^{n-2} - c_3 x^{n-3} + \cdots + (-1)^{n-1}c_{n-1}x + (-1)^n c_n \tag{49}$$

are real numbers, then the coefficients c_1 and c_2 satisfy the inequality

$$(n-1)c_1^2 \geq 2nc_2.$$

SOLUTION.   If $x_1, x_2, \ldots, x_n \in \mathbb{R}$ is the n-tuple of zeros of (49), then by Vieta's relations (see Chapter 1, Section 3) we have

$$c_1 = x_1 + x_2 + \cdots + x_n,$$
$$c_2 = x_1 x_2 + x_1 x_3 + \cdots + x_1 x_n + x_2 x_3 + \cdots + x_{n-1} x_n.$$

This implies $x_1^2 + x_2^2 + \cdots + x_n^2 = c_1^2 - 2c_2$, and by (48) with $u_k = x_k$ we therefore have $c_1^2 \leq n(c_1^2 - 2c_2)$, which gives $(n-1)c_1^2 \geq 2nc_2$.   □

(x) Show that the inequality $\sqrt{x+1} + \sqrt{2x-3} + \sqrt{50-3x} \leq 12$ holds for all values $x \in \mathbb{R}$ for which the left-hand side is defined.

SOLUTION.   By (48) with $n = 3$, $u_1 = \sqrt{x+1}$, $u_2 = \sqrt{2x-3}$, and $u_3 = \sqrt{50-3x}$ we have

$$(\sqrt{x+1} + \sqrt{2x-3} + \sqrt{50-3x})^2 \leq 3(x+1+2x-3+50-3x) = 144,$$

from which, upon taking the square root, we obtain the result.   □

(xi) Suppose that the sum of the n positive numbers $a_1, a_2, \ldots, a_n$ is equal to 1. Prove the inequality

$$\left(a_1 + \frac{1}{a_1}\right)^2 + \left(a_2 + \frac{1}{a_2}\right)^2 + \cdots + \left(a_n + \frac{1}{a_n}\right)^2 \geq \frac{(n^2+1)^2}{n}. \tag{50}$$

SOLUTION.   Using (48), we estimate the left-hand side L from below:

$$L \geq \frac{1}{n}\left(a_1 + \frac{1}{a_1} + a_2 + \frac{1}{a_2} + \cdots + a_n + \frac{1}{a_n}\right)^2$$
$$= \frac{1}{n}\left(1 + \frac{1}{a_1} + \frac{1}{a_2} + \cdots + \frac{1}{a_n}\right)^2.$$

Hence it suffices to show that

$$\frac{1}{a_1} + \frac{1}{a_2} + \cdots + \frac{1}{a_n} \geq n^2.$$

But this is (46), with $a_1 + a_2 + \cdots + a_n = 1$.   □

(xii) Let $n > 1$ and suppose that the numbers $x, y_1, y_2, \ldots, y_n \in \mathbb{R}$ satisfy

$$(n-1)(x + y_1^2 + y_2^2 + \cdots + y_n^2) < (y_1 + y_2 + \cdots + y_n)^2.$$

Show that if $1 \leq i < j \leq n$, then $x < 2y_i y_j$.

SOLUTION. In view of the symmetry we will prove only the inequality $x < 2y_1 y_2$. By Cauchy's inequality (48) for the $(n-1)$-tuple of numbers $y_1 + y_2, y_3, y_4, \ldots, y_n$ we have

$$(y_1 + y_2 + \cdots + y_n)^2 \leq (n-1)[(y_1 + y_2)^2 + y_3^2 + \cdots + y_n^2]$$
$$= 2(n-1)y_1 y_2 + (n-1)(y_1^2 + y_2^2 + \cdots + y_n^2).$$

From this and the hypothesis we obtain upon dividing by $n-1$,

$$x + y_1^2 + y_2^2 + \cdots + y_n^2 < 2y_1 y_2 + y_1^2 + y_2^2 + \cdots + y_n^2,$$

and thus $x < 2y_1 y_2$. □

(xiii) Show that if the numbers a, b, $c \in \mathbb{R}^+$ satisfy $abc = 1$, then

$$\frac{1}{a^3(b+c)} + \frac{1}{b^3(a+c)} + \frac{1}{c^3(a+b)} \geq \frac{1}{2}(ab + bc + ac).$$

SOLUTION. Since $abc = 1$, we have

$$L = \frac{x^2}{y+z} + \frac{y^2}{x+z} + \frac{z^2}{x+y}$$

and $R = \frac{1}{2}(x + y + z)$, where $x = \frac{1}{a}$, $y = \frac{1}{b}$ and $z = \frac{1}{c}$. By Cauchy's inequality (39) with $n = 3$ and the triples

$$(u_1, u_2, u_3) = (\sqrt{y+z}, \sqrt{x+z}, \sqrt{x+y}),$$
$$(v_1, v_2, v_3) = \left(\frac{x}{\sqrt{y+z}}, \frac{y}{\sqrt{x+z}}, \frac{z}{\sqrt{x+y}}\right)$$

we have $(x + y + z)^2 \leq 2(x + y + z) \cdot L$, which upon division by $2(x + y + z)$ leads to the conclusion $R \leq L$. □

5.6 Exercises

(i) Show that if $a, b, c, d \in \mathbb{R}_0^+$, then $\sqrt{(a+b)(c+d)} \geq \sqrt{ac} + \sqrt{bd}$.

(ii) Find the smallest value of the sum $x_1^2 + x_2^2 + \cdots + x_n^2$ under the condition that $x_1, x_2, \ldots, x_n \in \mathbb{R}$ satisfy $a_1 x_1 + a_2 x_2 + \cdots + a_n x_n = b$, where $a_1, a_2, \ldots, a_n, b$ are given real numbers with $a_k \neq 0$ for some $k = 1, 2, \ldots, n$.

Show that (iii) and (iv) hold for $a_1, a_2, \ldots, a_n \in \mathbb{R}^+$ and $x_1, x_2, \ldots, x_n \in \mathbb{R}$:

(iii) $\left(\sum_{k=1}^n a_k x_k\right)^2 \leq \left(\sum_{k=1}^n a_k\right)\left(\sum_{k=1}^n a_k x_k^2\right)$.

(iv) $\dfrac{x_1^2}{a_1} + \dfrac{x_2^2}{a_2} + \cdots + \dfrac{x_n^2}{a_n} \geq \dfrac{(x_1 + x_2 + \cdots + x_n)^2}{a_1 + a_2 + \cdots + a_n}.$

(v) Show that if $p, q, r, s \in \mathbb{R}$, then

$$(1 + p^4)(1 + q^4)(1 + r^4)(1 + s^4) \geq (1 + pqrs)^4.$$

(vi) Using the result of (v), show that if $a, b, c \in \mathbb{R}_0^+$, then

$$(1 + a^3)(1 + b^3)(1 + c^3) \geq (1 + abc)^3.$$

(vii) Show that if $p_k, q_k, r_k, s_k \in \mathbb{R}$, $1 \leq k \leq n$, then

$$\left(\sum_{k=1}^{n} p_k q_k r_k s_k \right)^4 \leq \left(\sum_{k=1}^{n} p_k^4 \right) \left(\sum_{k=1}^{n} q_k^4 \right) \left(\sum_{k=1}^{n} r_k^4 \right) \left(\sum_{k=1}^{n} s_k^4 \right).$$

(viii) Show that if $a, b, c, d \in \mathbb{R}^+$, then

$$\frac{1}{a+b+c} + \frac{1}{a+b+d} + \frac{1}{a+c+d} + \frac{1}{b+c+d} \geq \frac{16/3}{a+b+c+d}.$$

(ix) Show that if $x_1, x_2, \ldots, x_n \in \mathbb{R}$, then

$$(x_1 + x_2 + \cdots + x_n)^4 \leq n^3(x_1^4 + x_2^4 + \cdots + x_n^4).$$

(x) In Example 5.5.(ix), prove the inequality $(n-1)c_{n-1}^2 \geq 2nc_{n-2}c_n$.

(xi) Suppose that the real numbers $x_1, x_2, \ldots, x_n$ satisfy the inequality

$$x_1 + x_2 + \cdots + x_n \geq \sqrt{(n-1)(x_1^2 + x_2^2 + \cdots + x_n^2)}.$$

Show that $x_k \geq 0$ for each $k = 1, 2, \ldots, n$.

(xii) Show that if $a_k, b_k, c_k \in \mathbb{R}_0^+$, $1 \leq k \leq n$, then

$$\left(\sum_{k=1}^{n} a_k b_k c_k \right)^3 \leq \left(\sum_{k=1}^{n} a_k^3 \right) \left(\sum_{k=1}^{n} b_k^3 \right) \left(\sum_{k=1}^{n} c_k^3 \right).$$

Proceed in a way similar to Proof 2 in 5.4 and use 4.1.(ix).

Show that if $a_1, a_2, \ldots, a_n \in \mathbb{R}_0^+$, then (xiii) and (xiv) hold:

(xiii) $(a_1 + a_2 + \cdots + a_n)^3 \leq n^2(a_1^3 + a_2^3 + \cdots + a_n^3).$

(xiv) $(a_1^2 + a_2^2 + \cdots + a_n^2)^3 \leq n(a_1^3 + a_2^3 + \cdots + a_n^3)^2.$

*(xv) Show that for $a_1, a_2, \ldots, a_n \in \mathbb{R}^+$ we have

$$\frac{(a_1 + a_2 + \cdots + a_n)^2}{2(a_1^2 + a_2^2 + \cdots + a_n^2)} \leq \frac{a_1}{a_2 + a_3} + \frac{a_2}{a_3 + a_4} + \cdots + \frac{a_n}{a_1 + a_2}.$$

*(xvi) Let $a_1, a_2, \ldots, a_n \in \mathbb{R}_0^+$. Show that any numbers $x_1, x_2, \ldots, x_n \in \mathbb{R}$ satisfy

$$(a_1 x_1^2 + a_2 x_2^2 + \cdots + a_n x_n^2)^2 \leq a_1 x_1^4 + a_2 x_2^4 + \cdots + a_n x_n^4$$

if and only if $a_1 + a_2 + \cdots + a_n \leq 1$.

6   The Induction Principle

For the solution of inequality problems we can use induction methods in two different situations. The first one deals with inequalities of the type $L(n) \geq R(n)$, that is, inequalities where both sides are functions of the integer variable n. It is also assumed that these inequalities are composed of finite sums or products and possibly powers with exponents depending on n. This will be done in 6.1–6.5. The second situation concerns inequalities of the type $L(x_1, x_2, \ldots, x_n) \geq R(x_1, x_2, \ldots, x_n)$, where induction is used on the number n of the variables $x_1, x_2, \ldots, x_n$ (Sections 6.6–6.7). In the entire section, i, j, k, n are integers.

6.1   Examples

We begin with two examples to explain the two most common approaches to proofs of inequalities $L(n) \geq R(n)$ by mathematical induction. Instead of the usual concise solutions, we offer here some rather more informal remarks.

(i) Let us prove that for each $n \geq 6$ we have $2^n > (n+1)^2$. If we substitute the smallest value $n = 6$, we get $L(6) = 2^6 = 64$ and $R(6) = 7^2 = 49$. Hence the inequality $L(6) > R(6)$ is true. Instead of also verifying the next inequality $L(7) > P(7)$ by direct computation, we note that $L(7) = 2^7 = 2L(6)$, while $R(7)/R(6) = 64/49 < 2$. This shows that in the transition from $L(6) > P(6)$ to $L(7) > P(7)$ the larger side L increases by the factor 2, while the smaller side R increases only by the factor $64/49$; hence $L(7) > P(7)$ holds. Next, the relations $L(8)/L(7) = 2$ and $P(8)/P(7) = 81/64 < 2$ show that $L(8) > P(8)$ holds, etc. All these particular verifications have to be carried out *simultaneously*: we must show that for all $n \geq 6$ we have

$$\frac{L(n+1)}{L(n)} > \frac{P(n+1)}{P(n)}.$$

But this is quite easy:

$$\frac{L(n+1)}{L(n)} > \frac{P(n+1)}{P(n)} \iff 2 > \frac{(n+2)^2}{(n+1)^2}$$

$$\iff 2(n+1)^2 > (n+2)^2 \iff n^2 > 2.$$

We stress that such a "quotient" criterion can be used only for inequalities between positive expressions.

(ii) We explain the second, "difference," criterion by way of the inequality

$$\frac{1}{2^2} + \frac{1}{3^2} + \cdots + \frac{1}{n^2} \geq \frac{7}{12} - \frac{1}{n+1}, \tag{51}$$

which we will prove for each $n \geq 2$. For the smallest $n = 2$, equality occurs
in (51), since $L(2) = R(2) = \frac{1}{4}$. We note further that the left-hand side of
(51) is a finite sum; if we increase n by 1, there will be an additional term
on the left:

$$L(n+1) - L(n) = \frac{1}{(n+1)^2}.$$

It is also easy to determine the increase on the right-hand side of (51):

$$R(n+1) - R(n) = \left(\frac{7}{12} - \frac{1}{n+2}\right) - \left(\frac{7}{12} - \frac{1}{n+1}\right)$$
$$= \frac{1}{(n+1)(n+2)} < \frac{1}{(n+1)^2}.$$

Thus we see that the increase on the left is greater than that on the right:

$$L(n+1) - L(n) > R(n+1) - R(n) \quad \text{for all } n \geq 2.$$

The equality $L(2) = R(2)$ therefore implies $L(3) > R(3)$, and from this
follows $L(4) > R(4)$, etc. Hence we have $L(n) > R(n)$ for *all* $n \geq 3$.

These examples illustrate the fact that in induction proofs of the inequality $L(n) \geq R(n)$ we most often compare the *absolute*, resp. *relative*,
growths of both sides L, R as the value of n is increased by 1. We express
this in the following principle.

6.2   The Induction Principle

Theorem. *Let $L(n_0) \geq R(n_0)$ and suppose that for each $n \geq n_0$ one of
the two conditions* (i) *and* (ii) *holds:*

(i) *The numbers $L(n)$, $R(n)$, $L(n+1)$, and $R(n+1)$ are positive and
satisfy*

$$\frac{L(n+1)}{L(n)} \geq \frac{R(n+1)}{R(n)}.$$

(ii) $L(n+1) - L(n) \geq R(n+1) - R(n)$.

*Then the inequality $L(n) \geq R(n)$ holds for all $n \geq n_0$. If furthermore
$L(n_1) > R(n_1)$ for some $n_1 \geq n_0$, then $L(n) > R(n)$ for all $n \geq n_1$.*

Remark 1. In order to prove a finite number of inequalities $L(n) \geq R(n)$,
$n_0 \leq n \leq N$, we require that (i) or (ii) hold only for $n = n_0, n_0 + 1, \ldots,$
$N - 1$. In this case we talk about so-called *finite induction*.

Remark 2. The conditions (i), resp. (ii), often enter in a "hidden" manner
when from the assumption that for some $n \geq n_0$ the inequality $L(n) \geq R(n)$
holds, we derive the truth of the inequality $L(n+1) \geq R(n+1)$. Thus,
in Example 6.1.(i) we could have proceeded as follows: Multiplying the

inequality $2^n > (n+1)^2$ by 2, we obtain $2^{n+1} > 2(n+1)^2$. Hence it suffices to verify that $2(n+1)^2 \geq (n+2)^2$. In Example 6.1.(ii), by adding $1/(n+1)^2$ to both sides of (51), we obtain

$$\frac{1}{2^2} + \frac{1}{3^2} + \cdots + \frac{1}{n^2} + \frac{1}{(n+1)^2} \geq \frac{7}{12} - \frac{1}{n+1} + \frac{1}{(n+1)^2} = \frac{7}{12} - \frac{n}{(n+1)^2}.$$

Hence it suffices to verify the inequality

$$\frac{7}{12} - \frac{n}{(n+1)^2} > \frac{7}{12} - \frac{1}{n+2}.$$

6.3 Exercises

Show that the inequalities (i)–(viii) hold for all $n \geq n_0$:

(i) $(1-a)^n < \dfrac{1}{1+na}$ $(n_0 = 1,\ 0 < a < 1)$.

(ii) $\dfrac{4^n}{n+1} < \dfrac{(2n)!}{(n!)^2} < 4^n$ $(n_0 = 2)$.

(iii) $\dfrac{1}{n+1} + \dfrac{1}{n+2} + \cdots + \dfrac{1}{3n+1} > 1$ $(n_0 = 1)$.

(iv) $2^{\frac{1}{2}n(n-1)} > n!$ $(n_0 = 3)$.

(v) $x^{2^n} + y^{2^n} + n \geq (xy)^{2^{n-1}} + (xy)^{2^{n-2}} + \cdots + xy + x + y$
$(x, y \in \mathbb{R}, n_0 = 1)$.

(vi) $\dfrac{n}{2} < 1 + \dfrac{1}{2} + \dfrac{1}{3} + \cdots + \dfrac{1}{2^n - 1} < n$ $(n_0 = 2)$.

*(vii) $(1+a)^n > (1+a^{n+1})(1+a^{n+2}) \cdots (1+a^{2n})$ $(n_0 = 1,\ 0 < a \leq \frac{7}{12})$.

(viii) $2!4! \cdots (2n)! > [(n+1)!]^n$ $(n_0 = 2)$.

(ix) Show that if $1 \leq k \leq n$, then we have

$$1 + \frac{k}{n} \leq \left(1 + \frac{1}{n}\right)^k < 1 + \frac{k}{n} + \frac{k^2}{n^2}. \tag{52}$$

Remark. For $k = n$, (52) implies (compare with 3.8.(iv))

$$2 \leq \left(1 + \frac{1}{n}\right)^n < 3 \quad \text{for all } n \geq 1. \tag{53}$$

(x) Show that for all $n \geq 6$ we have

$$\left(\frac{n}{2}\right)^n > n! > \left(\frac{n}{3}\right)^n.$$

(xi) Show that if $k \geq 2$ and $n \geq 2$, then at least one of the numbers $\sqrt[k]{n}$ and $\sqrt[n]{k}$ is less than 2.

(xii) Given the expression

$$Q(n) = \frac{1}{2} \cdot \frac{3}{4} \cdots \frac{2n-1}{2n},$$

prove the estimates

$$Q(p)\sqrt{\frac{4p+1}{4n+1}} \leq Q(n) \leq Q(p)\sqrt{\frac{3p+1}{3n+1}} \qquad (n \geq p) \qquad (54)$$

for all $p \in \mathbb{N}$. Show that the inequalities (54) are stronger than those in 2.6.(x).

(xiii) Show that for all $n \geq 1$ we have

$$\sqrt{n + \sqrt{(n-1) + \sqrt{\cdots + \sqrt{2 + \sqrt{1}}}}} < \sqrt{n} + 1.$$

6.4 Examples

We will show now that in the course of using the induction principle it is often necessary to transform the desired inequality, or even change it to a stronger inequality.

(i) Out of n threes and n fours we form the numbers

$$A_n = 3^{3^{\cdot^{\cdot^{\cdot^3}}}} \quad \text{and} \quad B_n = 4^{4^{\cdot^{\cdot^{\cdot^4}}}}.$$

Which of the two numbers A_n and B_{n-1} is larger?

SOLUTION. We show that for each $n \geq 2$ we have $A_n > B_{n-1}$. With the induction method it is easier to show the stronger inequality $A_n > 2B_{n-1}$. By direct verification we find that $A_2 > 2B_1$. If the inequality $A_n > 2B_{n-1}$ holds for some $n \geq 2$, then

$$A_{n+1} = 3^{A_n} > 3^{2B_{n-1}} = 9^{B_{n-1}} = \left(\frac{9}{4}\right)^{B_{n-1}} \cdot 4^{B_{n-1}} > 2 \cdot 4^{B_{n-1}} = 2B_n. \quad \square$$

(ii) Find the smallest number A for which one can prove

$$\left(1 + \frac{1}{2^2}\right)\left(1 + \frac{1}{3^2}\right)\cdots\left(1 + \frac{1}{n^2}\right) \leq A \qquad (55)$$

by using the following approach: The sharper estimate

$$\left(1 + \frac{1}{2^2}\right)\left(1 + \frac{1}{3^2}\right)\cdots\left(1 + \frac{1}{n^2}\right) \leq A - \frac{B}{n+1} \qquad (56)$$

with appropriate $B > 0$ can be verified by way of the principle 6.2 with the "division" condition (i).

SOLUTION. Let us first point out that the inequality (55) cannot be directly proven by induction, since its left-hand side grows with increasing n, while the right-hand side remains constant. We therefore consider the stronger inequality (56); condition (i) in 6.2 means that for each $n \geq 2$ we need

$$\left(A - \frac{B}{n+1}\right)\left(1 + \frac{1}{(n+1)^2}\right) \leq A - \frac{B}{n+2},$$

which, upon equivalent transformations, can be written as

$$A(n^2 + 3n + 2) \leq B(n^2 + 3n + 3).$$

This last inequality holds for each $n \geq 2$ if and only if $A \leq B$. This implies with (56) for $n = 2$ that

$$1 + \frac{1}{4} \leq A - \frac{B}{3} \leq A - \frac{A}{3} = \frac{2A}{3}.$$

Thus we obtain $A \geq 15/8$. It is also clear that (56) with $A = B = 15/8$ holds for all $n \geq 2$; hence we may set $A = 15/8$ in (55). Can we use this approach to determine whether (55) is true also for some $A < 15/8$? We note that when (55) holds for some integer $n = p > 2$, then it holds also for all preceding $n = 2, 3, \ldots, p - 1$. Therefore, it suffices to show that the stronger inequality (56) is true for all $n \geq p$, where p is a certain (arbitrarily large) number. This occurs if and only if $A \leq B$ and

$$\left(1 + \frac{1}{2^2}\right)\left(1 + \frac{1}{3^2}\right)\cdots\left(1 + \frac{1}{p^2}\right) \leq A - \frac{B}{p+1}.$$

We can therefore set

$$A = B = \frac{p+1}{p} \cdot \left(1 + \frac{1}{2^2}\right)\left(1 + \frac{1}{3^2}\right)\cdots\left(1 + \frac{1}{p^2}\right).$$

We have thus shown that the estimate (55) holds with the constant

$$A = \frac{p+1}{p}\left(1 + \frac{1}{2^2}\right)\left(1 + \frac{1}{3^2}\right)\cdots\left(1 + \frac{1}{p^2}\right), \tag{57}$$

where $p \geq 2$ is an arbitrary integer. If we denote the right-hand side of (57) by A_p, then it is easy to see that

$$A_{p+1} = \left[1 - \frac{1}{(p+1)^4}\right] A_p.$$

This means that the value of A_p decreases as p grows. Actual computations give, for instance, $A_2 = 15/8 = 1.875$, $A_{13} \approx 1.83828$, and $A_{30} \approx 1.83806$.

We finally remark that the sequence of numbers $A_2, A_3, A_4, \ldots$ converges to a minimal value A that satisfies (55); this value is equal to the *infinite product*

$$\left(1 + \frac{1}{2^2}\right)\left(1 + \frac{1}{3^2}\right)\cdots\left(1 + \frac{1}{n^2}\right)\cdots.$$

By means of complex analysis one can show that the value of this product is equal to the number $(4\pi)^{-1} \cdot (e^{\pi} - e^{-\pi}) \approx 1.83804$ (compare this with the value of A_{30} above). □

6.5 Exercises

(i) Eight fives and five eights are used to form the numbers

$$5^{5^{\cdot^{\cdot^{\cdot^5}}}} \quad \text{and} \quad 8^{8^{\cdot^{\cdot^{\cdot^8}}}}.$$

Which one is larger?

(ii) Show that if $0 < a < 1$ and $0 < b < 1$, then for each $n \geq 1$ we have

$$(a + b - ab)^n + (1 - a^n)(1 - b^n) \geq 1.$$

(iii) Show that for each $n \geq 1$ we have

$$\left(1 + \frac{1}{1 \cdot 2}\right)\left(1 + \frac{1}{2 \cdot 3}\right)\cdots\left(1 + \frac{1}{n(n+1)}\right) \leq \frac{5}{2}.$$

To do this, replace the right-hand side by the expression

$$A - \frac{B}{n + C}$$

with appropriate constants A, B, C.

*(iv) Try to improve the upper bound in (54) in the form

$$Q(n) \leq \frac{Q(p)}{\sqrt{An + B}} \qquad (n \geq p),$$

where A and B are appropriate constants (depending on p).

(v) Show that for each $n \geq 1$ we have

$$\left(1 + \frac{1}{2}\right)\left(1 + \frac{1}{4}\right)\cdots\left(1 + \frac{1}{2^n}\right) \leq 3.$$

*(vi) Let $S(n) = 1^1 + 2^2 + \cdots + n^n$. Show that the inequality

$$\frac{1}{n^n} > \frac{1}{S(n)} + \frac{1}{S(n+1)} + \cdots + \frac{1}{S(n+k)}$$

holds for arbitrary integers $n \geq 2$ and $k \geq 0$. (To do this, use the fact that the numbers $c_k = (1 + \frac{1}{k})^k$, where $k = 1, 2, \ldots$, form an increasing sequence; see 3.4.(vi) and 3.8.(iv).)

6.6 Induction on the Number of Variables

We will now use the induction method to solve problems concerning inequalities of the form $Q(x_1, x_2, \ldots, x_n) \geq 0$. As a rule one can derive an inequality $Q(x_1, x_2, \ldots, x_{n+1}) \geq 0$ from the preceding inequality $Q(y_1, y_2, \ldots, y_n) \geq 0$, where $y_1, y_2, \ldots, y_n$ is some n-tuple, appropriately chosen and dependent on the given $(n+1)$-tuple $x_1, x_2, \ldots, x_{n+1}$. In the simplest case one can choose $y_1 = x_1, y_2 = x_2, \ldots, y_n = x_n$; in more complicated cases where the variables $x_1, x_2, \ldots, x_{n+1}$ are dependent on each other through conditions that have to be satisfied, the chosen variables $y_1, y_2, \ldots, y_n$ have to satisfy the same conditions. We will now consider five examples.

(i) Suppose that the real numbers $a_1, a_2, \ldots, a_n$ satisfy either the inequalities $a_k \geq 0$ $(k = 1, 2, \ldots, n)$, or $-1 \leq a_k < 0$ $(k = 1, 2, \ldots, n)$. Prove that

$$(1 + a_1)(1 + a_2) \cdots (1 + a_n) \geq 1 + a_1 + a_2 + \cdots + a_n .$$

(In the case $a_1 = a_2 = \cdots = a_n$ we obtain Bernoulli's inequality (17).)

SOLUTION.   Using the method of finite induction we show that for each index $k = 1, 2, \ldots, n$ we have

$$(1 + a_1)(1 + a_2) \cdots (1 + a_k) \geq 1 + a_1 + a_2 + \cdots + a_k. \qquad (58)$$

For $k = 1$, equality occurs in (58). If (58) holds for some $k < n$, then upon multiplication of both sides by the nonnegative number $1 + a_{k+1}$ we get

$$(1 + a_1)(1 + a_2) \cdots (1 + a_{k+1}) \geq (1 + a_1 + a_2 + \cdots + a_k)(1 + a_{k+1})$$
$$= 1 + a_1 + a_2 + \cdots + a_{k+1} + b_k,$$

where $b_k = a_1 a_{k+1} + a_2 a_{k+1} + \cdots + a_k a_{k+1} \geq 0$, since by assumption all the numbers $a_j a_{k+1}$ are nonnegative. Therefore, we have

$$(1 + a_1)(1 + a_2) \cdots (1 + a_{k+1}) \geq 1 + a_1 + a_2 + \cdots + a_{k+1},$$

which completes the proof by induction.   □

(ii) Suppose that $a_1, a_2, a_3, \ldots$ is a sequence of real numbers satisfying

$$a_{n+1} + a_{n-1} \geq 2a_n \qquad (59)$$

for each $n \geq 2$. Show that for each $n \geq 1$ we have

$$\frac{a_1 + a_3 + \cdots + a_{2n+1}}{n+1} \geq \frac{a_2 + a_4 + \cdots + a_{2n}}{n}. \qquad (60)$$

SOLUTION.   We rewrite the inequality (60) in the equivalent form

$$n(a_1 + a_3 + \cdots + a_{2n+1}) \geq (n+1)(a_2 + a_4 + \cdots + a_{2n}).$$

We denote this inequality by $L(n) \geq R(n)$ and prove it by induction. For $n = 1$ we have $L(1) = a_1 + a_3 \geq 2a_2 = R(1)$, by (59) with $n = 2$. The inequality $L(n+1) - L(n) \geq R(n+1) - R(n)$, which is of the form

$$(n+1)a_{2n+3} + a_1 + a_3 + \cdots + a_{2n+1} \geq (n+2)a_{2n+2} + a_2 + a_4 + \cdots + a_{2n},$$

can be rewritten as

$$S(n) = a_1 - a_2 + a_3 - \cdots - a_{2n} + a_{2n+1} - (n+2)a_{2n+2} + (n+1)a_{2n+3} \geq 0$$

and will also be proved by induction. For $n = 0$ we have $S(0) = a_1 - 2a_2 + a_3 \geq 0$, by (59) with $n = 2$. Since, again by (59), $a_{2n+5} \geq 2a_{2n+4} - a_{2n+3}$, we estimate the difference $S(n+1) - S(n)$ as follows:

$$
\begin{aligned}
S(n+1) &- S(n) \\
&= (n+2)a_{2n+5} - (n+3)a_{2n+4} - na_{2n+3} + (n+1)a_{2n+2} \\
&\geq (n+2)(2a_{2n+4} - a_{2n+3}) - (n+3)a_{2n+4} - na_{2n+3} + (n+1)a_{2n+2} \\
&= (n+1)(a_{2n+4} - 2a_{2n+3} + a_{2n+2}).
\end{aligned}
$$

By (59) the last term is nonnegative. Hence $S(n) \geq 0$ for all $n \geq 0$.   □

(iii) Show that if the numbers $a_1, a_2, \ldots, a_n \in \mathbb{R}^+, n \geq 3$, satisfy the inequality

$$(a_1^2 + a_2^2 + \cdots + a_n^2)^2 > (n-1)(a_1^4 + a_2^4 + \cdots + a_n^4), \qquad (61)$$

then any three numbers a_i, a_j, a_k $(1 \leq i < j < k \leq n)$ are side lengths of some triangle.

SOLUTION.   Since (61) for $n = 3$ is actually identical with (35), the assertion for $n = 3$ follows from 5.2.(i). Now let $n \geq 3$. In view of the symmetry it suffices to show that (61) is a consequence of the inequality

$$(a_1^2 + a_2^2 + \cdots + a_n^2 + a_{n+1}^2)^2 > n(a_1^4 + a_2^4 + \cdots + a_n^4 + a_{n+1}^4) \qquad (62)$$

for arbitrary $a_1, a_2, \ldots, a_{n+1} \in \mathbb{R}^+$. If we set $S_k = a_1^k + a_2^k + \cdots + a_n^k$ for $k = 2$ and $k = 4$, then (62) can be rewritten as $(S_2 + a_{n+1}^2)^2 > n(S_4 + a_{n+1}^4)$, which is equivalent to

$$(n-1)\left(a_{n+1}^2 - \frac{S_2}{n-1}\right)^2 < \frac{n}{n-1}(S_2)^2 - nS_4.$$

(Here we have used the method of completing the squares.) If (62) holds, then the right-hand side of the last inequality is positive, and thus we have $(S_2)^2 > (n-1)S_4$. But this, by the definition of S_2 and S_4, is just the inequality (61).   □

(iv) Suppose that the two n-tuples of real numbers $u_1, u_2, \ldots, u_n$ and $v_1, v_2, \ldots, v_n$ satisfy

$$u_1 \geq v_1, u_1 + u_2 \geq v_1 + v_2, \ldots, \quad u_1 + u_2 + \cdots + u_n \geq v_1 + v_2 + \cdots + v_n.$$

Show that if $x_1 \geq x_2 \geq \cdots \geq x_n \geq 0$, then also

$$u_1 x_1 + u_2 x_2 + \cdots + u_n x_n \geq v_1 x_1 + v_2 x_2 + \cdots + v_n x_n.$$

SOLUTION.   We show by finite induction that for $k = 1, 2, \ldots, n$ we have

$$u_1 x_1 + u_2 x_2 + \cdots + u_k x_k \geq v_1 x_1 + v_2 x_2 + \cdots + v_k x_k. \tag{63}$$

For $k = 1$, (63) is of the form $u_1 x_1 \geq v_1 x_1$, which follows from $u_1 \geq v_1$ and $x_1 \geq 0$. We will now consider the question: Is it possible to derive (63) for a given k, $1 < k \leq n$, from the inequality

$$u_1 y_1 + u_2 y_2 + \cdots + u_{k-1} y_{k-1} \geq v_1 y_1 + v_2 y_2 + \cdots + v_{k-1} y_{k-1}, \tag{64}$$

where $y_1 \geq y_2 \geq \cdots \geq y_{k-1} \geq 0$ is an appropriate $(k-1)$-tuple? To begin with, it is clear that we cannot choose $y_j = x_j$ $(1 \leq j \leq k-1)$; this, in order to derive (63) from (64), would require $u_k x_k \geq v_k x_k$, which does not follow from the assumptions. However, we have

$$(u_1 + u_2 + \cdots + u_k) x_k \geq (v_1 + v_2 + \cdots + v_k) x_k.$$

Now it suffices to note that we obtain (63) by adding this last inequality to (64) with $y_j = x_j - x_k$ $(1 \leq j \leq k-1)$. Since for these numbers we have $y_1 \geq y_2 \geq \cdots \geq y_{k-1} \geq 0$ (this follows from $x_1 \geq x_2 \geq \cdots \geq x_k$), the use of (64) is justified, and the proof by induction is complete.   □

(v) Show that if $x_1, x_2, \ldots, x_n \in \mathbb{R}^+$ and $x_1 x_2 \cdots x_n = 1$, then we have $x_1 + x_2 + \cdots + x_n \geq n$, with equality if and only if $x_1 = x_2 = \cdots = x_n = 1$.

SOLUTION.   For $n = 1$ the assertion is true. We assume that we have $(n+1)$ positive numbers $x_1, x_2, \ldots, x_{n+1}$ that satisfy $x_1 x_2 \cdots x_{n+1} = 1$, and we assume further that the assertion holds for n. If we choose $y_1 = x_1, y_2 = x_2, \ldots, y_{n-1} = x_{n-1}$ and $y_n = x_n x_{n+1}$, then $y_1 y_2 \cdots y_n = 1$, which means, by our assumption, that

$$y_1 + y_2 + \cdots + y_n = x_1 + x_2 + \cdots + x_{n-1} + x_n x_{n+1} \geq n.$$

We have to show that $x_1 + x_2 + \cdots + x_{n+1} \geq n + 1$. To this end, it clearly suffices to verify that $x_n + x_{n+1} \geq x_n x_{n+1} + 1$, that is, $(x_n - 1)(x_{n+1} - 1) \leq 0$. This, in general, does not follow from the assumption $x_1 x_2 \cdots x_{n+1} = 1$ alone; however, in view of symmetry we can order the numbers $x_1, x_2, \ldots, x_{n+1}$ in advance such that $x_n \leq x_i \leq x_{n+1}$

for $i = 1, 2, \ldots, n-1$. Then we have $(x_n)^{n+1} \leq x_1 x_2 \cdots x_{n+1} \leq (x_{n+1})^{n+1}$, that is, $x_n \leq 1 \leq x_{n+1}$, which implies $(x_n - 1)(x_{n+1} - 1) \leq 0$. This proves the inequality $x_1 + x_2 + \cdots + x_{n+1} \geq n + 1$; equality means that $x_1 = x_2 = \cdots = x_{n-1} = x_n x_{n+1} = 1$ and $(x_n - 1)(x_{n+1} - 1) = 0$, which implies $x_1 = x_2 = \cdots = x_{n+1} = 1$. □

Remark. The result of (v) follows immediately from the so-called AM–GM inequality, which will be studied in detail in Section 8. Since that inequality is homogeneous, we make use of the opposite direction: The proof of the AM–GM inequality in 8.1 will be based on the above result of (v).

6.7 Exercises

(i) Show that for each $n \geq 2$ we have

$$\left(1 - \frac{1}{2^2}\right)\left(1 - \frac{1}{2^3}\right)\cdots\left(1 - \frac{1}{2^n}\right) > \frac{1}{2}.$$

(ii) Suppose that the sequence of real numbers $x_1, x_2, x_3, \ldots$ satisfies

$$x_{n+1} \geq \frac{1}{n}(x_1 + x_2 + \cdots + x_n)$$

for each $n \geq 1$. Show that for all $n \geq 1$ we have

$$x_1 + 2x_2 + \cdots + nx_n \geq \frac{n+1}{2}(x_1 + x_2 + \cdots + x_n).$$

(iii) Suppose that the sequence of real numbers $x_1, x_2, x_3, \ldots$ is nondecreasing, that is, $x_{n+1} \geq x_n$ for each $n \geq 1$. Show that for all $n \geq 1$ we have the inequality

$$n(x_1 + 2x_2 + 4x_3 + \cdots + 2^{n-1}x_n) \geq (2^n - 1)(x_1 + x_2 + \cdots + x_n).$$

(iv) Suppose that the sequence of real numbers $a_1, a_2, a_3, \ldots$ satisfies the inequality $a_{n+1} \cdot a_{n-1} \geq a_n^2$ for each $n \geq 2$. Show that if we set $b_n = \sqrt[n]{a_1 a_2 \cdots a_n}$, then for each $n \geq 2$ the inequality $b_{n+1} \cdot b_{n-1} \geq b_n^2$ holds as well.

*(v) Suppose that the two n-tuples of positive numbers $a_1 \geq a_2 \geq \cdots \geq a_n$ and $b_1 \geq b_2 \geq \cdots \geq b_n$ satisfy $a_1 \geq b_1$, $a_1 + a_2 \geq b_1 + b_2, \ldots$, $a_1 + a_2 + \cdots + a_n \geq b_1 + b_2 + \cdots + b_n$. Show that for each $k \in \mathbb{N}$ we have $a_1^k + a_2^k + \cdots + a_n^k \geq b_1^k + b_2^k + \cdots + b_n^k$.

(vi) Use a method similar to 6.6.(v) to show that if the positive numbers $x_1, x_2, \ldots, x_n$ satisfy $x_1 + x_2 + \cdots + x_n = n$, then $x_1 x_2 \cdots x_n \leq 1$.

(vii) Show that if $n \geq 4$, then all $x_1, x_2, \ldots, x_n \in \mathbb{R}_0^+$ satisfy

$$(x_1 + x_2 + \cdots + x_n)^2 \geq 4(x_1 x_2 + x_2 x_3 + \cdots + x_n x_1).$$

*(viii) Show that if $a_1, a_2, \ldots, a_n \in \mathbb{N}$ are distinct, then

$$(a_1^7 + \cdots + a_n^7) + (a_1^5 + \cdots + a_n^5) \geq 2(a_1^3 + \cdots + a_n^3)^2.$$

7   Chebyshev's Inequality

7.1   Estimating a Sum Using Permutations

We will be concerned with the following problem: Given two n-tuples of real numbers $x_1, x_2, \ldots, x_n$ and $y_1, y_2, \ldots, y_n$, find the smallest and the largest value of the sum

$$S = x_1 z_1 + x_2 z_2 + \cdots + x_n z_n, \qquad (65)$$

where $z_1, z_2, \ldots, z_n$ is an arbitrary ordering of the numbers $y_1, y_2, \ldots, y_n$. A special case of this problem is the following situation: On a table there are four piles of coins, a pile each of one-crown, two-crown, five-crown, and ten-crown coins. From two piles you may take eight coins each, from the next one five, and from the last one three coins. How will you decide?

Obviously, it will be convenient to first order both the given n-tuples x_j and y_j. Let us therefore assume that $x_1 \leq x_2 \leq \cdots \leq x_n$ and $y_1 \leq y_2 \leq \cdots \leq y_n$. First we show that for each $k = 1, 2, \ldots, n$ the inequality

$$x_1 z_1 + x_2 z_2 + \cdots + x_k z_k \leq x_1 y_1 + x_2 y_2 + \cdots + x_k y_k \qquad (66)$$

holds, where $z_1, z_2, \ldots, z_k$ is an arbitrary ordering of the numbers $y_1, y_2, \ldots, y_k$; also, equality occurs in (66) if and only if the condition

$$(x_j - x_i)(z_j - z_i) \geq 0 \quad (1 \leq i < j \leq k) \qquad (67)$$

is satisfied. The reader should carefully consider what the condition (67) means; it may be surprising that equality in (66) can also occur for a k-tuple $z_1, z_2, \ldots, z_k$ different from the one for which $z_1 \leq z_2 \leq \cdots \leq z_k$, that is, $z_i = y_i$ $(1 \leq i \leq k)$. Note, however, that in the case where $x_i = x_j$ for some $i \neq j$, switching the numbers z_i and z_j does not change the value of the left-hand side of (66).

We prove the inequality (66) by finite induction. The case $k = 1$ is trivial. Let us now assume that the assertion holds for some $k < n$; we will show that for an arbitrary ordering $z_1, z_2, \ldots, z_{k+1}$ of the numbers $y_1, y_2, \ldots, y_{k+1}$ we have the inequality

$$x_1 z_1 + x_2 z_2 + \cdots + x_{k+1} z_{k+1} \leq x_1 y_1 + x_2 y_2 + \cdots + x_{k+1} y_{k+1}. \qquad (68)$$

If $z_{k+1} = y_{k+1}$, then $z_1, z_2, \ldots, z_k$ is an ordering of the numbers $y_1, y_2, \ldots, y_k$; hence the inequality (66) follows, and upon adding $x_{k+1} z_{k+1} = x_{k+1} y_{k+1}$ we obtain (68). If $z_{k+1} = y_{k+1}$ does not hold, then $z_{k+1} < y_{k+1}$ and $y_{k+1} = z_i$ for some $i \leq k$. We denote by $z_1', z_2', \ldots, z_{k+1}'$ the ordering that is obtained from $z_1, z_2, \ldots, z_{k+1}$ by switching the terms z_i and z_{k+1}. Obviously,

$$\begin{aligned}
(x_1 z_1' + \cdots + x_{k+1} z_{k+1}') &- (x_1 z_1 + \cdots + x_{k+1} z_{k+1}) \\
&= (x_i z_{k+1} + x_{k+1} y_{k+1}) - (x_i y_{k+1} + x_{k+1} z_{k+1}) \\
&= (x_{k+1} - x_i)(y_{k+1} - z_{k+1}) \geq 0,
\end{aligned}$$

since $x_{k+1} \geq x_i$ and $y_{k+1} > z_{k+1}$; to prove (68), it therefore suffices to verify that

$$x_1 z_1' + x_2 z_2' + \cdots + x_{k+1} z_{k+1}' \leq x_1 y_1 + x_2 y_2 + \cdots + x_{k+1} y_{k+1}. \qquad (69)$$

However, since $z_1', z_2', \ldots, z_{k+1}'$ is the ordering of the numbers $y_1, y_2, \ldots, y_{k+1}$ for which $z_{k+1}' = y_{k+1}$, the inequality (69) has already been proven. Through an easy examination of our approach we can verify the truth of criterion (67) for equality in (66); in the process it is determined that the left-hand side of (69) is larger than the left-hand side of (68) as long as $x_i < x_{k+1}$ (since $z_{k+1} < y_{k+1}$). This proves (66) for all $k = 1, 2, \ldots, n$.

To obtain a lower bound

$$x_1 z_1 + x_2 z_2 + \cdots + x_n z_n \geq x_1 y_n + x_2 y_{n-1} + \cdots + x_n y_1 \qquad (70)$$

we now use a simple trick; $z_1, z_2, \ldots, z_n$ is some ordering of the numbers $y_1 \leq y_2 \leq \cdots \leq y_n$ if and only if $-z_1, -z_2, \ldots, -z_n$ is an ordering of the numbers $-y_n \leq -y_{n-1} \leq \cdots \leq -y_1$. Therefore, by the previous discussion we have

$$x_1(-z_1) + x_2(-z_2) + \cdots + x_n(-z_n) \leq x_1(-y_n) + x_2(-y_{n-1}) + \cdots + x_n(-y_1),$$

which is just (70); furthermore, equality occurs in (70) if and only if

$$(x_j - x_i)(-z_j + z_i) \geq 0 \qquad (1 \leq i < j \leq n). \qquad (71)$$

We note that the conditions (67) and (71) are formulated in such a way that for their verification in specific situations the n-tuple $x_1, x_2, \ldots, x_n$ need not be previously ordered.

To summarize everything concisely (Theorem 7.3), we need the following definition.

7.2 Orderings

Definition. We say that the two n-tuples $u_1, u_2, \ldots, u_n$ and $v_1, v_2, \ldots, v_n$ of real numbers *have the same ordering*, resp. the *opposite ordering* if

$$(u_j - u_i)(v_j - v_i) \geq 0 \qquad (1 \leq i \leq n,\ 1 \leq j \leq n),$$

respectively

$$(u_j - u_i)(v_j - v_i) \leq 0 \qquad (1 \leq i \leq n,\ 1 \leq j \leq n).$$

(*Caution.* The fact that the n-tuples $u_1, u_2, \ldots, u_n$ and $v_1, v_2, \ldots, v_n$ have the same ordering does not necessarily mean that the n-tuples $u_1, u_2, \ldots, u_n$ and $v_n, v_{n-1}, \ldots, v_1$ have opposite orderings—consider, for instance, the quadruples 1, 1, 2, 3 and 2, 1, 3, 4. However, such a conclusion is true if, for instance, the numbers $u_1, u_2, \ldots, u_n$ are distinct.)

7.3   Extremal Values of a Sum

Theorem. *Let* $x_1, x_2, \ldots, x_n$ *and* $y_1, y_2, \ldots, y_n$ *be two n-tuples of real numbers. The sum* $S = x_1 z_1 + x_2 z_2 + \cdots + x_n z_n$, *where* $z_1, z_2, \ldots, z_n$ *is an arbitrary ordering of the numbers* $y_1, y_2, \ldots, y_n$, *is maximal, resp. minimal, if* $x_1, x_2, \ldots, x_n$ *and* $z_1, z_2, \ldots, z_n$ *have the same, resp. opposite, orderings. In particular, if the n-tuples* $x_1, x_2, \ldots, x_n$ *and* $y_1, y_2, \ldots, y_n$ *have the same ordering, and at the same time* $x_1, x_2, \ldots, x_n$ *and* $y_n, y_{n-1}, \ldots, y_1$ *have opposite orderings, then each of the n! sums S satisfies the inequalities*

$$x_1 y_n + x_2 y_{n-1} + \cdots + x_n y_1 \leq S \leq x_1 y_1 + x_2 y_2 + \cdots + x_n y_n. \qquad (72)$$

The proof was already carried out in Section 7.1.

7.4   Examples

The above theorem will be applied in the following two examples.

(i) Show that for arbitrary numbers $a_1, a_2, \ldots, a_n, p \in \mathbb{R}^+$ the inequality

$$a_1^p a_2 + a_2^p a_3 + \cdots + a_{n-1}^p a_n + a_n^p a_1 \leq a_1^{p+1} + a_2^{p+1} + \cdots + a_n^{p+1} \qquad (73)$$

holds, with equality if and only if $a_1 = a_2 = \cdots = a_n$.

SOLUTION.   For $p > 0$ the n-tuples $a_1, a_2, \ldots, a_n$ and $a_1^p, a_2^p, \ldots, a_n^p$ have the same ordering, by 1.10. Then by Theorem 7.3 the ordering $a_2, a_3, \ldots, a_n, a_1$ satisfies

$$a_1^p \cdot a_2 + a_2^p \cdot a_3 + \cdots + a_n^p \cdot a_1 \leq a_1^p \cdot a_1 + a_2^p \cdot a_2 + \cdots + a_n^p \cdot a_n,$$

and this is (73). Here we obtain equality only in the case where the n-tuples $a_1, a_2, \ldots, a_n$ and $a_2, a_3, \ldots, a_n, a_1$ have the same ordering. Let us assume that this is the case and that we do not at the same time have $a_1 = a_2 = \cdots = a_n$; if we set $M = \max(a_1, a_2, \ldots, a_n)$, then there exist indices i and j such that $a_{i-1} < a_i = M$ and $a_{j+1} < a_j = M$, where $a_0 = a_n$ and $a_{n+1} = a_1$. The inequality $(a_{i-1} - a_j)(a_i - a_{j+1}) < 0$ means that the original n-tuples $a_1, a_2, \ldots, a_n$ and $a_2, a_3, \ldots, a_n, a_1$ do not have the same ordering, and this is a contradiction.   □

(ii) Show that for arbitrary $a, b, c \in \mathbb{R}^+$ we have

$$a + b + c \leq \frac{a^2 + b^2}{2c} + \frac{b^2 + c^2}{2a} + \frac{c^2 + a^2}{2b}.$$

SOLUTION.   By 1.9 and 1.10, the triples (a^2, b^2, c^2) and $(\frac{1}{a}, \frac{1}{b}, \frac{1}{c})$ have opposite orderings, and thus Theorem 7.3 gives the inequalities

$$a^2 \cdot \frac{1}{a} + b^2 \cdot \frac{1}{b} + c^2 \cdot \frac{1}{c} \leq a^2 \cdot \frac{1}{b} + b^2 \cdot \frac{1}{c} + c^2 \cdot \frac{1}{a},$$

$$a^2 \cdot \frac{1}{a} + b^2 \cdot \frac{1}{b} + c^2 \cdot \frac{1}{c} \leq a^2 \cdot \frac{1}{c} + b^2 \cdot \frac{1}{a} + c^2 \cdot \frac{1}{b}.$$

If we add these two inequalities and then divide by two, we obtain the desired inequality.   □

7.5 Exercises

(i) Show that if $a_1, a_2, \ldots, a_n, p \in \mathbb{R}^+$, then the inequality

$$\frac{a_1^p}{a_2} + \frac{a_2^p}{a_3} + \cdots + \frac{a_{n-1}^p}{a_n} + \frac{a_n^p}{a_1} \geq a_1^{p-1} + a_2^{p-1} + \cdots + a_n^{p-1} \tag{74}$$

holds, with equality if and only if $a_1 = a_2 = \cdots = a_n$.

(ii) Give a new proof of the result of 6.6.(v): If $x_1, x_2, \ldots, x_n \in \mathbb{R}^+$ and if $x_1 x_2 \cdots x_n = 1$, then $x_1 + x_2 + \cdots + x_n > n$, with the exception of the case $x_1 = x_2 = \cdots = x_n = 1$. (Hint: Use (74) with $p = 1$ and an appropriate n-tuple $a_1, a_2, \ldots, a_n$.)

(iii) Show that for arbitrary $a, b, c \in \mathbb{R}^+$ we have

$$\frac{a^2 + b^2}{2c} + \frac{b^2 + c^2}{2a} + \frac{c^2 + a^2}{2b} \leq \frac{a^3}{bc} + \frac{b^3}{ac} + \frac{c^3}{ab}.$$

7.6 Chebyshev's Inequality

Theorem. *Suppose that we are given two ordered n-tuples of real numbers $x_1 \leq x_2 \leq \cdots \leq x_n$ and $y_1 \leq y_2 \leq \cdots \leq y_n$. Set*

$$\begin{aligned} S_{\min} &= x_1 y_n + x_2 y_{n-1} + \cdots + x_n y_1, \\ S_{\max} &= x_1 y_1 + x_2 y_2 + \cdots + x_n y_n \end{aligned} \tag{75}$$

(this notation is consistent with Theorem 7.3). Then the inequalities

$$n S_{\min} \leq (x_1 + x_2 + \cdots + x_n)(y_1 + y_2 + \cdots + y_n) \leq n S_{\max} \tag{76}$$

hold, with equality in the right and left parts of (76) occurring only simultaneously, and if and only if $x_1 = x_2 = \cdots = x_n$ or $y_1 = y_2 = \cdots = y_n$.

PROOF. By Theorem 7.3 and definition (75) we have

$$\begin{aligned} S_{\max} &= x_1 y_1 + x_2 y_2 + \cdots + x_n y_n, \\ S_{\max} &\geq x_1 y_2 + x_2 y_3 + \cdots + x_n y_1, \\ S_{\max} &\geq x_1 y_3 + x_2 y_4 + \cdots + x_n y_2, \\ &\vdots \\ S_{\max} &\geq x_1 y_n + x_2 y_1 + \cdots + x_n y_{n-1}. \end{aligned} \tag{77}$$

Adding all the relations in (77), we obtain

$$n S_{\max} \geq (x_1 + x_2 + \cdots + x_n)(y_1 + y_2 + \cdots + y_n),$$

and this is the right-hand part of (76). The left part of (76) can be proved in a similar way. If equality holds, for example in the left part of (76), then we have equality everywhere in (77). Let us assume that $x_1 = x_2 = \cdots = x_n$ does not hold, that is, that $x_1 < x_n$. Then according to 7.1 the equalities in (77) mean that $y_2 \leq y_1, y_3 \leq y_2, \ldots, y_n \leq y_{n-1}$, which implies $y_1 \geq y_2 \geq \cdots \geq y_n$. However, by the theorem's hypothesis we have $y_1 \leq y_2 \leq \cdots \leq y_n$, thus $y_1 = y_2 = \cdots = y_n$, and the proof is complete. $\qquad\square$

7.7 Example

We now consider a typical instance of the use of Chebyshev's inequality: We will prove that arbitrary positive numbers a, b, c satisfy the inequalities

$$(a^2 + b^2 + c^2)(a^3 + b^3 + c^3) \leq 3(a^5 + b^5 + c^5) \tag{78}$$

and

$$(a^4 + b^4 + c^4)\left(\frac{1}{a} + \frac{1}{b} + \frac{1}{c}\right) \geq 3(a^3 + b^3 + c^3), \tag{79}$$

with equality in both (78) and (79) if and only if $a = b = c$. In view of the symmetry we may assume that $0 < a \leq b \leq c$. Then clearly $a^n \leq b^n \leq c^n$ for $n = 2, 3, 4$, and $c^{-1} \leq b^{-1} \leq a^{-1}$. Hence (78) is the right part of (76) with $n = 3$, $x_1 = a^2$, $x_2 = b^2$, $x_3 = c^2$, $y_1 = a^3$, $y_2 = b^3$, and $y_3 = c^3$; similarly, (79) is the left part of (76) with $n = 3$, $x_1 = a^4$, $x_2 = b^4$, $x_3 = c^4$, $y_1 = c^{-1}$, $y_2 = b^{-1}$, and $y_3 = a^{-1}$. In both cases the triples x_1, x_2, x_3 and y_1, y_2, y_3 have the same ordering. This completes the proof.

Further consequences of Chebyshev's inequality will be derived in the following exercises. Especially the result of (v) deserves some attention; it is an interesting generalization of Cauchy's inequality (48).

7.8 Exercises

(i) Another proof of Chebyshev's inequality: Show that the expression

$$n(x_1y_1 + x_2y_2 + \cdots + x_ny_n) - (x_1 + x_2 + \cdots + x_n)(y_1 + y_2 + \cdots + y_n)$$

can be written as the sum

$$\frac{1}{2} \sum_{i,j=1}^{n} s_{ij},$$

where $s_{ij} = (x_i - x_j)(y_i - y_j)$.

(ii) Show that the inequality (46) is a consequence of Chebyshev's inequality.

(iii) How can inequalities (78) and (79) be generalized?

(iv) Show that for each integer $n \geq 2$ we have the inequality

$$\frac{1}{n} + \frac{2}{n-1} + \frac{3}{n-2} + \cdots + \frac{n}{1} > \frac{n+1}{2}\left(\frac{1}{n} + \frac{1}{n-1} + \frac{1}{n-2} + \cdots + \frac{1}{1}\right).$$

(v) Show that for $a_1, a_2, \ldots, a_n \in \mathbb{R}_0^+$ and each integer $k \geq 2$ the strict inequality

$$(a_1 + a_2 + \cdots + a_n)^k < n^{k-1}(a_1^k + a_2^k + \cdots + a_n^k) \qquad (80)$$

holds, with the exception of the case $a_1 = a_2 = \cdots = a_n$.

(vi) Show that for arbitrary $a_1, a_2, \ldots, a_n \in \mathbb{R}^+$ we have

$$(a_1 + \cdots + a_n)(a_1^3 + \cdots + a_n^3) \leq (a_1^5 + \cdots + a_n^5)\left(\frac{1}{a_1} + \cdots + \frac{1}{a_n}\right).$$

(vii) Show that the result of 6.7.(iii) is a certain Chebyshev inequality.

(viii) Suppose that the numbers $x, y \in \mathbb{R}$ satisfy $xy + 1 \geq x + y$. Show that for each integer $n \geq 1$ we have the inequality

$$(n-1)xy + 1 \geq \frac{(n-1)^2 xy + (n-1)(x+y) + 1}{n}.$$

(ix) Show that if $0 \leq a_{j1} \leq a_{j2} \leq \cdots \leq a_{jn}$ $(1 \leq j \leq k)$, then

$$\left(\sum_{i=1}^n a_{1i}\right)\left(\sum_{i=1}^n a_{2i}\right) \cdots \left(\sum_{i=1}^n a_{ki}\right) \leq n^{k-1} \sum_{i=1}^n a_{1i} a_{2i} \cdots a_{ki}.$$

(x) Show that the inequality

$$(x_1 + x_2 + \cdots + x_n)(y_1 + y_2 + \cdots + y_n) \leq n\,(x_1 y_1 + x_2 y_2 + \cdots + x_n y_n)$$

holds also in more general situations than in 7.6: The inequality $y_i \leq y_j$ holds for any indices $i, j \in \{1, 2, \ldots, n\}$ for which $x_i < A < x_j$, where $A = (x_1 + x_2 + \cdots + x_n)/n$.

8 Inequalities Between Means

We encounter the term *mean value* in many different situations; in most cases it refers to the *arithmetic mean* of the values $x_1, x_2, \ldots, x_n$, determined by the relation

$$\mathcal{A}_n(x_1, x_2, \ldots, x_n) = \frac{x_1 + x_2 + \cdots + x_n}{n}. \qquad (81)$$

However, there are also other kinds of means; the best known among them is the *geometric mean* of nonnegative numbers $x_1, x_2, \ldots, x_n$, which is defined by

$$\mathcal{G}_n(x_1, x_2, \ldots, x_n) = \sqrt[n]{x_1 x_2 \cdots x_n}. \tag{82}$$

Between the various kinds of means there are inequalities that can be successfully used to solve a variety of problems. Especially the inequality between the arithmetic and geometric mean, more concisely known as the *AM–GM inequality*, is among the most important results of the entire theory of inequalities, with numerous applications (see 8.1–8.5). A generalization of the AM–GM inequality to the case of *weighted means* will be discussed and applied in Sections 8.6–8.9; the remaining sections are then devoted to the theory of *power means* with arbitrary real degrees.

8.1   The AM–GM Inequality

Theorem. *For arbitrary numbers $a_1, a_2, \ldots, a_n \in \mathbb{R}_0^+$ the inequality*

$$\frac{a_1 + a_2 + \cdots + a_n}{n} \geq \sqrt[n]{a_1 a_2 \cdots a_n} \tag{83}$$

holds, with equality if and only if $a_1 = a_2 = \cdots = a_n$.

PROOF.   If $a_k = 0$ for some $k = 1, 2, \ldots, n$, then the statement is clear, since the right-hand side of (83) vanishes. Let therefore $a_k > 0$, $1 \leq k \leq n$. Since the inequality (83) is homogeneous (see 2.7), the n-tuple $a_1, a_2, \ldots, a_n$ can be normalized such that $a_1 a_2 \cdots a_n = 1$. But then the assertion (83) is identical with the result that was derived in Section 6.6.(v). This completes the proof of the AM–GM inequality.   □

Remark. Other proofs of (83) follow from the results of 3.4.(iv) and 6.7.(vi). The significance of the inequality (83) has stimulated the interest of mathematicians in finding proofs by different approaches and different methods (see, e.g., [1], [3]).

8.2   Applications of the AM–GM Inequality

The proofs of numerous inequalities follow from comparing the sum and the product of appropriate n-tuples of numbers, such as in the AM–GM inequality (83). In some situations (examples (i)–(vii)) we require a *lower bound for the sum*

$$A_1 + A_2 + \cdots + A_n \geq n \cdot \sqrt[n]{A_1 A_2 \cdots A_n}, \tag{84}$$

while in others (examples (viii)–(xi)) we need an *upper bound for the product*

$$A_1 A_2 \cdots A_n \leq \left(\frac{A_1 + A_2 + \cdots + A_n}{n} \right)^n. \tag{85}$$

We emphasize that the inequalities (84) and (85) hold for an arbitrary n-tuple of *nonnegative* numbers $A_1, A_2, \ldots, A_n$, with equality if and only if the numbers $A_1, A_2, \ldots, A_n$ are all equal. We have already seen the significance of (84) and (85) for $n = 2$ in Sections 4.3 and 4.5.

(i) Show that if $a \geq 0$, then $a^{11} - 3a^5 + a^4 + 1 \geq 0$, and that this inequality is sharp when $a \neq 1$.

SOLUTION.   By (84), the triple of numbers $a^{11}, a^4, 1$ satisfies

$$a^{11} + a^4 + 1 \geq 3 \cdot \sqrt[3]{a^{11} \cdot a^4 \cdot 1} = 3a^5,$$

and equality occurs if and only if $a^{11} = a^4 = 1$, that is, $a = 1$. This implies the assertion.   □

(ii) Show that for each $a \geq 0$ and for each integer $n \geq 1$ we have

$$\frac{a^n}{1 + a + a^2 + \cdots + a^{2n}} \leq \frac{1}{2n + 1}.$$

SOLUTION.   By (84) the numbers $1, a, a^2, \ldots, a^{2n}$ satisfy

$$1 + a + a^2 + \cdots + a^{2n} \geq (2n + 1) \cdot \sqrt[2n+1]{1 \cdot a^2 \cdots a^{2n}}$$
$$= (2n + 1) \cdot \sqrt[2n+1]{a^{n(2n+1)}} = (2n + 1)a^n,$$

which is equivalent to the desired inequality.   □

(iii) Determine the largest value $p \in \mathbb{R}$ for which the inequality $a^2 b^2 c^2 + ab + bc + ca \geq pabc$ holds for arbitrary numbers $a, b, c \in \mathbb{R}^+$.

SOLUTION.   If we set $a = b = c = 1$, we obtain the condition $p \leq 4$. On the other hand, by (84) the four numbers $a^2 b^2 c^2, ab, bc, ca$ satisfy

$$a^2 b^2 c^2 + ab + bc + ca \geq 4 \cdot \sqrt[4]{a^2 b^2 c^2 abbcca} = 4abc.$$

Hence the desired largest value of p is equal to 4.   □

(iv) Show that if $0 < b < a$, then we have the inequality

$$a + \frac{1}{(a - b)b} \geq 3.$$

SOLUTION.   By (84), applied to the number triple $b, a - b$, and $[(a-b)b]^{-1}$, we have

$$a + \frac{1}{(a - b)b} = b + (a - b) + \frac{1}{(a - b)b} \geq 3 \cdot \sqrt[3]{b(a - b) \cdot \frac{1}{(a - b)b}} = 3. \quad \square$$

(v) Show that each $a > 0$ satisfies the inequality

$$\frac{a^4 + 9}{10a} > \frac{4}{5}.$$

SOLUTION. We apply (84) with an appropriate quadruple of numbers:

$$\frac{a^4 + 9}{10a} = \frac{a^3}{10} + \frac{3}{10a} + \frac{3}{10a} + \frac{3}{10a} \geq 4 \cdot \sqrt[4]{\frac{a^3}{10} \cdot \frac{3}{10a} \cdot \frac{3}{10a} \cdot \frac{3}{10a}} = \frac{2}{5} \cdot \sqrt[4]{27}.$$

Now it suffices to note that $\sqrt[4]{27} > 2$, since $27 > 16 = 2^4$. $\qquad\square$

(vi) Show that if $a, b \in \mathbb{R}^+$, then $3a^3 + 7b^3 > 9ab^2$.

SOLUTION. The term ab^2 on the right is the geometric mean of the three numbers a^3, b^3, b^3. We therefore try to use the estimate

$$3a^3 + p_1 b^3 + p_2 b^3 \geq 3 \cdot \sqrt[3]{3p_1 p_2 a^3 b^6} = 3 \cdot \sqrt[3]{3p_1 p_2} \cdot ab^2 \qquad (86)$$

for appropriate positive numbers p_1, p_2. We require that $p_1 + p_2 < 7$ and $3 \cdot \sqrt[3]{3p_1 p_2} \geq 9$, that is, $p_1 p_2 \geq 9$. We easily see that both conditions are satisfied by $p_1 = p_2 = 3$. This solves the problem. It is left to the reader to explain why the estimate (86) is "most valuable" in the case $p_1 = p_2$. $\qquad\square$

(vii) Show that for any $a, b, c \in \mathbb{R}_0^+$ we have the inequality

$$2(a + b + c)(a^2 + b^2 + c^2) \geq a^3 + b^3 + c^3 + 15abc.$$

SOLUTION. Upon expanding the left-hand side and subtracting the sum $a^3 + b^3 + c^3$ from both sides of the inequality we obtain

$$a^3 + b^3 + c^3 + 2(a^2b + a^2c + b^2a + b^2c + c^2a + c^2b) \geq 15abc.$$

The left-hand side L of this last inequality is a sum of $3 + 2 \cdot 6 = 15$ terms; by (84) we have

$$L \geq 15 \cdot \sqrt[15]{a^3 b^3 c^3 (a^2 b)^2 (a^2 c)^2 (b^2 a)^2 (b^2 c)^2 (c^2 a)^2 (c^2 b)^2} = 15abc. \qquad\square$$

(viii) Prove the inequality $1 \cdot 3 \cdot 5 \cdots (2n - 1) < n^n$ for all $n > 1$.

SOLUTION. Applying (85) to the numbers $1, 3, 5, \ldots, 2n - 1$, we obtain

$$1 \cdot 3 \cdot 5 \cdots (2n - 1) < \left[\frac{1 + 3 + 5 + \cdots + (2n - 1)}{n} \right]^n = \left(\frac{n^2}{n} \right)^n = n^n,$$

since $1 + 3 + 5 + \cdots + (2n - 1) = n^2$ (see Chapter 1, Section 2). $\qquad\square$

(ix) Show that if $0 < b < 2a$, then $16b(2a - b)^3 \leq 27a^4$.

SOLUTION.   We use (85) with the four numbers $3b, 2a-b, 2a-b, 2a-b$ (the coefficient 3 of b is chosen such that the sum of all 4 numbers is independent of b):

$$(3b)(2a - b)^3 \leq \left[\frac{3b + 3(2a - b)}{4} \right]^4 = \left(\frac{3a}{2} \right)^4,$$

which, upon multiplying by $16/3$, gives the desired inequality.   □

(x) Find the largest value of the expression $V = (a - x)(b - y)(cx + dy)$, where $a, b, c, d \in \mathbb{R}^+$ are given numbers, and the real variables x and y satisfy $x < a$, $y < b$ and $cx + dy > 0$.

SOLUTION.   The expression V is the product of the three terms in parentheses, whose sum, however, is not generally constant. But we have $c(a - x) + d(b - y) + (cx + dy) = ac + bd$; hence we apply (85) with the three numbers $c(a - x), d(b - y), (cx + dy)$:

$$cdV = c(a - x)d(b - y)(cx + dy) \leq \left(\frac{ac + bd}{3} \right)^3.$$

This gives the bound

$$V \leq \frac{(ac + bd)^3}{27cd}, \tag{87}$$

with equality if and only if $c(a-x) = d(b-y) = cx+dy$; an easy computation gives

$$x = \frac{2ac - bd}{3c} \quad \text{and} \quad y = \frac{2bd - ac}{3d}.$$

The reader should verify that these values are admissible. The number on the right-hand side of (87) is therefore the largest value of V.   □

(xi) Show that for each $k \in \mathbb{N}$ we have the inequality

$$1^1 \cdot 2^2 \cdot 3^3 \cdots k^k \geq \left(\frac{k + 1}{2} \right)^{\frac{k(k+1)}{2}}. \tag{88}$$

SOLUTION.   We consider the system of the $n_k = \frac{1}{2}(k^2 + k)$ numbers

$$1, \frac{1}{2}, \frac{1}{2}, \frac{1}{3}, \frac{1}{3}, \frac{1}{3}, \ldots, \frac{1}{k}, \frac{1}{k}, \ldots, \frac{1}{k}.$$

Since their sum is equal to k, by (85) we have

$$1 \cdot \frac{1}{2^2} \cdot \frac{1}{3^3} \cdots \frac{1}{k^k} \leq \left(\frac{k}{n_k} \right)^{n_k} = \left(\frac{2}{k + 1} \right)^{n_k}.$$

By writing down the opposite inequality for the reciprocals, we obtain the inequality (88).   □

8.3   Exercises

Using (84), prove the inequalities (i)–(vii), where $a, b, c \in \mathbb{R}^+$:

(i) $a^4 + a^3 - 8a^2 + 4a + 4 > 0$.

(ii) $a^2(b + c) + b^2(c + a) + c^2(a + b) \geq 6abc$.

(iii) $a^4 + 2a^3b + 2ab^3 + b^4 \geq 6a^2b^2$.

(iv) $2\sqrt{a} + 3 \cdot \sqrt[3]{b} \geq 5 \cdot \sqrt[5]{ab}$.

(v) $\dfrac{a^3 + b^6}{2} \geq 3ab^2 - 4$.

(vi) $a + \dfrac{4}{a^2} \geq 3$.

(vii) $a^n - 1 \geq n\left(a^{\frac{n+1}{2}} - a^{\frac{n-1}{2}}\right)$, if $a > 1$ and $n \in \mathbb{N}$.

Using (85), prove the inequalities (viii)–(xv), where $a, b, c, d \in \mathbb{R}^+$ and $k, m \in \mathbb{N}$:

(viii) $ab^2c^3d^4 \leq \left(\dfrac{a + 2b + 3c + 4d}{10}\right)^{10}$.

(ix) $(k^2 + 3k)(k^2 + 5k + 6)^{k+2} < (k^2 + 5k + 4)^{k+3}$.

(x) $27a^2b \leq 4(a + b)^3$.

(xi) $432ab^2c^3 \leq (a + b + c)^6$.

(xii) $(1 - a)^5(1 + a)(1 + 2a)^2 < 1$.

(xiii) $(b - a)^k(b + ka) < b^{k+1}$, if $a < b$.

(xiv) $1^1 \cdot 2^2 \cdot 3^3 \cdots k^k \leq \left(\dfrac{2k + 1}{3}\right)^{\frac{k(k+1)}{2}}$.

(xv) $k^m \cdot m^k \leq \left(\dfrac{2km}{k + m}\right)^{k+m}$.

(xvi) Show that for arbitrary $a_1, a_2, \ldots, a_k \in \mathbb{R}_0^+$ we have

$$(1 + a_1)(1 + a_2) \cdots (1 + a_k) \leq \sum_{j=0}^{k} \frac{(a_1 + a_2 + \cdots + a_k)^j}{j!}.$$

8.4   Further Applications of the AM–GM Inequality

In order to solve the following problems, which are somewhat more difficult than those in 8.2, we will usually have to apply the AM–GM inequality several times over. For clarity we will use the symbols $\mathcal{A}_n$ and $\mathcal{G}_n$ for the two means introduced in (81) and (82). It is important that the reader be well acquainted with these symbols, for instance through verifying the following relations (which hold for arbitrary $a_k, b_k, c \in \mathbb{R}_0^+$):

$$a_1 + a_2 + \cdots + a_n = n \cdot \mathcal{A}_n(a_1, a_2, \ldots, a_n),$$
$$a_1 a_2 \cdots a_n = \mathcal{G}_n(a_1^n, a_2^n, \ldots, a_n^n),$$
$$\mathcal{A}_n(ca_1, \ldots, ca_n) = c \cdot \mathcal{A}_n(a_1, \ldots, a_n),$$
$$\mathcal{G}_n(ca_1, \ldots, ca_n) = c \cdot \mathcal{G}_n(a_1, \ldots, a_n),$$
$$\mathcal{G}_n(a_1, \ldots, a_n)\mathcal{G}_n(b_1, \ldots, b_n) = \mathcal{G}_n(a_1 b_1, \ldots, a_n b_n),$$
$$\mathcal{G}_n\left(\frac{1}{a_1}, \ldots, \frac{1}{a_n}\right) = \frac{1}{\mathcal{G}_n(a_1, \ldots, a_n)} \quad (a_k > 0, \ 1 \le k \le n).$$

These will be used repeatedly in the solutions to the problems below. We also remark that the lower index n with the symbols $\mathcal{A}$ and $\mathcal{G}$ helps avoid misunderstandings in some ambiguous expressions of the type $\mathcal{A}_n(a, b, b, \ldots, b)$. In this example we are dealing with the arithmetic mean of n numbers: the number a and $n - 1$ numbers equal to b.

(i) Show that for arbitrary $a, b, c \in \mathbb{R}_0^+$ we have the inequality

$$(a + b + c)(a^2 + b^2 + c^2) \ge 9abc.$$

SOLUTION.   We prove the statement as follows:

$$L = 9\mathcal{A}_3(a, b, c)\mathcal{A}_3(a^2, b^2, c^2) \ge 9\mathcal{G}_3(a, b, c)\mathcal{G}_3(a^2, b^2, c^2)$$
$$= 9\mathcal{G}_3(a^3, b^3, c^3) = 9abc = R.$$

A different proof follows from the AM–GM inequality applied to the 9 numbers that are obtained by multiplying out the left-hand side of the given inequality.   □

(ii) The *harmonic mean* of the numbers $a_1, a_2, \ldots, a_n \in \mathbb{R}^+$ is defined as follows:

$$\mathcal{H}_n(a_1, a_2, \ldots, a_n) = \frac{n}{\frac{1}{a_1} + \frac{1}{a_2} + \cdots + \frac{1}{a_n}}. \tag{89}$$

Show that the harmonic and geometric means satisfy the inequality

$$\mathcal{H}_n(a_1, a_2, \ldots, a_n) \le \mathcal{G}_n(a_1, a_2, \ldots, a_n), \tag{90}$$

with equality if and only if $a_1 = a_2 = \cdots = a_n$.

SOLUTION.   By definition (89) and by the AM–GM inequality we have

$$\mathcal{H}_n(a_1, \ldots, a_n) = \frac{1}{\mathcal{A}_n(\frac{1}{a_1}, \ldots, \frac{1}{a_n})} \le \frac{1}{\mathcal{G}_n(\frac{1}{a_1}, \ldots, \frac{1}{a_n})} = \mathcal{G}_n(a_1, \ldots, a_n),$$

with equality if and only if $1/a_1 = 1/a_2 = \cdots = 1/a_n$, that is, $a_1 = a_2 = \cdots = a_n$.

As a consequence of (83) and (90) we obtain an inequality between the arithmetic and the harmonic means; this also follows from the inequality (46).   □

(iii) Show that for arbitrary $a, b, c \in \mathbb{R}_0^+$ we have

$$\sqrt[3]{ab} + \sqrt[3]{bc} + \sqrt[3]{ca} - 1 \le \frac{2}{3}(a + b + c).$$

SOLUTION. The desired inequality is the sum of the three inequalities

$$\sqrt[3]{ab} - \frac{1}{3} \le \frac{a + b}{3}, \quad \sqrt[3]{bc} - \frac{1}{3} \le \frac{b + c}{3}, \quad \text{and} \quad \sqrt[3]{ca} - \frac{1}{3} \le \frac{c + a}{3}.$$

In view of symmetry it suffices to prove the first one of these,

$$\sqrt[3]{ab} = \mathcal{G}_3(a, b, 1) \le \mathcal{A}_3(a, b, 1) = \frac{a + b + 1}{3},$$

and subtract $\frac{1}{3}$.   □

(iv) Show that for arbitrary $a_1, a_2, \ldots, a_n \in \mathbb{R}^+$ we have

$$\frac{S}{S - a_1} + \frac{S}{S - a_2} + \cdots + \frac{S}{S - a_n} \ge \frac{n^2}{n - 1},$$

where $n \ge 2$ and $S = a_1 + a_2 + \cdots + a_n$.

SOLUTION. To obtain a lower bound for the left-hand side, we apply the AM–GM inequality twice:

$$L = n \cdot \mathcal{A}_n \left(\frac{S}{S - a_1}, \ldots, \frac{S}{S - a_n} \right) \ge n \mathcal{G}_n \left(\frac{S}{S - a_1}, \ldots, \frac{S}{S - a_n} \right)$$

$$= nS \cdot \mathcal{G}_n \left(\frac{1}{S - a_1}, \ldots, \frac{1}{S - a_n} \right) = \frac{nS}{\mathcal{G}_n(S - a_1, \ldots, S - a_n)}$$

$$\ge \frac{nS}{\mathcal{A}_n(S - a_1, \ldots, S - a_n)} = \frac{n^2 S}{(S - a_1) + \cdots + (S - a_n)}$$

$$= \frac{n^2 S}{nS - (a_1 + \cdots + a_n)} = \frac{n^2 S}{nS - S} = \frac{n^2}{n - 1}.$$ □

(v) Suppose that the numbers $x_1, x_2, \ldots, x_n$ are all greater than -1, and that their sum is equal to zero. Prove the inequality

$$\frac{x_1}{x_1 + 1} + \frac{x_2}{x_2 + 1} + \cdots + \frac{x_n}{x_n + 1} \le 0.$$

SOLUTION. After rewriting the right-hand side we apply the AM–GM inequality twice:

$$L = \frac{(x_1+1)-1}{x_1+1} + \cdots + \frac{(x_n+1)-1}{x_n+1} = n - \left(\frac{1}{x_1+1} + \cdots + \frac{1}{x_n+1}\right)$$

$$= n - n \cdot \mathcal{A}_n\left(\frac{1}{x_1+1}, \ldots, \frac{1}{x_n+1}\right) \le n - n \cdot \mathcal{G}_n\left(\frac{1}{x_1+1}, \ldots, \frac{1}{x_n+1}\right)$$

$$= n - \frac{n}{\mathcal{G}_n(x_1+1, \ldots, x_n+1)} \le n - \frac{n}{\mathcal{A}_n(x_1+1, \ldots, x_n+1)}$$

$$= n - \frac{n^2}{(x_1+1) + \cdots + (x_n+1)} = n - \frac{n^2}{(x_1 + \cdots + x_n) + n}$$

$$= n - \frac{n^2}{0+n} = 0. \qquad \square$$

(vi) Show that for each integer $n \ge 2$ we have the inequality

$$1 + \frac{1}{2} + \frac{1}{3} + \cdots + \frac{1}{n} > n(\sqrt[n]{n+1} - 1).$$

SOLUTION. If we denote the left-hand side by L_n, then

$$L_n + n = (1+1) + \left(\frac{1}{2}+1\right) + \left(\frac{1}{3}+1\right) + \cdots + \left(\frac{1}{n}+1\right)$$

$$= \frac{2}{1} + \frac{3}{2} + \frac{4}{3} + \cdots + \frac{n+1}{n} = n\mathcal{A}_n\left(\frac{2}{1}, \frac{3}{2}, \ldots, \frac{n+1}{n}\right)$$

$$> n\mathcal{G}_n\left(\frac{2}{1}, \frac{3}{2}, \ldots, \frac{n+1}{n}\right) = n \cdot \sqrt[n]{\frac{(n+1)!}{n!}} = n \cdot \sqrt[n]{n+1},$$

which, after subtracting the number n, gives the desired inequality. $\square$

(vii) Show that if $a_1, a_2, \ldots, a_n, b_1, b_2, \ldots, b_n \in \mathbb{R}^+$, then

$$\mathcal{G}_n(a_1, a_2, \ldots, a_n) + \mathcal{G}_n(b_1, b_2, \ldots, b_n) \le \mathcal{G}_n(a_1+b_1, a_2+b_2, \ldots, a_n+b_n). \tag{91}$$

Furthermore, determine when equality occurs in (91).

SOLUTION. Dividing by the right-hand side, we rewrite (91) as follows:

$$\mathcal{G}_n\left(\frac{a_1}{a_1+b_1}, \ldots, \frac{a_1}{a_n+b_n}\right) + \mathcal{G}_n\left(\frac{b_1}{a_1+b_1}, \ldots, \frac{b_n}{a_n+b_n}\right) \le 1,$$

which is the sum of the two AM–GM inequalities

$$\mathcal{G}_n\left(\frac{a_1}{a_1+b_1}, \ldots, \frac{a_n}{a_n+b_n}\right) \le \frac{1}{n}\left(\frac{a_1}{a_1+b_1} + \cdots + \frac{a_n}{a_n+b_n}\right)$$

and

$$\mathcal{G}_n \left(\frac{b_1}{a_1 + b_1}, \ldots, \frac{b_n}{a_n + b_n} \right) \leq \frac{1}{n} \left(\frac{b_1}{a_1 + b_1} + \cdots + \frac{b_n}{a_n + b_n} \right).$$

Hence (91) holds, with equality if and only if

$$\frac{a_1}{a_1 + b_1} = \frac{a_2}{a_2 + b_2} = \cdots = \frac{a_n}{a_n + b_n} \quad \text{and} \quad \frac{b_1}{a_1 + b_1} = \cdots = \frac{b_n}{a_n + b_n},$$

which means that $a_1/b_1 = a_2/b_2 = \cdots = a_n/b_n$, that is, $a_k = tb_k$ ($1 \leq k \leq n$) for an appropriate $t > 0$. It is easy to verify (for instance by direct substitution into (91)) that this last condition for equality in (91) is also sufficient.   □

Remark. It is easy to verify that if we replace the symbol $\mathcal{G}_n$ everywhere in (91) by $\mathcal{A}_n$, then we obtain a trivial equality. This fact has an interesting "dual": The inequality

$$\mathcal{A}_n(a_1, a_2, \ldots, a_n) \cdot \mathcal{A}_n(b_1, b_2, \ldots, b_n) \leq \mathcal{A}_n(a_1 b_1, a_2 b_2, \ldots, a_n b_n)$$

holds by Chebyshev's inequality (76) in 7.6 if the n-tuples $a_1, a_2, \ldots, a_n$ and $b_1, b_2, \ldots, b_n$ have the same ordering; if the n-tuples have opposite orderings, then the opposite inequality holds. If we replace the symbol $\mathcal{A}_n$ in this inequality by $\mathcal{G}_n$, we get again a trivial equality.

(viii) Suppose that the sum of the positive numbers $a_1, a_2, \ldots, a_n$ is equal to 1. Prove the inequality

$$\left(1 + \frac{1}{a_1} \right) \left(1 + \frac{1}{a_2} \right) \cdots \left(1 + \frac{1}{a_n} \right) \geq (n+1)^n.$$

SOLUTION.   For each $k = 1, 2, \ldots, n$ we have by the AM–GM inequality

$$1 + \frac{1}{a_k} = (n+1)\mathcal{A}_{n+1}\left(1, \frac{1}{na_k}, \frac{1}{na_k}, \ldots, \frac{1}{na_k} \right)$$
$$\geq (n+1)\mathcal{G}_{n+1}\left(1, \frac{1}{na_k}, \frac{1}{na_k}, \ldots, \frac{1}{na_k} \right).$$

Multiplying these inequalities, we obtain

$$\left(1 + \frac{1}{a_1} \right) \cdots \left(1 + \frac{1}{a_n} \right) \geq (n+1)^n \cdot \sqrt[n+1]{\frac{1}{(na_1)^n} \cdots \frac{1}{(na_n)^n}}.$$

The proof is complete if we verify the inequality $(na_1)^n (na_2)^n \cdots (na_n)^n \leq 1$, which is equivalent to $\mathcal{G}_n(a_1, a_2, \ldots, a_n) \leq \frac{1}{n}$. But this is the AM–GM inequality for the n-tuple $a_1, a_2, \ldots, a_n$, since $a_1 + a_2 + \cdots + a_n = 1$.   □

(ix) Suppose that the nonnegative numbers $a_1, a_2, \ldots, a_n$ $(n \geq 2)$ have the following property: If their sum is denoted by S, i.e., $S = a_1 + a_2 + \cdots + a_n$, then the numbers $b_k = S - (n-1)a_k$, $1 \leq k \leq n$, are nonnegative as well. Prove the inequality $a_1 a_2 \cdots a_n \geq b_1 b_2 \cdots b_n$.

SOLUTION.   First we note that

$$b_2 + b_3 + \cdots + b_n$$
$$= [S - (n-1)a_2] + [S - (n-1)a_3] + \cdots + [S - (n-1)a_n]$$
$$= (n-1)S - (n-1)(a_2 + a_3 + \cdots + a_n)$$
$$= (n-1)(S - a_2 - a_3 - \cdots - a_n) = (n-1)a_1.$$

Hence the AM–GM inequality for the $(n-1)$ numbers $b_2, b_3, \ldots, b_n$ can be written as

$$a_1 \geq \sqrt[n-1]{b_2 b_3 \cdots b_n}\,.$$

Similarly, we have $a_k \geq \mathcal{G}_{n-1}(b_1, \ldots, b_{k-1}, b_{k+1}, \ldots, b_n)$ for $k = 2, 3, \ldots, n$. Multiplying these n inequalities together, we obtain

$$a_1 a_2 \cdots a_n \geq \sqrt[n-1]{b_1^{n-1} b_2^{n-1} \cdots b_n^{n-1}} = b_1 b_2 \cdots b_n\,. \qquad \square$$

(x) Show that for each $n \in \mathbb{N}$ we have the inequality

$$\sqrt{2} \cdot \sqrt[4]{4} \cdot \sqrt[8]{8} \cdots \sqrt[2^n]{2^n} < n + 1.$$

SOLUTION.   We rewrite the left-hand side as one root:

$$\sqrt[2^n]{2^{2^{n-1}} \cdot 4^{2^{n-2}} \cdot 8^{2^{n-3}} \cdots (2^{n-1})^{2^1} \cdot (2^n)^{2^0}}\,.$$

Underneath the radical sign we have the product of $2^n - 1$ numbers: 2^{n-1} times the number 2, 2^{n-2} times 4, $\ldots$, 2 times the number 2^{n-1}, and once the number 2^n. If we take the additional factor 1, the product will not change. The root in question is then the geometric mean of this 2^n-tuple; by the AM–GM inequality it is then at most equal to

$$\frac{1}{2^n}(2^{n-1} \cdot 2 + 2^{n-2} \cdot 4 + \cdots + 2 \cdot 2^{n-1} + 2^n + 1) = \frac{1}{2^n}(n2^n + 1)$$

$$= n + \frac{1}{2^n} < n + 1.$$

However, it is possible to prove a sharper bound without the use of the AM–GM inequality:

$$\sqrt[2]{2} \cdot \sqrt[4]{4} \cdot \sqrt[8]{8} \cdots \sqrt[2^n]{2^n} = 2^{\frac{1}{2} + \frac{2}{4} + \frac{3}{8} + \cdots + \frac{n}{2^n}} = 2^{2 - \frac{n+2}{2^n}} < 4\,,$$

where the sum in the exponent was determined in Chapter 1, Example 2.37. $\qquad \square$

8.5 Exercises

(i) Show that for arbitrary $a_1, a_2, \ldots, a_n \in \mathbb{R}_0^+$ we have the inequality
$$(a_1 + a_2 + \cdots + a_n)(a_1^{n-1} + a_2^{n-1} + \cdots + a_n^{n-1}) \geq n^2 a_1 a_2 \cdots a_n.$$

(ii) Use (84) to give another proof of the consequence (46) of Cauchy's inequality.

(iii) Show that if $a_1, a_2, \ldots, a_n \in \mathbb{R}_0^+$ and $b_1, b_2, \ldots, b_n \in \mathbb{R}^+$, then
$$\frac{1}{n}\sqrt{\sum_{k=1}^{n} a_k^2 \sum_{k=1}^{n} \frac{1}{b_k^2}} \geq \sqrt[n]{\frac{a_1 a_2 \cdots a_n}{b_1 b_2 \cdots b_n}}.$$

(iv) Show that if $a, b, c \in \mathbb{R}_0^+$, then
$$\sqrt{a^2 + b^2 + c^2}\left(ab + bc + ca + \sqrt{a^2 b^2 + b^2 c^2 + c^2 a^2}\right)$$
$$\geq 3\left(1 + \sqrt{3}\right) abc.$$

(v) Show that for arbitrary $a_1, a_2, \ldots, a_n \in \mathbb{R}^+, n \geq 3$, we have
$$\frac{a_1 - a_3}{a_2 + a_3} + \frac{a_2 - a_4}{a_3 + a_4} + \cdots + \frac{a_{n-1} - a_1}{a_n + a_1} + \frac{a_n - a_2}{a_1 + a_2} \geq 0.$$

(vi) Show that for each integer $n \geq 2$ we have the inequality
$$1 + \frac{1}{2} + \frac{1}{3} + \cdots + \frac{1}{n} < 1 + n\left(1 - \frac{1}{n\sqrt{n}}\right).$$

(vii) Show that for arbitrary numbers $a, b_1, b_2, \ldots, b_n \in \mathbb{R}^+$ we have
$$\frac{a^{b_1 - b_2}}{b_1 + b_2} + \frac{a^{b_2 - b_3}}{b_2 + b_3} + \cdots + \frac{a^{b_n - b_1}}{b_n + b_1} \geq \frac{n^2}{2(b_1 + b_2 + \cdots + b_n)}.$$

(viii) Suppose that the product of the positive numbers $a_1, a_2, \ldots, a_k$ is equal to 1. Show that for each $n \in \mathbb{N}$ we have
$$(n + a_1)(n + a_2) \cdots (n + a_k) \geq (n + 1)^k.$$

(ix) Suppose that the sum of the positive numbers $a_1, a_2, \ldots, a_n$ is equal to 1. Show that for each real number $b \geq 1$ we have the inequality
$$\frac{a_1}{b - a_1} + \frac{a_2}{b - a_2} + \cdots + \frac{a_n}{b - a_n} \geq \frac{n}{bn - 1}.$$

(x) Suppose that the sum of the positive numbers $a_1, a_2, \ldots, a_n$ is equal to 1. Prove the inequality
$$\left(\frac{1}{a_1} - 1\right)\left(\frac{1}{a_2} - 1\right) \cdots \left(\frac{1}{a_n} - 1\right) \geq (n - 1)^n.$$

Prove the inequalities (xi)–(xii) for arbitrary $a_1, a_2, \ldots, a_n \in \mathbb{R}_0^+$:

(xi) $a_1 a_2 \cdots a_n \leq \dfrac{a_1^2}{2} + \dfrac{a_2^4}{4} + \dfrac{a_3^8}{8} + \cdots + \dfrac{a_n^{2^n}}{2^n} + \dfrac{1}{2^n}.$

*(xii) $\sqrt[n]{(1+a_1)(1+a_2)\cdots(1+a_n)} \geq 1 + \sqrt[n]{a_1 a_2 \cdots a_n}.$

8.6 Weighted Means

Sometimes, in determining the mean value of numbers $x_1, x_2, \ldots, x_n$, we associate with different numbers x_j different levels of "importance," indicated by *weight coefficients* $p_j \in \mathbb{R}^+$. The *weighted arithmetic mean* is then defined by the expression

$$\mathcal{A}_n^{\mathrm{W}}(x_1, x_2, \ldots, x_n) = \frac{p_1 x_1 + p_2 x_2 + \cdots + p_n x_n}{p_1 + p_2 + \cdots + p_n}.$$

It is convenient to normalize the weight coefficients p_j in such a way that their sum is equal to 1. We achieve this by introducing new coefficients

$$v_j = \frac{p_j}{p_1 + p_2 + \cdots + p_n} \qquad (1 \leq j \leq n),$$

since clearly $v_1 + v_2 + \cdots + v_n = 1$; the relation for the mean then simplifies to

$$\mathcal{A}_n^{\mathrm{W}}(x_1, x_2, \ldots, x_n) = v_1 x_1 + v_2 x_2 + \cdots + v_n x_n. \tag{92}$$

We also define the *weighted geometric mean*

$$\mathcal{G}_n^{\mathrm{W}}(x_1, x_2, \ldots, x_n) = x_1^{v_1} x_2^{v_2} \cdots x_n^{v_n}; \tag{93}$$

then the question arises whether the AM–GM inequality also holds in this more general situation. (If in (92) and (93) we set $v_1 = v_2 = \cdots = v_n = \frac{1}{n}$, we get formulas (81) and (82) for the original nonweighted means $\mathcal{A}_n$ and $\mathcal{G}_n$.) The following theorem gives an affirmative answer.

8.7 The AM–GM Inequality for Weighted Means

Theorem. *Let the sum of the positive numbers $v_1, v_2, \ldots, v_n$ be equal to 1. Then for arbitrary numbers $a_1, a_2, \ldots, a_n \in \mathbb{R}^+$ we have the inequality*

$$v_1 a_1 + v_2 a_2 + \cdots + v_n a_n \geq a_1^{v_1} a_2^{v_2} \cdots a_n^{v_n}, \tag{94}$$

with equality if and only if $a_1 = a_2 = \cdots = a_n$.

PROOF.   We restrict ourselves to the case where the numbers $v_1, v_2, \ldots, v_n$ are rational (the general case will be dealt with in Section 9). Then there exists a number $k \in \mathbb{N}$ such that all numbers $m_j = k v_j$ $(1 \leq j \leq n)$ are positive integers. Furthermore,

$$m_1 + m_2 + \cdots + m_n = k(v_1 + v_2 + \cdots + v_n) = k \cdot 1 = k.$$

We now use the AM–GM inequality for the following set of k numbers: m_1 times the number a_1, m_2 times $a_2, \ldots, m_n$ times a_n. We thus obtain

$$\frac{m_1 a_1 + m_2 a_2 + \cdots + m_n a_n}{k} \geq \sqrt[k]{a_1^{m_1} a_2^{m_2} \cdots a_n^{m_n}}, \qquad (95)$$

which proves (94), since $m_j/k = v_j$ ($1 \leq j \leq n$). By the result 8.1 we have equality if and only if the above set of k numbers consists only of identical elements, that is, when $a_1 = a_2 = \cdots = a_n$. This implies the assertion about equality in (94), and the proof (for the case $v_j \in \mathbb{Q}$, $1 \leq j \leq n$) is complete. $\qquad \square$

8.8 Examples

We now give six examples for the use of inequality (94).

(i) We call the positive numbers p, q *associated* if they satisfy $\frac{1}{p} + \frac{1}{q} = 1$. (Note that both numbers are greater than 1.) Prove the so-called *Young inequality*: If p, q are associated numbers, then the inequality

$$xy \leq \frac{x^p}{p} + \frac{y^q}{q} \qquad (96)$$

holds for arbitrary $x, y \in \mathbb{R}^+$.

SOLUTION. If p, q are associated numbers, then we can use (94) with $n = 2$ and $v_1 = 1/p$, $v_2 = 1/q$. Then with $a_1 = x^p$ and $a_2 = y^q$ we immediately obtain (96).

We remark that the Young inequality is in fact equivalent to (94) for $n = 2$: If v_1, v_2 are weight coefficients in (94) for $n = 2$, then the numbers $p = 1/v_1, q = 1/v_2$ are associated. $\qquad \square$

(ii) Show that if $a, b, c \in \mathbb{R}^+$ and $c \neq 1$, then

$$ac^b + \frac{b}{c^a} > a + b.$$

SOLUTION. If in (94) with $n = 2$ we set

$$v_1 = \frac{a}{a+b}, \quad v_2 = \frac{b}{a+b}, \quad a_1 = c^b, \quad \text{and} \quad a_2 = c^{-a},$$

we obtain (in view of the fact that $a_1 \neq a_2$) the strict inequality

$$\frac{a}{a+b} \cdot c^b + \frac{b}{a+b} \cdot c^{-a} > (c^b)^{a/(a+b)} \cdot (c^{-a})^{b/(a+b)} = 1,$$

which upon multiplying by $a + b$ gives the desired inequality. $\qquad \square$

(iii) Show that if $0 < x < 1$, then

$$\frac{1}{2-x} < x^x < x^2 - x + 1. \tag{97}$$

SOLUTION. If $0 < x < 1$, we can use (94) with $n = 2$ and set $v_1 = a_1 = x$, $v_2 = 1 - x$, and $a_2 = 1$. Since $a_1 < a_2$, we obtain the strict inequality $x \cdot x + (1 - x) \cdot 1 > x^x \cdot 1^{1-x}$, which is the right-hand part of (97). If we interchange the numbers a_1 and a_2, we get $x \cdot 1 + (1 - x) \cdot x > 1^x \cdot x^{1-x}$, that is, $2x - x^2 > x^{1-x}$, and this gives the left part of (97).   □

(iv) Show that if a, b, c are side lengths of a triangle, then

$$\left(1 + \frac{b-c}{a}\right)^a \cdot \left(1 + \frac{c-a}{b}\right)^b \cdot \left(1 + \frac{a-b}{c}\right)^c \leq 1.$$

SOLUTION. We use the inequality (94) with $n = 3$ for the coefficients

$$v_1 = \frac{a}{a+b+c}, \quad v_2 = \frac{b}{a+b+c}, \quad \text{and} \quad v_3 = \frac{c}{a+b+c}. \tag{98}$$

Then for the (positive!) numbers

$$a_1 = 1 + \frac{b-c}{a}, \quad a_2 = 1 + \frac{c-a}{b}, \quad \text{and} \quad a_3 = 1 + \frac{a-b}{c}$$

we obtain the inequality

$$a_1^{v_1} a_2^{v_2} a_3^{v_3} \leq v_1 a_1 + v_2 a_2 + v_3 a_3 = \frac{a+b-c+b+c-a+c+a-b}{a+b+c} = 1,$$

which upon raising to the power $a+b+c$ gives the desired inequality.   □

(v) Show that for arbitrary numbers $a, b, c \in \mathbb{R}^+$ we have the inequality

$$a^a b^b c^c \geq \left(\frac{a+b+c}{3}\right)^{a+b+c}.$$

SOLUTION. We use again (94) with $n = 3$ and the coefficients (98). However, this time we set $a_1 = 1/a$, $a_2 = 1/b$, and $a_3 = 1/c$; we thus obtain

$$\left(\frac{1}{a}\right)^{v_1} \left(\frac{1}{b}\right)^{v_2} \left(\frac{1}{c}\right)^{v_3} \leq \frac{v_1}{a} + \frac{v_2}{b} + \frac{v_3}{c} = \frac{3}{a+b+c},$$

since

$$\frac{v_1}{a} = \frac{v_2}{b} = \frac{v_3}{c} = \frac{1}{a+b+c}.$$

If we write the opposite inequality for the reciprocals and then raise both sides to the power $a + b + c$, we obtain the desired inequality.   □

(vi) Prove the general *Bernoulli inequality*: If $x \in \mathbb{R}$, $p \in \mathbb{R}^+$, $x > -1$, and $p \neq 1$, then

$$(1+x)^p \geq 1 + px \qquad (p > 1), \qquad\qquad (99)$$

respectively

$$(1+x)^p \leq 1 + px \qquad (0 < p < 1), \qquad\qquad (100)$$

with equality in (99) or (100) if and only if $x = 0$ (compare with 3.3.(iii)).

SOLUTION. If $0 < p < 1$, we substitute in (94) $n = 2$, $v_1 = p$, $v_2 = 1 - p$, $a_1 = 1 + x > 0$, and $a_2 = 1$; we thus obtain

$$p(1+x) + (1-p) \cdot 1 \geq (1+x)^p \cdot 1^{1-p},$$

that is, inequality (100), with equality if and only if $a_1 = a_2$, which means $x = 0$.

Now let $p > 1$. If $1 + px \leq 0$, we obviously have a strict inequality in (99). If $1 + px > 0$, then by the already proven inequality (100) we have

$$(1 + px)^{\frac{1}{p}} \leq 1 + \frac{1}{p} \cdot px = 1 + x.$$

(Equality occurs only for $x = 0$.) From this we obtain (99) after raising both sides to the power p. $\qquad\qquad\square$

8.9 Exercises

(i) Show that the weighted means (92) and (93) of the positive numbers $x_1, \ldots, x_n$ have the *fundamental mean property*. Their value is at most (at least) equal to the largest (smallest) of the numbers $x_1, \ldots, x_n$.

(ii) Show that if $a, b, p \in \mathbb{R}^+$ and $p < 1$, then $(a+b)^p a^{1-p} < a + pb$.

Show that (iii)–(v) hold for arbitrary $a, b, c, d \in \mathbb{R}^+$:

(iii) $ad^{b-c} + bd^{c-a} + cd^{a-b} \geq a + b + c$.

(iv) $a^c b^d (c+d)^{c+d} \leq c^c d^d (a+b)^{c+d}$.

(v) $\left(\dfrac{a+b+c}{3} \right)^{a+b+c} \geq \dfrac{(b+c)^a (c+a)^b (a+b)^c}{2^{a+b+c}}$.

(vi) Show that if the sum of the positive numbers a, b equals 1, then the inequalities $a^b b^a \leq 2ab$ and $\frac{1}{2} \leq a^a b^b \leq a^2 + b^2$ hold. When does equality occur in these inequalities?

(vii) Prove the inequality $a^c b^{1-c} + (1-a)^c (1-b)^{1-c} \leq 1$ for arbitrary positive numbers a, b, c less than 1.

Prove (viii) and (ix) with the help of Bernoulli's inequality in 8.8.(vi):

(viii) $(1+x)^{a+1} \geq \left(1+x+\dfrac{x}{a}\right)^a$ $(x > -1, a > 0)$.

(ix) $\left(1+\dfrac{1}{a}\right)^a < \left(1+\dfrac{1}{b}\right)^b$ $(0 < a < b)$.

(The inequality (ix) was derived for integers $a \geq 2$ and $b = a+1$ in 3.4.(vi) and 3.8.(iv).)

(x) Prove *Hölder's inequality*: If p, q are associated numbers (see 8.8.(i)), then for arbitrary numbers $x_1, x_2, \ldots, x_n, y_1, y_2, \ldots, y_n \in \mathbb{R}^+$ we have

$$\sum_{k=1}^{n} x_k y_k \leq \left(\sum_{k=1}^{n} x_k^p\right)^{1/p} \left(\sum_{k=1}^{n} y_k^q\right)^{1/q}. \tag{101}$$

Use (96) and proceed in a similar fashion as in the second proof of 5.4. (Note that for $p = q = 2$, (101) becomes Cauchy's inequality (42).)

(xi) Use (101) to show that if $a, b, c, d \in \mathbb{R}^+$, $a^5 + b^5 \leq 1$, and $c^5 + d^5 \leq 1$, then $a^2 c^3 + b^2 d^3 \leq 1$.

*(xii) Use (101) to prove *Minkowski's inequality*: If $p > 1$, then for arbitrary numbers $x_1, x_2, \ldots, x_n, y_1, y_2, \ldots, y_n \in \mathbb{R}^+$ we have

$$\left(\sum_{k=1}^{n} (x_k + y_k)^p\right)^{1/p} \leq \left(\sum_{k=1}^{n} x_k^p\right)^{1/p} + \left(\sum_{k=1}^{n} y_k^p\right)^{1/p}.$$

Proceed as in 5.5.(iv) for the proof of the triangle inequality (45).

(xiii) Let $c, p_1, p_2, \ldots, p_n \in \mathbb{R}^+$ and suppose that the positive variables $x_1, x_2, \ldots, x_n$ satisfy the condition $x_1 + x_2 + \cdots + x_n = c$. Show that the expression

$$x_1^{p_1} x_2^{p_2} \cdots x_n^{p_n}$$

attains its largest value if and only if

$$\frac{x_1}{p_1} = \frac{x_2}{p_2} = \cdots = \frac{x_n}{p_n}.$$

8.10   Power Means

Up to the end of Section 8 we will assume that $a_1, a_2, \ldots, a_n$ are arbitrary positive numbers. We will derive several useful inequalities between the sums $a_1^r + a_2^r + \cdots + a_n^r$ for various $r \in \mathbb{R}$. For this purpose it is convenient to introduce the so-called *power means*. By a mean of degree r of the numbers $a_1, a_2, \ldots, a_n$ we understand the value

$$\mathcal{M}_n^r(a_1, a_2, \ldots, a_n) = \left(\frac{a_1^r + a_2^r + \cdots + a_n^r}{n} \right)^{\frac{1}{r}}, \qquad (102)$$

which is meaningful for each $r \in \mathbb{R} \setminus \{0\}$. We note that for $r = 1$ (resp. $r = -1$) this is the arithmetic (resp. harmonic) mean (compare with (81) and (89)). The following theorem deals with the comparison of the power means and the geometric mean.

8.11 Power Means of Positive and Negative Degrees

Theorem. *If $r < 0 < s$, then*

$$\mathcal{M}_n^r(a_1, a_2, \ldots, a_n) \leq \mathcal{G}_n(a_1, a_2, \ldots, a_n) \leq \mathcal{M}_n^s(a_1, a_2, \ldots, a_n), \qquad (103)$$

with equality anywhere in (103) if and only if $a_1 = a_2 = \cdots = a_n$.

PROOF. We obtain the right part of (103) if we raise the AM–GM inequality

$$\mathcal{G}_n(a_1^s, a_2^s, \ldots, a_n^s) \leq \mathcal{A}_n(a_1^s, a_2^s, \ldots, a_n^s)$$

to the positive power $1/s$; the left part is obtained by raising the AM–GM inequality $\mathcal{G}_n(a_1^r, a_2^r, \ldots, a_n^r) \leq \mathcal{A}_n(a_1^r, a_2^r, \ldots, a_n^r)$ to the negative power $1/r$. This also implies the statement about equality in (103). $\qquad \square$

Remark. Using mathematical analysis one can show that

$$\mathcal{G}_n(a_1, a_2, \ldots, a_n) = \lim_{r \to 0} \mathcal{M}_n^r(a_1, a_2, \ldots, a_n).$$

Therefore, the geometric mean is often called *mean of degree zero*.
We now state the main result on power means.

8.12 Inequalities Between Power Means

Theorem. *Let $r, s \in \mathbb{R}$, $r < s$, and $rs \neq 0$. Then*

$$\mathcal{M}_n^r(a_1, a_2, \ldots, a_n) \leq \mathcal{M}_n^s(a_1, a_2, \ldots, a_n), \qquad (104)$$

with equality if and only if $a_1 = a_2 = \cdots = a_n$.

The inequality (104) for integers $r = 1$ and $s > 1$ is equivalent to the earlier inequality (80) in Exercise 7.8.(v).

PROOF. By Theorem 8.11 it suffices to examine only two cases: $0 < r < s$ and $r < s < 0$. First let $0 < r < s$. If we set $p = s/r > 1$, then $a_k^s = b_k^p$, where $b_k = a_k^r$ $(1 \le k \le n)$. Therefore, upon raising to the power r, the inequality (104) has the form

$$\frac{b_1 + b_2 + \cdots + b_n}{n} \le \left(\frac{b_1^p + b_2^p + \cdots + b_n^p}{n}\right)^{1/p}. \tag{105}$$

It suffices to show that (105) with $p > 1$ holds for arbitrary positive numbers b_k, with equality if and only if $b_1 = b_2 = \cdots = b_n$. Since we are dealing with a homogeneous inequality in the variables $b_1, b_2, \ldots, b_n$ (see 2.7), we may assume that

$$b_1 + b_2 + \cdots + b_n = n. \tag{106}$$

This means that for the numbers $x_k = b_k - 1$ we have

$$x_1 + x_2 + \cdots + x_n = (b_1 + b_2 + \cdots + b_n) - n = 0. \tag{107}$$

Since $x_k > -1$ and $p > 1$, Bernoulli's inequality (99) implies

$$b_k^p = (1 + x_k)^p \ge 1 + p x_k, \qquad 1 \le k \le n. \tag{108}$$

Adding the inequalities (108), we obtain in view of (107),

$$b_1^p + b_2^p + \cdots + b_n^p \ge n + p(x_1 + x_2 + \cdots + x_n) = n,$$

which implies (105) under the assumption (106); equality occurs in (105) if and only if there is equality in (108) for each k, and by 8.8.(vi) this means that $x_k = 0$ $(1 \le k \le n)$, that is, $b_1 = b_2 = \cdots = b_n = 1$.

In the second case, when $r < s < 0$, (104) follows from the inequality

$$\mathcal{M}_n^{-s}\left(\frac{1}{a_1}, \frac{1}{a_2}, \ldots, \frac{1}{a_n}\right) \le \mathcal{M}_n^{-r}\left(\frac{1}{a_1}, \frac{1}{a_2}, \ldots, \frac{1}{a_n}\right),$$

which has already been proven, since for each $t \ne 0$ we clearly have

$$\mathcal{M}_n^{-t}\left(\frac{1}{a_1}, \frac{1}{a_2}, \ldots, \frac{1}{a_n}\right) = \frac{1}{\mathcal{M}_n^t(a_1, a_2, \ldots, a_n)}.$$

This completes the proof of the theorem.   □

8.13 Examples

We show how inequalities between power means can be applied.

(i) Find the smallest value of the sum $a^3 + b^3 + c^3$ for positive real numbers a, b, c satisfying the condition $a^2 + b^2 + c^2 = 27$.

SOLUTION. By Theorem 8.12, the power means of degrees 2 and 3 for the numbers a, b, c satisfy the inequality

$$\sqrt[2]{\frac{a^2 + b^2 + c^2}{3}} \leq \sqrt[3]{\frac{a^3 + b^3 + c^3}{3}}.$$

Upon substituting $a^2 + b^2 + c^2 = 27$ and some simplification we obtain the bound $a^3 + b^3 + c^3 \geq 81$. Using Theorem 8.12 again, we see that the equation $a^3 + b^3 + c^3 = 81$ occurs if and only if $a = b = c\,(= 3)$. The desired smallest value of the sum $a^3 + b^3 + c^3$ is therefore equal to 81.

If we wanted to find its *largest* value (under the same condition), Jensen's inequality from Section 2.7 (with exponents $r = 2$ and $s = 3$) would give the bound $a^3 + b^3 + c^3 < (\sqrt{27})^3 = 81 \cdot \sqrt{3}$. It is also clear that the value of the sum $a^3 + b^3 + c^3$ can get arbitrarily close to $81 \cdot \sqrt{3}$ (since, for instance, the number a can be arbitrarily close to $\sqrt{27}$). A largest value of the given sum therefore does not exist (in the set of triples of *positive* numbers a, b, c satisfying the condition $a^2 + b^2 + c^2 = 27$). □

(ii) Show that for arbitrary numbers $a, b, c \in \mathbb{R}^+$ we have the inequality

$$8(a^3 + b^3 + c^3)^2 \geq 9(a^2 + bc)(b^2 + ca)(c^2 + ab).$$

SOLUTION. With the help of (85) we first estimate from above the product of the three terms on the right,

$$R \leq 9 \left[\frac{(a^2 + bc) + (b^2 + ca) + (c^2 + ab)}{3} \right]^3 = \frac{1}{3}(a^2 + b^2 + c^2 + ab + bc + ca)^3,$$

which together with the inequality $ab + bc + ca \leq a^2 + b^2 + c^2$ (see (27)) leads to the conclusion that

$$R \leq \frac{1}{3}(2a^2 + 2b^2 + 2c^2)^3 = \frac{8}{3}(a^2 + b^2 + c^2)^3.$$

Therefore, the proof is complete if we can verify the inequality

$$\frac{8}{3}(a^2 + b^2 + c^2)^3 \leq 8(a^3 + b^3 + c^3)^2 \,(= L);$$

this, however, is clearly equivalent to the inequality between the means of degrees 2 and 3 of the numbers a, b, c, which is a special case of Theorem 8.12. □

9   Appendix on Irrational Numbers

Let us first recall that the general power a^r with real exponent r is for each $a > 0$ defined as the limit

$$a^r = \lim_{k \to \infty} a^{r(k)}, \tag{109}$$

where $\{r(k)\}$ is an arbitrary sequence of rational numbers converging to r. Of course, it needs to be shown that the limit on the right of (109) exists and is independent of the choice of the sequence $\{r(k)\}$ (see [5], p. 61).

ADDITION TO 1.10.   We need to show that $a^r > b^r$ under the assumption that $a > b > 0$ and $r \in \mathbb{R}^+ \setminus \mathbb{Q}$. Some sequence $\{r(k)\}$ of numbers in $\mathbb{Q}^+$ converges to this number r. By the proof in 1.10 we have $a^{r(k)} > b^{r(k)}$ for all k, and by taking the limit as $k \to \infty$ we obtain by (109) the *no longer strict* inequality $a^r \geq b^r$. However, in the case $a^r = b^r$ we get

$$a = (a^r)^{1/r} = (b^r)^{1/r} = b,$$

which contradicts the assumption $a > b$. Hence $a^r > b^r$.   □

ADDITION TO 8.6.   We have to prove the assertion concerning inequality (94) in the case that some of the numbers $v_1, v_2, \ldots, v_n$ are irrational. Given $v_1, v_2, \ldots, v_n$, we find sequences $\{v_j(k)\}$ of positive numbers in $\mathbb{Q}$ such that $v_j(k) \to v_j$ as $k \to \infty$ $(1 \leq j \leq n)$, and then we define numbers

$$v'_j(k) = \frac{v_j(k)}{v_1(k) + v_2(k) + \cdots + v_n(k)} \in \mathbb{Q}^+ \qquad (1 \leq j \leq n, \ k \in \mathbb{N}).$$

By the proof of 8.6 we have for each k,

$$v'_1(k)a_1 + v'_2(k)a_2 + \cdots + v'_n(k)a_n \geq a_1^{v'_1(k)} a_2^{v'_2(k)} \cdots a_n^{v'_n(k)},$$

which, upon taking the limit as $k \to \infty$, implies (94), since $v'_j(k) \to v_j$ $(1 \leq j \leq n)$. It remains to show that (94) is a strict inequality unless $a_1 = a_2 = \cdots = a_n$. In that case we choose $u_1, u_2, \ldots, u_n \in \mathbb{Q}^+$ such that $u_j < v_j$ $(1 \leq j \leq n)$, and we define the positive numbers

$$u = \sum_{j=1}^{n} u_j, \quad u'_j = \frac{u_j}{u}, \quad w_j = v_j - u_j, \quad w = \sum_{j=1}^{n} w_j, \quad w'_j = \frac{w_j}{w}. \quad (110)$$

Since $u'_j \in \mathbb{Q}^+$, by the proof of 8.6 we have the sharp inequality

$$u'_1 a_1 + u'_2 a_2 + \cdots + u'_n a_n > a_1^{u'_1} a_2^{u'_2} \cdots a_n^{u'_n}; \quad (111)$$

by the above we also have

$$w'_1 a_1 + w'_2 a_2 + \cdots + w'_n a_n \geq a_1^{w'_1} a_2^{w'_2} \cdots a_n^{w'_n}. \quad (112)$$

If we raise (111), resp. (112), to the power u, resp. w, and multiply the resulting inequalities together, we obtain (in view of the fact that $u_j = uu'_j$, $w_j = ww'_j$, and $v_j = u_j + w_j$)

$$\left(\sum_{j=1}^{n} u'_j a_j \right)^u \left(\sum_{j=1}^{n} w'_j a_j \right)^w > a_1^{u_1+w_1} \cdots a_n^{u_n w_n} = a_1^{v_1} a_2^{v_2} \cdots a_n^{v_n}.$$

The inequality (94) is therefore strict if

$$\sum_{j=1}^{n} v_j a_j \geq \left(\sum_{j=1}^{n} u_j' a_j\right)^u \left(\sum_{j=1}^{n} w_j' a_j\right)^w. \qquad (113)$$

We note that the identity

$$u + w = \sum_{j=1}^{n}(u_j + w_j) = \sum_{j=1}^{n} v_j = 1$$

holds; hence for arbitrary $x, y \in \mathbb{R}^+$ we have $ux + wy \geq x^u \cdot y^w$. If we choose

$$x = \sum_{j=1}^{n} u_j' a_j \quad \text{and} \quad y = \sum_{j=1}^{n} w_j' a_j,$$

we obtain (113), since by (110) we have

$$ux + wy = \sum_{j=1}^{n}(uu_j' + ww_j')a_j = \sum_{j=1}^{n}(u_j + w_j)a_j = \sum_{j=1}^{n} v_j a_j. \qquad \square$$

3
Number Theory

In this chapter we deal with problems involving integers. We will be mainly concerned with questions of divisibility of integers, and also with solving equations in integers or natural numbers.

Although the natural numbers and more generally the integers represent, in a certain sense, the simplest of mathematical structures, the investigation of their properties has presented generations of mathematicians with a large number of very difficult problems. Often these are problems that are easy to formulate, but are still not solved. Let us mention a few of the most famous among them: The *twin prime conjecture* (are there infinitely many primes p such that $p + 2$ is also a prime?), the *Goldbach conjecture* (can every even number greater than 2 be written as a sum of two primes?), or the crown among all the problems of number theory, namely *Fermat's last theorem* (are there natural numbers n, x, y, z such that $n > 2$ and $x^n + y^n = z^n$?)

We note that in the six years between the first and the second Czech editions of this book, there have been very significant developments on this last problem. In June of 1993, *Andrew Wiles*, a professor at Princeton University, announced during a lecture at a conference in Cambridge that he had proved the so-called *Shimura–Taniyama conjecture*; this conjecture was already known to imply Fermat's last theorem (that is, the nonexistence of natural numbers n, x, y, z, as mentioned above). After initial problems with Wiles's proof, Fermat's last theorem is now settled beyond any doubt.

1   Basic Concepts

The concept of *divisibility of numbers* is encountered already in elementary school; we are therefore dealing with a rather familiar notion. However, since it is very important for this chapter, we will briefly recall the definition and some basic properties.

1.1   Divisibility

Definition. We say that the integer a *divides* the integer b (or b is *divisible* by a, or b is a *multiple* of a) if there exists an integer c such that $a \cdot c = b$. In this case we write $a \mid b$.

The following easy statements follow directly from the definition; we leave the proofs to the reader as exercises, with hints given in the appendix.

The number 0 is divisible by every integer; the only integer divisible by 0 is 0; for any number a we have $a \mid a$; for arbitrary numbers a, b, c the following four implications hold:

$$a \mid b \wedge b \mid c \Longrightarrow a \mid c, \tag{1}$$

$$a \mid b \wedge a \mid c \Longrightarrow a \mid b + c \wedge a \mid b - c, \tag{2}$$

$$c \neq 0 \Longrightarrow (a \mid b \Longleftrightarrow ac \mid bc), \tag{3}$$

$$a \mid b \wedge b > 0 \Longrightarrow a \leq b. \tag{4}$$

We also recall that numbers that are (resp., are not) divisible by 2, are called *even* (resp., *odd*).

We are now ready to solve a few problems.

(i) Determine for which natural number n the number $n^2 + 1$ is divisible by $n + 1$.

SOLUTION.   We have $n^2 - 1 = (n+1)(n-1)$, and thus $n+1$ divides $n^2 - 1$. Let us assume that $n+1$ also divides $n^2 + 1$. Then it also has to divide the difference $(n^2 + 1) - (n^2 - 1) = 2$. Since $n \in \mathbb{N}$, we have $n+1 \geq 2$, and so from $n+1 \mid 2$ it follows that $n+1 = 2$; hence $n = 1$. Thus the only natural number with the given property is 1. $\square$

(ii) Find all integers $a \neq 3$ for which $a - 3 \mid a^3 - 3$.

SOLUTION.   By equation (5) in Chapter 1 we have

$$a^3 - 27 = (a - 3)(a^2 + 3a + 9),$$

and therefore $a - 3$ divides $a^3 - 27$. Let us assume that $a - 3 \mid a^3 - 3$. Then $a - 3$ also divides the difference $(a^3 - 3) - (a^3 - 27) = 24$, and conversely, if $a - 3 \mid 24$, then $a - 3$ also divides the sum $(a^3 - 27) + 24 = a^3 - 3$.

The number 24 is divisible only by ± 1, ± 2, ± 3, ± 4, ± 6, ± 8, ± 12, ± 24; therefore, $a - 3$ must be equal to one of these numbers. This means that $a \in \{-21, -9, -5, -3, -1, 0, 1, 2, 4, 5, 6, 7, 9, 11, 15, 27\}$. □

(iii) Show that for each $n \in \mathbb{N}$,

$$169 \mid 3^{3n+3} - 26n - 27.$$

SOLUTION.   We set $A(n) = 3^{3n+3} - 26n - 27$ and rewrite 3^{3n+3} using the binomial theorem:

$$3^{3n+3} = 27^{n+1} = (26 + 1)^{n+1} = \sum_{i=0}^{n+1} \binom{n+1}{i} 26^i$$

$$= 1 + 26(n + 1) + \sum_{i=2}^{n+1} \binom{n+1}{i} 26^i.$$

Since for $i \geq 2$ we have $26^i = 169 \cdot 4 \cdot 26^{i-2}$, all the summands in the summation above are divisible by 169, and we have therefore

$$\sum_{i=2}^{n+1} \binom{n+1}{i} 26^i = 169t$$

for some natural number t. Finally, $A(n) = 1 + 26(n+1) + 169t - 26n - 27 = 169t$ and $169 \mid A(n)$. □

(iv) Show that there are infinitely many natural numbers n such that $2^n + 1$ is divisible by n.

SOLUTION.   We will show that all numbers of the form $n = 3^k$, where $k \in \mathbb{N}$ (i.e., infinitely many numbers), have the desired property. We prove this by induction. If $k = 1$, then $n = 3$ and clearly $3 \mid 2^3 + 1$. Let us now assume that for some $k \in \mathbb{N}$ we have $3^k \mid 2^{3^k} + 1$. Then there exists an integer m such that $m \cdot 3^k = 2^{3^k} + 1$. The binomial theorem gives

$$2^{3^{k+1}} = (2^{3^k})^3 = (m \cdot 3^k - 1)^3$$
$$= m^3 \cdot 3^{3k} - m^2 \cdot 3^{2k+1} + m \cdot 3^{k+1} - 1 = t \cdot 3^{k+1} - 1,$$

where $t = m^3 \cdot 3^{2k-1} - m^2 \cdot 3^k + m$ is an integer. Thus,

$$3^{k+1} \mid 2^{3^{k+1}} + 1,$$

which was to be shown. □

1.2 Exercises

(i) Determine the integers n for which $7n + 1$ is divisible by $3n + 4$.

(ii) For which natural numbers n do we have $3n + 2 \mid 5n^2 + 2n + 4$?

Show that (iii)–(vi) hold for arbitrary $n \in \mathbb{N}$:

(iii) $9 \mid 4^n + 15n - 1$.

(iv) $64 \mid 3^{2n+3} + 40n - 27$.

(v) $n^2 \mid (n + 1)^n - 1$.

(vi) $(2^n - 1)^2 \mid 2^{(2^n-1)n} - 1$.

(vii) Show that for arbitrary $a, b \in \mathbb{Z}$ we have $17 \mid 2a + 3b$ if and only if $17 \mid 9a + 5b$.

(viii) Show that if $n \mid 2^n - 2$ for some integer $n \geq 2$, then also $m \mid 2^m - 2$ for $m = 2^n - 1$.

1.3 The Division Theorem

Theorem. *Given arbitrary numbers $a \in \mathbb{Z}$ and $m \in \mathbb{N}$, there exist unique numbers $q \in \mathbb{Z}$ and $r \in \{0, 1, \ldots, m - 1\}$ such that $a = qm + r$.*

PROOF. First we prove the existence of the numbers q and r. We assume that the positive integer m is fixed, and we prove the result for arbitrary $a \in \mathbb{Z}$. To this end, we first assume that $a \in \mathbb{N}_0$, and prove the existence of q and r by induction:

If $0 \leq a < m$, we choose $q = 0$, $r = a$, and the equality $a = qm + r$ clearly holds.

Now we assume that $a \geq m$ and that we have already shown the existence of the numbers q, r for all $a' \in \{0, 1, 2, \ldots, a - 1\}$. Then in particular, for $a' = a - m$ there exist q', r' such that $a' = q'm + r'$ with $r' \in \{0, 1, \ldots, m - 1\}$. If we choose $q = q' + 1$ and $r = r'$, then we have $a = a' + m = (q' + 1)m + r' = qm + r$, which was to be shown.

We have thus shown the existence of the numbers q, r for arbitrary $a \geq 0$. If, on the other hand, $a < 0$, then we may use the fact that for the positive integer $-a$ there exist numbers $q' \in \mathbb{Z}$, $r' \in \{0, 1, \ldots, m - 1\}$ such that $-a = q'm + r'$; hence $a = -q'm - r'$. If $r' = 0$, we set $r = 0$, $q = -q'$; if $r > 0$, we set $r = m - r'$, $q = -q' - 1$. In both cases we have $a = q \cdot m + r$, which proves the existence of numbers q, r with the desired properties, for all $a \in \mathbb{Z}$, $m \in \mathbb{N}$.

We will now prove uniqueness. We assume that the numbers $q_1, q_2 \in \mathbb{Z}$ and $r_1, r_2 \in \{0, 1, \ldots, m - 1\}$ satisfy $a = q_1 m + r_1 = q_2 m + r_2$. Rewriting, we obtain $r_1 - r_2 = (q_2 - q_1)m$, and thus $m \mid r_1 - r_2$. Since $0 \leq r_1 < m$ and $0 \leq r_2 < m$, we have $-m < r_1 - r_2 < m$, and by (4) we get $r_1 - r_2 = 0$. But then also $(q_2 - q_1)m = 0$, and therefore $q_1 = q_2$, $r_1 = r_2$. Hence the numbers q, r are uniquely determined, and this completes the proof. □

The numbers q and r in Theorem 1.3 are called the (incomplete) *quotient* and *remainder*, respectively, in the division of a by m with remainder. It becomes clear that this terminology is appropriate if we write the equation $a = mq + r$ in the form

$$\frac{a}{m} = q + \frac{r}{m}, \quad \text{where} \quad 0 \le \frac{r}{m} < 1.$$

It is also useful to convince oneself that m divides a if and only if the remainder r is zero; this follows from Theorem 1.3 as well.

We will now give some examples to illustrate the usefulness of Theorem 1.3 for solving number theoretic problems.

1.4 Examples

(i) Show that if the remainders of $a, b \in \mathbb{Z}$ upon dividing by $m \in \mathbb{N}$ are 1, then the remainder of ab upon dividing by m is also 1.

SOLUTION. By Theorem 1.3 there exist $s, t \in \mathbb{Z}$ such that $a = sm + 1$, $b = tm + 1$. Multiplying, we obtain the expression

$$ab = (sm + 1)(tm + 1) = (stm + s + t)m + 1 = qm + r,$$

where $q = stm + s + t$ and $r = 1$ are, by Theorem 1.3, uniquely determined; the remainder in the division of ab by m is therefore equal to 1. □

(ii) Show that among m consecutive integers there is exactly one that is divisible by m.

SOLUTION. We denote the given numbers by $a + 1, a + 2, \ldots, a + m$. According to Theorem 1.3 there exist uniquely determined numbers $q \in \mathbb{Z}$, $r \in \{0, 1, \ldots, m - 1\}$ such that $a = qm + r$, which implies

$$a + (m - r) = qm + r + m - r = (q + 1)m,$$

and thus m divides $a + (m - r)$, which is one of our m numbers. If, on the other hand, $a + k$, $k \in \{1, 2, \ldots, m\}$, is divisible by m, then we have $a + k = c \cdot m$ for an appropriate $c \in \mathbb{Z}$, and thus $a = (c - 1)m + (m - k)$. The uniqueness of the number r now implies that $k = m - r$, and therefore only $a + (m - r)$ is divisible by m. □

1.5 Exercises

Show that

(i) the sum of the squares of two consecutive natural numbers, divided by 4, leaves the remainder 1;

(ii) the remainder of the square of an odd number, divided by 8, is 1;

(iii) for an odd k and a positive integer n, the number $k^{2^n} - 1$ is divisible by 2^{n+2}.

1.6 The Greatest Common Divisor and the Least Common Multiple

Definition. Let the integers a_1, a_2 be given. Any integer m such that $m \mid a_1$ and $m \mid a_2$ (resp. $a_1 \mid m$ and $a_2 \mid m$) is called a *common divisor* (resp. *common multiple*) of a_1 and a_2. A common divisor (resp. multiple) $m \geq 0$ of a_1 and a_2 that is divisible by any common divisor (resp. divides any common multiple) of a_1, a_2, is called *greatest common divisor* (resp. *least common multiple*) of a_1 and a_2 and is denoted by (a_1, a_2) (resp. $[a_1, a_2]$).

It follows directly from the definitions that for any $a, b \in \mathbb{Z}$ we have $(a, b) = (b, a)$, $[a, b] = [b, a]$, $(a, 1) = 1$, $[a, 1] = |a|$, $(a, 0) = |a|$, $[a, 0] = 0$. However, it is not yet clear whether the numbers (a, b) and $[a, b]$ exist at all for every pair $a, b \in \mathbb{Z}$. But if they do exist, they are unique, since for any two numbers $m_1, m_2 \in \mathbb{N}_0$ we know by (4) that if both $m_1 \mid m_2$ and $m_2 \mid m_1$, then necessarily $m_1 = m_2$. We prove the existence of (a, b) (along with an algorithm for finding it) in Section 1.7, while an existence proof of $[a, b]$ and a method of determining its value will be described in 1.9.

1.7 The Euclidean Algorithm

Theorem. *Let a_1, a_2 be natural numbers. For each $n \geq 3$ such that $a_{n-1} \neq 0$, we denote by a_n the remainder when a_{n-2} is divided by a_{n-1}. Then we will eventually get $a_k = 0$, and in this case $a_{k-1} = (a_1, a_2)$.*

PROOF. By 1.3 we have $a_2 > a_3 > a_4 > \cdots$. Since we are dealing with nonnegative integers, each number is at least one less than the previous one, and therefore we necessarily arrive at the final step $a_k = 0$, in which case $a_{k-1} \neq 0$. From the definitions of the numbers a_n it follows that there exist integers $q_1, q_2, \ldots, q_{k-2}$ such that

$$a_1 = q_1 \cdot a_2 + a_3,$$
$$a_2 = q_2 \cdot a_3 + a_4,$$
$$\vdots \tag{5}$$
$$a_{k-3} = q_{k-3} \cdot a_{k-2} + a_{k-1},$$
$$a_{k-2} = q_{k-2} \cdot a_{k-1}.$$

From the last equation it follows that $a_{k-1} \mid a_{k-2}$, from the second-to-last that $a_{k-1} \mid a_{k-3}$, etc., until in the end from the second equation we obtain $a_{k-1} \mid a_2$, and finally the first one gives $a_{k-1} \mid a_1$. Hence a_{k-1} is a common divisor of a_1, a_2. On the other hand, an arbitrary common divisor of a_1, a_2 also divides $a_3 = a_1 - q_1 a_2$, and therefore $a_4 = a_2 - q_2 a_3, \ldots$, and finally also $a_{k-1} = a_{k-3} - q_{k-3} a_{k-2}$. We have therefore shown that a_{k-1} is the greatest common divisor of a_1 and a_2. $\square$

1.8   Bézout's Equality

Theorem. *Any two integers a_1, a_2 have a greatest common divisor (a_1, a_2), which is unique by 1.6, and there exist integers k_1, k_2 such that $(a_1, a_2) = k_1 a_1 + k_2 a_2$.*

PROOF.   It clearly suffices to prove this theorem for $a_1, a_2 \in \mathbb{N}$. We note that if it is possible to express some numbers $r, s \in \mathbb{Z}$ in the form $r = r_1 a_1 + r_2 a_2$, $s = s_1 a_1 + s_2 a_2$, where $r_1, r_2, s_1, s_2 \in \mathbb{Z}$, then we can also express in this way

$$r + s = (r_1 + s_1)a_1 + (r_2 + s_2)a_2$$

and

$$c \cdot r = (c \cdot r_1)a_1 + (c \cdot r_2)a_2,$$

for arbitrary $c \in \mathbb{Z}$. Since $a_1 = 1 \cdot a_1 + 0 \cdot a_2$ and $a_2 = 0 \cdot a_1 + 1 \cdot a_2$, it follows from (5) that we can also express in this way $a_3 = a_1 - q_1 a_2$, $a_4 = a_2 - q_2 a_3$, ..., and finally $a_{k-1} = a_{k-3} - q_{k-3} a_{k-2}$, which of course is (a_1, a_2).   □

1.9   Existence of the Least Common Multiple

Theorem. *For arbitrary integers a_1, a_2 there exists the least common multiple $[a_1, a_2]$, and we have $(a_1, a_2) \cdot [a_1, a_2] = |a_1 \cdot a_2|$.*

PROOF.   The theorem is certainly true when one of the numbers a_1, a_2 is zero. We may furthermore assume that both nonzero numbers a_1, a_2 are positive, since a change in sign does not change the formula. We will be done if we can show that $q = \frac{a_1 \cdot a_2}{(a_1, a_2)}$ is the least common multiple of a_1, a_2. Since (a_1, a_2) is a common divisor of a_1 and a_2, the numbers $\frac{a_1}{(a_1, a_2)}$ and $\frac{a_2}{(a_1, a_2)}$ are integers, and therefore

$$q = \frac{a_1 a_2}{(a_1, a_2)} = \frac{a_1}{(a_1, a_2)} \cdot a_2 = \frac{a_2}{(a_1, a_2)} \cdot a_1$$

is a common multiple of a_1 and a_2. By 1.8 there exist $k_1, k_2 \in \mathbb{Z}$ such that $(a_1, a_2) = k_1 a_1 + k_2 a_2$. We now assume that $n \in \mathbb{Z}$ is an arbitrary common multiple of a_1, a_2 and we show that it is divisible by q. Indeed, we have $n/a_1, n/a_2 \in \mathbb{Z}$, and therefore the expression

$$\frac{n}{a_2} \cdot k_1 + \frac{n}{a_1} \cdot k_2 = \frac{n(k_1 a_1 + k_2 a_2)}{a_1 a_2} = \frac{n(a_1, a_2)}{a_1 a_2} = \frac{n}{q}$$

is also an integer. But this means that $q \mid n$, which remained to be shown.   □

1.10 Divisors and Multiples of More Than Two Numbers

We will define the greatest common divisor and the least common multiple of n numbers $a_1, a_2, \ldots, a_n \in \mathbb{Z}$ in the same way as in 1.6. Any $m \in \mathbb{Z}$ such that $m \mid a_1$, $m \mid a_2$, $\ldots$, $m \mid a_n$ (resp. $a_1 \mid m$, $a_2 \mid m$, $\ldots$, $a_n \mid m$) is called a *common divisor* (resp. *common multiple*) of the numbers $a_1, a_2, \ldots, a_n$. A common divisor (resp. multiple) $m \geq 0$ of $a_1, a_2, \ldots, a_n$ that is divisible by any common divisor (resp. divides any common multiple) is called the *greatest common divisor* (resp. *least common multiple*) of the numbers $a_1, a_2, \ldots, a_n$ and is denoted by $(a_1, a_2, \ldots, a_n)$ (resp. $[a_1, a_2, \ldots, a_n]$).

We can easily convince ourselves that we have

$$(a_1, \ldots, a_{n-1}, a_n) = ((a_1, \ldots, a_{n-1}), a_n), \tag{6}$$

$$[a_1, \ldots, a_{n-1}, a_n] = [[a_1, \ldots, a_{n-1}], a_n]. \tag{7}$$

Indeed, the greatest common divisor $(a_1, \ldots, a_n)$ divides all numbers $a_1, \ldots, a_n$; hence it is a common divisor of $a_1, \ldots, a_{n-1}$, and therefore it divides the greatest common divisor $(a_1, \ldots, a_{n-1})$, that is, $(a_1, \ldots, a_n) \mid ((a_1, \ldots, a_{n-1}), a_n)$. Conversely, the greatest common divisor of the numbers $(a_1, \ldots, a_{n-1})$ and a_n divides a_n, but also all numbers $a_1, \ldots, a_{n-1}$; hence it divides their greatest common divisor, which means that $((a_1, \ldots, a_{n-1}), a_n) \mid (a_1, \ldots, a_n)$. Together, this implies equation (6), and (7) can be obtained in complete analogy.

With the help of (6) and (7) we can easily obtain the existence of the greatest common divisor and the least common multiple of any n integers. Indeed, we use induction on n: For $n = 2$ the existence is guaranteed by 1.7 and 1.9, and if for some $n > 2$ we know that the greatest common divisor and the least common multiple exist for any $n - 1$ numbers, then by (6) and (7) they exist also for any n integers.

1.11 Relative Primality

Definition. The numbers $a_1, a_2, \ldots, a_n \in \mathbb{Z}$ are called *relatively prime* (or *coprime*) if they satisfy $(a_1, a_2, \ldots, a_n) = 1$. The numbers $a_1, a_2, \ldots, a_n \in \mathbb{Z}$ are called *pairwise relatively prime* if for any i, j with $1 \leq i < j \leq n$, we have $(a_i, a_j) = 1$.

In the case $n = 2$, these two concepts coincide, and for $n > 2$ pairwise relative primality implies relative primality, but not conversely: For example, 6, 10, and 15 are relatively prime, but they are not pairwise relatively prime, since no two of them are relatively prime: $(6, 10) = 2$, $(6, 15) = 3$, $(10, 15) = 5$.

1.12 Examples

(i) Find the greatest common divisor of $2^{63} - 1$ and $2^{91} - 1$.

SOLUTION.   Using the Euclidean algorithm:

$$2^{91} - 1 = 2^{28}(2^{63} - 1) + 2^{28} - 1,$$
$$2^{63} - 1 = (2^{35} + 2^7)(2^{28} - 1) + 2^7 - 1,$$
$$2^{28} - 1 = (2^{21} + 2^{14} + 2^7 + 1)(2^7 - 1).$$

The desired greatest common divisor is therefore $2^7 - 1 = 127$.   □

(ii) Suppose that the natural numbers m, n satisfy $m < n$. Show that any set of n consecutive integers contains two distinct elements whose product is divisible by $m \cdot n$.

SOLUTION.   Let M be the given set, $M = \{a + 1, a + 2, \ldots, a + n\}$, with some $a \in \mathbb{Z}$. By 1.4.(ii) there exists exactly one $i \in \{1, 2, \ldots, n\}$ such that $n \mid a + i$, and exactly one $j \in \{1, 2, \ldots, m\}$ such that $m \mid a + j$. In the case $i \neq j$ the solution is easy, since we have $m \cdot n \mid (a + i)(a + j)$, where $a + i, a + j$ are distinct elements of M. Therefore, we now assume $i = j$. We set $(m, n) = d$, $[m, n] = q$; since by 1.9 we have $mn = dq$, the problem will be solved if we can find two distinct elements of M such that one is divisible by q and the other by d. The first one is easily found: Since the number $a + i$ is a common multiple of m and n, we have $q \mid a + i$. Next we show that M also contains the number $a + i + d$, which is divisible by d. Indeed, since $d \mid m$ and $d \mid n$, there exist $r, s \in \mathbb{N}$ such that $m = dr$, $n = ds$. The inequality $m < n$ implies $r < s$; thus $r + 1 \leq s$ and $m + d = d(r + 1) \leq ds = n$. Since $i \leq m$, we also have $i + d \leq m + d \leq n$; hence $a + i + d \in M$. It is now easy to verify that $d \mid a + i + d$: From $d \mid m$ and $m \mid a + i$ it follows that $d \mid a + i$, and thus $d \mid a + i + d$.   □

(iii) Show that there are infinitely many numbers of the form $t_n = \frac{n(n+1)}{2}$, $n \in \mathbb{N}$, that are pairwise relatively prime.

SOLUTION.   We will show that whenever we have k numbers $t_{i_1}, t_{i_2}, \ldots, t_{i_k}$ that are pairwise relatively prime, we can always find another index i_{k+1} such that $t_{i_{k+1}}$ is relatively prime to all of them. Let c denote the product of $t_{i_1}, t_{i_2}, \ldots, t_{i_k}$ and set $i_{k+1} = 2c + 1$. Then

$$t_{i_{k+1}} = \frac{1}{2}(2c + 1)(2c + 2) = (c + 1)(2c + 1) = 2c^2 + 3c + 1$$

has the desired property. Indeed, for $(t_{i_j}, t_{i_{k+1}}) = d > 1$ to hold for some index $j \in \{1, 2, \ldots, k\}$, we would need to have $d \mid t_{i_j}$, which would imply $d \mid c = t_{i_1} \cdots t_{i_k}$, and thus d would be a common divisor of c and $2c^2 + 3c + 1$; this, however, is impossible, since $(c, 2c^2 + 3c + 1) = 1$.   □

1.13   Exercises

In (i)–(iv), find the greatest common divisor of the given numbers for an arbitrary $n \in \mathbb{N}$:

(i) $(2n+1, 9n+4)$.   (ii) $(2n-1, 9n+4)$.

(iii) $(36n+3, 90n+6)$.   (iv) $(2n+3, n+7)$.

(v) Show that there are infinitely many numbers $r_n = n(n+1)(n+2)/6$, $n \in \mathbb{N}$, that are pairwise relatively prime.

(vi) Show that any integer $n > 6$ is the sum of two relatively prime integers greater than 1.

(vii) Show that for $m, n \in \mathbb{N}_0$, $m > n$, the numbers $2^{2^m} + 1$ and $2^{2^n} + 1$ are relatively prime.

(viii) Given any $a, b \in \mathbb{Z}$, show that if $d = ka + lb > 0$, where $k, l \in \mathbb{Z}$, satisfies $d \mid a$ and $d \mid b$, then $d = (a, b)$.

*(ix) For any $m, n \in \mathbb{N}$, determine the greatest common divisor of $2^m - 1$ and $2^n - 1$.

1.14 Further Properties of the Greatest Common Divisor

Theorem. *For arbitrary natural numbers a, b, c we have*

(i) $(ac, bc) = (a, b) \cdot c$,

(ii) *if* $(a, b) = 1$ *and* $a \mid bc$, *then* $a \mid c$,

(iii) $d = (a, b)$ *if and only if there exist* $q_1, q_2 \in \mathbb{N}$ *such that* $a = dq_1$, $b = dq_2$, *and* $(q_1, q_2) = 1$.

PROOF.   (i) Since (a, b) is a common divisor of a and b, then $(a, b) \cdot c$ is a common divisor of ac and bc; hence $(a, b) \cdot c \mid (ac, bc)$. By 1.8 there exist $k, l \in \mathbb{Z}$ such that $(a, b) = ka + lb$. Since (ac, bc) is a common divisor of ac and bc, it also divides $kac + lbc = (a, b) \cdot c$. We have thus shown that $(a, b) \cdot c$ and (ac, bc) are two natural numbers that divide one another, and so by (4) they are equal.

(ii) We assume that $(a, b) = 1$ and $a \mid bc$. By 1.8 there exist $k, l \in \mathbb{Z}$ such that $ka + lb = 1$, which implies that $c = c(ka + lb) = kca + lbc$. Since $a \mid bc$, it follows that also $a \mid c$.

(iii) Let $d = (a, b)$; then there exist $q_1, q_2 \in \mathbb{N}$ with $a = dq_1$, $b = dq_2$. Then by (i) we have $d = (a, b) = (dq_1, dq_2) = d \cdot (q_1, q_2)$, and thus $(q_1, q_2) = 1$. Conversely, if $a = dq_1$, $b = dq_2$, and $(q_1, q_2) = 1$, then $(a, b) = (dq_1, dq_2) = d(q_1, q_2) = d \cdot 1 = d$ (again using (i)).   □

1.15 Exercises

Prove (i)–(v) for arbitrary natural numbers a, b, c, d:

(i) $(a, b) \cdot (c, d) = (ac, ad, bc, bd)$.

(ii) if $(a, b) = (a, c) = 1$, then $(a, bc) = 1$.

(iii) if $(a, b) = 1$, then $(ac, b) = (c, b)$.

(iv) $abc = [a, b, c] \cdot (ab, ac, bc)$.

(v) $abc = (a, b, c) \cdot [ab, ac, bc]$.

Solve the systems of equations (vi)–(ix) in positive integers:

(vi) $x + y = 150$,
$\quad\;\; (x, y) = 30$.

(vii) $(x, y) = 45$,
$\quad\;\;\; 7x = 11y$.

(viii) $xy = 8400$,
$\quad\quad\; (x, y) = 20$.

(ix) $xy = 20$,
$\quad\;\; [x, y] = 10$.

2   Prime Numbers

Prime numbers are among the most important objects in elementary number theory. Their importance comes mainly from the theorem on unique factorization of a natural number into a product of primes, a strong and efficient tool for solving many problems in number theory.

2.1   Primes and Composite Numbers

Definition. Any integer $n \geq 2$ has at least two positive divisors: 1 and n. If it has no other positive divisors, it is called a *prime number* (or simply a *prime*). Otherwise, it is called a *composite number*.

In the remainder of this book we will, as a rule, denote a prime by the letter p. The smallest primes are 2, 3, 5, 7, 11, 13, 17, 19, 23, 29, 31.

2.2   An Equivalent Condition

Theorem. *The integer $p \geq 2$ is a prime if and only if the following holds: For any two integers a and b, the condition $p \mid ab$ implies $p \mid a$ or $p \mid b$.*

PROOF.   (i) We assume that p is a prime and $p \mid ab$, where $a, b \in \mathbb{Z}$. Since (p, a) is a positive divisor of p, we have $(p, a) = p$ or $(p, a) = 1$. In the first case $p \mid a$, while in the second case $p \mid b$, by 1.14.(ii).

(ii) If p is not a prime, it must have a positive divisor different from 1 and p. We denote it by a; then obviously $b = p/a \in \mathbb{N}$ and $p = ab$, which implies $1 < a < p$, $1 < b < p$. We have thus found integers a, b such that $p \mid ab$ but p divides neither a nor b.   □

2.3   Examples

(i) Find all primes that are at the same time the sum and the difference of appropriate pairs of primes.

SOLUTION. We assume that the prime p is, at the same time, the sum and the difference of two primes. Then, of course, $p > 2$, and thus p is odd. Since p is both the sum and the difference of two primes, one of them must always be even, namely 2. Thus we require primes p, p_1, p_2 such that $p = p_1 + 2 = p_2 - 2$, therefore p_1, p, p_2 are three consecutive odd numbers, which means that exactly one of them is divisible by 3 (this follows, for example, from 1.4.(ii); of the three consecutive integers $p_2 - 3$, p, $p_1 + 3$ exactly one is divisible by 3). Of course, 3 itself is the only prime divisible by 3, which in view of $p_1 > 1$ implies that $p_1 = 3$, $p = 5$, $p_2 = 7$. The only prime number that satisfies the condition is therefore $p = 5$.   □

(ii) Find the numbers $k \in \mathbb{N}_0$ for which among the ten consecutive integers $k + 1, k + 2, \ldots, k + 10$ there are the most prime numbers.

SOLUTION. For $k = 1$ the set contains five primes: 2, 3, 5, 7, 11. For $k = 0$ and $k = 2$ there are only four primes. If $k \geq 3$, the set in question does not contain the number 3. By 1.4.(ii), among ten consecutive integers there are five even and five odd numbers, and among the latter at least one is also divisible by 3. Among the integers $k + 1, k + 2, \ldots, k + 10$ we have therefore found at least six composite numbers; thus there are at most four primes in this set. The only solution to the problem is therefore $k = 1$.   □

(iii) Show that given an arbitrary natural number n, there exist n consecutive natural numbers none of which is a prime.

SOLUTION. We consider the numbers

$$(n + 1)! + 2, (n + 1)! + 3, \ldots, (n + 1)! + (n + 1).$$

Among these n consecutive integers there is no prime number because any $k \in \{2, 3, \ldots, n + 1\}$ satisfies $k \mid (n + 1)!$, and thus $k \mid (n + 1)! + k$, which means that $(n + 1)! + k$ cannot be a prime.   □

(iv) Show that for any prime p and any $k \in \mathbb{N}$, $k < p$, the binomial coefficient $\binom{p}{k}$ is divisible by p.

SOLUTION. By definition of the binomial coefficient,

$$\binom{p}{k} = \frac{p!}{k!(p-k)!} = \frac{p \cdot (p-1) \cdots (p-k+1)}{1 \cdot 2 \cdots k} \in \mathbb{N},$$

and thus $k! \mid p \cdot a$, where we have set $a = (p-1) \cdots (p-k+1)$. Since $k < p$, none of the numbers $1, 2, \ldots, k$ is divisible by the prime p, and so 2.2 shows that $k!$ is not divisible by p, which means that $(k!, p) = 1$. By 1.14.(ii) we have $k! \mid a$; hence $b = a/k!$ is an integer. Since $\binom{p}{k} = pa/k! = pb$, the number $\binom{p}{k}$ is divisible by p.   □

2.4   Exercises

(i) Show that there is no natural number n such that $6n + 5$ can be expressed as a sum of two primes.

(ii) Find all natural numbers n for which n, $n + 10$, and $n + 14$ are all primes.

(iii) Find all primes p such that $2p^2 + 1$ is also a prime.

(iv) Find all primes p such that $4p^2 + 1$ and $6p^2 + 1$ are also primes.

(v) Find all $n \in \mathbb{N}$ such that $2^n - 1$ and $2^n + 1$ are both primes.

(vi) Show that if p and $8p^2 + 1$ are both primes, then $8p^2 + 2p + 1$ is a prime.

(vii) Find all $n \in \mathbb{N}$ such that $n + 1$, $n + 3$, $n + 7$, $n + 9$, $n + 13$, and $n + 15$ are primes.

*(viii) Show that the number $5^{20} + 2^{30}$ is composite.

*(ix) Show that there exist infinitely many natural numbers n such that every number of the form $m^4 + n$, where $m \in \mathbb{N}$, is composite.

(x) Show that for any prime $p > 2$ the numerator m of the fraction

$$\frac{m}{n} = 1 + \frac{1}{2} + \frac{1}{3} + \cdots + \frac{1}{p-1},$$

where $m, n \in \mathbb{N}$, is divisible by p.

*(xi) For any $n \in \mathbb{N}$, find n pairwise relatively prime natural numbers $a_1, \ldots, a_n$ such that for every $1 < k \leq n$ the sum of any k of them is composite.

*(xii) Show that if $a, b, c, d \in \mathbb{N}$ satisfy $ab = cd$, then the sum $a^n + b^n + c^n + d^n$ is composite for any $n \in \mathbb{N}$.

2.5   Unique Factorization

Theorem. *Any natural number $n \geq 2$ can be expressed as a product of primes. This expression is unique up to the order of the factors. (If n itself is a prime, then by "product" we mean just this prime).*

PROOF.   First we prove by induction that every $n \geq 2$ can be decomposed into a product of primes.

If $n = 2$, then the product is just the prime 2.

We now assume that $n > 2$ and that we have already shown that any n', $2 \leq n' < n$, can be written as a product of primes. If n is a prime, then the product consists of just this one prime. If n is not prime, then it has a divisor d, $1 < d < n$. If we set $c = n/d$, then also $1 < c < n$. The induction hypothesis implies that c and d can be expressed as products of primes, and therefore their product $c \cdot d = n$ can also be expressed in this way.

To prove uniqueness, we assume that we have equality between the products $p_1 \cdot p_2 \cdots p_m = q_1 \cdot q_2 \cdots q_s$, where $p_1, \ldots, p_m, q_1, \ldots, q_s$ are primes such that $p_1 \leq p_2 \leq \cdots \leq p_m$, $q_1 \leq q_2 \leq \cdots \leq q_s$, and $1 \leq m \leq s$. We use induction on m to show that $m = s$, $p_1 = q_1, \ldots, p_m = q_m$.

In the case $m = 1$, $p_1 = q_1 \cdots q_s$ is a prime. If $s > 1$, then p_1 would have a divisor q_1 such that $1 < q_1 < p_1$ (since $q_2 q_3 \cdots q_s > 1$), which is impossible. Hence $s = 1$ and $p_1 = q_1$.

We now assume that $m \geq 2$ and that the assertion holds for $m - 1$. Since $p_1 \cdot p_2 \cdots p_m = q_1 \cdot q_2 \cdots q_s$, the prime p_m divides the product $q_1 \cdots q_s$, which by 2.2 is possible only when p_m divides a q_i for an appropriate $i \in \{1, 2, \ldots, s\}$. But q_i is a prime, so it follows that $p_m = q_i$ (since $p_m > 1$). In complete analogy it can be shown that $q_s = p_j$ for an appropriate $j \in \{1, 2, \ldots, m\}$. Hence

$$q_s = p_j \leq p_m = q_i \leq q_s,$$

and therefore $p_m = q_s$. Upon dividing we obtain $p_1 \cdot p_2 \cdots p_{m-1} = q_1 \cdot q_2 \cdots q_{s-1}$, and thus from the induction hypothesis, $m - 1 = s - 1$, $p_1 = q_1, \ldots, p_{m-1} = q_{m-1}$. In summary, $m = s$ and $p_1 = q_1, \ldots, p_{m-1} = q_{m-1}$, $p_m = q_m$. This proves uniqueness, and the proof of the theorem is complete. $\qquad\square$

2.6   Consequences

(i) If $p_1, \ldots, p_k$ are distinct primes and $n_1, \ldots, n_k \in \mathbb{N}_0$, then every positive divisor of the number $a = p_1^{n_1} \cdots p_k^{n_k}$ is of the form $p_1^{m_1} \cdots p_k^{m_k}$, with $m_1, \ldots, m_k \in \mathbb{N}_0$ and $m_1 \leq n_1$, $m_2 \leq n_2$, $\ldots$, $m_k \leq n_k$. The number a has therefore exactly $(n_1 + 1)(n_2 + 1) \cdots (n_k + 1)$ positive divisors.

(ii) Let $p_1, \ldots, p_k$ be distinct primes and $n_1, \ldots, n_k, m_1, \ldots, m_k \in \mathbb{N}_0$. If we set $r_i = \min\{n_i, m_i\}$, $t_i = \max\{n_i, m_i\}$ for all $i = 1, 2, \ldots, k$, then

$$(p_1^{n_1} \cdots p_k^{n_k}, p_1^{m_1} \cdots p_k^{m_k}) = p_1^{r_1} \cdots p_k^{r_k},$$
$$[p_1^{n_1} \cdots p_k^{n_k}, p_1^{m_1} \cdots p_k^{m_k}] = p_1^{t_1} \cdots p_k^{t_k}.$$

(iii) If a, b, c, m are natural numbers such that $(a, b) = 1$ and $ab = c^m$, then there exist relatively prime natural numbers x, y such that $a = x^m$, $b = y^m$.

The proofs of these consequences are left to the reader; hints can be found in the appendix.

2.7   Examples

(i) Show that for each integer $n > 2$ there is at least one prime between n and $n!$

SOLUTION.   Let p denote any prime dividing $n! - 1$ (it exists by 2.5, since $n! - 1 > 1$). If $p \leq n$, then p would have to divide $n!$ and could therefore not be a divisor of $n! - 1$. Hence $n < p$. Since $p \mid (n! - 1)$, we have $p \leq n! - 1$, and thus $p < n!$. The prime p therefore satisfies the given condition.   □

(ii) Show that there exist infinitely many primes of the form $3k + 2$, where $k \in \mathbb{N}_0$.

SOLUTION.   We assume the contrary, namely that there exist only finitely many primes of this form, and we denote them by $p_1 = 2$, $p_2 = 5$, $p_3 = 11$, $\ldots$, p_n. We set $N = 3p_2 \cdot p_3 \cdots p_n + 2$. If we write N as a product of primes according to Theorem 2.5, then this decomposition must contain at least one prime p of the form $3k + 2$, since otherwise N would be a product of primes of the form $3k + 1$ (note that N is not divisible by 3), and so by 1.4.(i), N would also be of the form $3k + 1$, which is not true. Now it follows from the particular form of N that the prime p cannot be among the primes $p_1, p_2, \ldots, p_n$, and this is a contradiction.   □

We note that the examples in 2.7 allow for considerable generalizations. In the book [11] *W. Sierpiński* proves the theorem: "For any natural number $n > 5$ there are at least two primes between n and $2n$," which generalizes *Chebyshev's theorem*: "For any integer $n > 3$ there is at least one prime between n and $2n - 2$." The proof uses only elementary methods, but is relatively long (pp. 131–137).

Example 2.7.(ii) is generalized by *Dirichlet's theorem on primes in arithmetic progressions*: "If a and m are relatively prime natural numbers, there exist infinitely many natural numbers k such that $mk + a$ is prime." This is a deep theorem, and its proof requires methods that go well beyond elementary number theory.

2.8   Exercises

(i) Find $(a + b, ab)$, where a, b are relatively prime integers.

(ii) Given integers a and b, find $(a + b, [a, b])$.

(iii) Show that if $a > 4$ is a composite integer, then $a \mid (a - 1)!$.

*(iv) Show that if $p > 5$ is a prime, then there is no $m \in \mathbb{N}$ such that $(p - 1)! + 1 = p^m$.

(v) Show that there exist infinitely many primes of the form $4k + 3$, where $k \in \mathbb{N}_0$.

(vi) Show that a prime of the form $2^{2^n} + 1$, where $n \in \mathbb{N}$, cannot be expressed as a difference of fifth powers of two natural numbers.

*(vii) Suppose that a and n are natural numbers with the property that there exists an $s \in \mathbb{N}$ such that $(a - 1)^s$ is divisible by n. Show that $1 + a + \cdots + a^{n-1}$ is also divisible by n.

(viii) Show that if $a, b \in \mathbb{Z}$ are such that $(a, b) = 1$ and $2 \mid ab$, then $(a + b, a^2 + b^2) = 1$.

2.9 Exponents in a Factorization

To any prime p and any natural number n there is, according to Theorem 2.5, a uniquely determined number of appearances of p in the decomposition of n into prime factors (if p does not divide n, we take this number to be zero). We denote this number by the symbol $v_p(n)$. For a negative integer n we set $v_p(n) = v_p(-n)$.

According to 2.6.(i) we can characterize this notation by saying that $v_p(n)$ is the exponent of the largest power of the prime p that divides n, or by $n = p^{v_p(n)} \cdot m$, where m is an integer not divisible by p. From this it easily follows that any nonzero integers a, b satisfy

$$v_p(ab) = v_p(a) + v_p(b), \tag{8}$$
$$v_p(a) \le v_p(b) \wedge a + b \ne 0 \implies v_p(a + b) \ge v_p(a), \tag{9}$$
$$v_p(a) < v_p(b) \implies v_p(a + b) = v_p(a), \tag{10}$$
$$v_p(a) \le v_p(b) \implies v_p((a, b)) = v_p(a) \wedge v_p([a, b]) = v_p(b). \tag{11}$$

2.10 Example

Show that any natural numbers a, b, c satisfy

$$([a, b], [a, c], [b, c]) = [(a, b), (a, c), (b, c)].$$

SOLUTION.   By 2.5, we are done if we can show that $v_p(L) = v_p(R)$ for any prime p, where L, resp. R, denotes the term on the left, resp. on the right. Now let p be any prime. In view of the symmetry on both sides, we may without loss of generality assume that $v_p(a) \le v_p(b) \le v_p(c)$. By (11) we have $v_p([a, b]) = v_p(b)$, $v_p([a, c]) = v_p([b, c]) = v_p(c)$; $v_p((a, b)) = v_p((a, c)) = v_p(a)$, $v_p((b, c)) = v_p(b)$, and this implies $v_p(L) = v_p(b) = v_p(R)$, which was to be shown. □

2.11 Exercises

(i) Give new proofs of 1.15.(i)–(v), using the method of 2.10.

(ii) Show that any natural numbers a, b, c, such that $(a, b) = 1$, satisfy $(ab, c) = (a, c) \cdot (b, c)$.

(iii) Show that if for $m, n \in \mathbb{N}$ the number $\sqrt[m]{n}$ is rational, then it is in fact a positive integer.

(iv) Show that if $n, r, s \in \mathbb{N}$, $(r, s) = 1$ and $\sqrt[s]{n^r} \in \mathbb{N}$, then $n = m^s$ for an appropriate $m \in \mathbb{N}$.

(v) Show that if for some positive rational numbers a and b the number $\sqrt{a} + \sqrt{b}$ is rational, then $\sqrt{a}$ and $\sqrt{b}$ are also rational.

*(vi) Let a, b, c, d be integers such that ac, bd, $bc + ad$ are all divisible by the same natural number m. Show that then the numbers bc and ad are also divisible by m.

3   Congruences

The concept of congruence was introduced by Gauss. Although it is a very simple notion, its importance and usefulness in number theory are enormous. This is particularly obvious in that even complicated arguments can be presented in a concise and clear manner.

3.1   Definition of Congruence

Definition. If the two integers a, b have the same remainder r upon division by the natural number m, where $0 \le r < m$, then a and b are called *congruent modulo m*, and we write

$$a \equiv b \pmod{m}.$$

Otherwise, we say that a and b are not congruent, or incongruent, modulo m,

$$a \not\equiv b \pmod{m}.$$

3.2   Equivalent conditions

Theorem. *For any numbers $a, b \in \mathbb{Z}$ and $m \in \mathbb{N}$ the following conditions are equivalent:*

(i) $a \equiv b \pmod{m}$,
(ii) $a = b + mt$ *for an appropriate* $t \in \mathbb{Z}$,
(iii) $m \mid a - b$.

PROOF.   I. If $a = q_1 m + r$ and $b = q_2 m + r$, then $a - b = (q_1 - q_2)m$, and therefore (i) implies (iii).

II. If $m \mid a - b$, then there exists a $t \in \mathbb{Z}$ such that $m \cdot t = a - b$, that is, $a = b + mt$, and so (iii) implies (ii).

III. If $a = b + mt$, then it follows from $b = mq + r$ that $a = m(q + t) + r$, so a and b have the same remainder r upon division by m, that is, $a \equiv b \pmod{m}$; hence (ii) implies (i).   □

3.3 Basic Properties of Congruences

It follows directly from Definition 3.1 that congruence modulo m is reflexive (i.e., $a \equiv a \pmod m$ for any $a \in \mathbb{Z}$), symmetric (i.e., for each $a, b \in \mathbb{Z}$, $a \equiv b$ $\pmod m$ implies $b \equiv a \pmod m$)) and transitive (i.e., for each $a, b, c \in \mathbb{Z}$, $a \equiv b \pmod m$ and $b \equiv c \pmod m$ imply $a \equiv c \pmod m$)). We now prove some further properties:

(i) Congruences with the same modulus can be added. Any summand can be moved, with the opposite sign, from one side of the congruence to the other. To any side of the congruence we can add an arbitrary multiple of the modulus.

PROOF.   If $a_1 \equiv b_1 \pmod m$ and $a_2 \equiv b_2 \pmod m$, then by 3.2 there exist $t_1, t_2 \in \mathbb{Z}$ such that $a_1 = b_1 + mt_1$, $a_2 = b_2 + mt_2$. Then $a_1 + a_2 = b_1 + b_2 + m(t_1 + t_2)$, and again by 3.2 we have $a_1 + a_2 \equiv b_1 + b_2 \pmod m$. If we add $a + b \equiv c \pmod m$ and the obviously valid congruence $-b \equiv -b$ $\pmod m$, we obtain $a \equiv c - b \pmod m$. If we add another obviously valid congruence, $mk \equiv 0 \pmod m$, to $a \equiv b \pmod m$, then we get $a + mk \equiv b$ $\pmod m$.   □

(ii) Congruences with the same modulus can be multiplied together. Both sides of a congruence can be raised to the same positive integer power. Both sides of a congruence can be multiplied by the same integer.

PROOF.   If $a_1 \equiv b_1 \pmod m$ and $a_2 \equiv b_2 \pmod m$, then by 3.2 there exist $t_1, t_2 \in \mathbb{Z}$ such that $a_1 = b_1 + mt_1$ and $a_2 = b_2 + mt_2$. Then

$$a_1 a_2 = (b_1 + mt_1)(b_2 + mt_2) = b_1 b_2 + m(t_1 b_2 + b_1 t_2 + mt_1 t_2),$$

and from 3.2 we obtain $a_1 a_2 \equiv b_1 b_2 \pmod m$.

   If $a \equiv b \pmod m$, we prove by induction on n that $a^n \equiv b^n \pmod m$. For $n = 1$ there is nothing to show. If $a^n \equiv b^n \pmod m$ for some fixed n, we multiply this congruence with $a \equiv b \pmod m$ and obtain $a^n \cdot a \equiv b^n \cdot b$ $\pmod m$, hence $a^{n+1} \equiv b^{n+1} \pmod m$, and this is the assertion for $n + 1$. The proof by induction is now complete.

   If we multiply the congruences $a \equiv b \pmod m$ and $c \equiv c \pmod m$, we obtain $ac \equiv bc \pmod m$.   □

(iii) Both sides of a congruence can be divided by a common divisor, as long as this divisor is relatively prime to the modulus.

PROOF.   We assume that $a \equiv b \pmod m$, $a = a_1 \cdot d$, $b = b_1 \cdot d$, and $(m, d) = 1$. By 3.2 the difference $a - b = (a_1 - b_1) \cdot d$ is divisible by m. Since $(m, d) = 1$, by 1.14.(ii) the number $a_1 - b_1$ is also divisible by m, and with 3.2 it follows that $a_1 \equiv b_1 \pmod m$.   □

(iv) Both sides of a congruence and its modulus can be simultaneously multiplied by the same natural number.

PROOF. If $a \equiv b \pmod{m}$, then by 3.2 there exists an integer t such that $a = b + mt$; hence for $c \in \mathbb{N}$ we have $ac = bc + mc \cdot t$, and using 3.2 again we get $ac \equiv bc \pmod{mc}$. □

(v) Both sides of a congruence and its modulus can be divided by a positive common divisor.

PROOF. We assume that $a \equiv b \pmod{m}$, $a = a_1 \cdot d$, $b = b_1 \cdot d$, $m = m_1 \cdot d$, where $d \in \mathbb{N}$. By 3.2 there exists a $t \in \mathbb{Z}$ such that $a = b + mt$, that is, $a_1 \cdot d = b_1 \cdot d + m_1 dt$, which implies $a_1 = b_1 + m_1 t$; this means, by 3.2, that $a_1 \equiv b_1 \pmod{m_1}$. □

(vi) If the congruence $a \equiv b$ holds with various moduli $m_1, \ldots, m_k$, then it also holds modulo $[m_1, \ldots, m_k]$, the least common multiple of these numbers.

PROOF. If $a \equiv b \pmod{m_1}, a \equiv b \pmod{m_2}, \ldots, a \equiv b \pmod{m_k}$, then by 3.2 the difference $a - b$ is a common multiple of the moduli $m_1, m_2, \ldots, m_k$, and is thus divisible by the least common multiple $[m_1, m_2, \ldots, m_k]$, which implies $a \equiv b \pmod{[m_1, \ldots, m_k]}$. □

(vii) If a congruence holds modulo m, then it also holds modulo d, where d is any divisor of m.

PROOF. If $a \equiv b \pmod{m}$, then $a - b$ is divisible by m, and therefore also by the divisor d of m, which means that $a \equiv b \pmod{d}$. □

(viii) If one side of a congruence and its modulus are divisible by some integer, then the other side will also be divisible by this integer.

PROOF. We assume that $a \equiv b \pmod{m}$, with $b = b_1 d$, $m = m_1 d$. Then by 3.2 there exists a $t \in \mathbb{Z}$ such that $a = b + mt = b_1 d + m_1 dt = (b_1 + m_1 t)d$, and thus $d \mid a$. □

We remark that we have already used several properties of congruences without having taken note of this fact. For instance, 1.4.(i) can be rewritten as "if $a \equiv 1 \pmod{m}$ and $b \equiv 1 \pmod{m}$, then also $ab \equiv 1 \pmod{m}$," which is a special case of the property in (ii). This is, of course, not a coincidence. Any statement using congruences can easily be rewritten in terms of divisibilities. The usefulness of congruences lies therefore not in the fact that one could solve problems that could not be solved otherwise, but in the fact that they provide a very convenient notation. By making consistent use of this notation we can significantly simplify not only the exposition, but also certain arguments. This is a typical situation: Appropriate notation plays a very important role in mathematics.

3.4 Examples

(i) Find the remainder when 5^{20} is divided by 26.

SOLUTION.   Since $5^2 = 25 \equiv -1 \pmod{26}$, we have by 3.3.(ii),

$$5^{20} \equiv (-1)^{10} = 1 \pmod{26},$$

so the desired remainder is 1.   □

(ii) Show that for any $n \in \mathbb{N}$, $37^{n+2} + 16^{n+1} + 23^n$ is divisible by 7.

SOLUTION.   Since $37 \equiv 16 \equiv 23 \equiv 2 \pmod 7$, by 3.3.(ii) and (i) we have

$$37^{n+2} + 16^{n+1} + 23^n \equiv 2^{n+2} + 2^{n+1} + 2^n = 2^n(4+2+1)$$
$$= 2^n \cdot 7 \equiv 0 \pmod 7,$$

which was to be shown.   □

(iii) Show that $n = (835^5 + 6)^{18} - 1$ is divisible by 112.

SOLUTION.   We factor $112 = 7 \cdot 16$. Since $(7, 16) = 1$, it suffices to show that $7 \mid n$ and $16 \mid n$. We have $835 \equiv 2 \pmod 7$, and thus by 3.3,

$$n \equiv (2^5 + 6)^{18} - 1 = 38^{18} - 1 \equiv 3^{18} - 1$$
$$= 27^6 - 1 \equiv (-1)^6 - 1 = 0 \pmod 7;$$

hence $7 \mid n$. Similarly, $835 \equiv 3 \pmod{16}$, and thus

$$n \equiv (3^5 + 6)^{18} - 1 = (3 \cdot 81 + 6)^{18} - 1 \equiv (3 \cdot 1 + 6)^{18} - 1$$
$$= 9^{18} - 1 = 81^9 - 1 \equiv 1^9 - 1 = 0 \pmod{16};$$

hence $16 \mid n$. Together, $112 \mid n$, which was to be shown.   □

(iv) Show that no number of the form $8k \pm 3$, $k \in \mathbb{N}$, can be expressed in the form $x^2 - 2y^2$ for any integers x, y.

SOLUTION.   Assume that $x^2 - 2y^2 = 8k \pm 3$ for some $x, y \in \mathbb{Z}$. Then x must be an odd number, and therefore $x^2 \equiv 1 \pmod 8$ by 1.5.(ii). If y is also odd, then $x^2 - 2y^2 \equiv 1 - 2 = -1 \pmod 8$. If y is even, then $2y^2 \equiv 0 \pmod 8$, and thus $x^2 - 2y^2 \equiv 1 \pmod 8$. In neither case do we have $x^2 - 2y^2 \equiv \pm 3 \pmod 8$, and this is a contradiction.   □

(v) Show that for any prime p and any $a, b \in \mathbb{Z}$ we have

$$a^p + b^p \equiv (a + b)^p \pmod p.$$

SOLUTION.   By the binomial theorem we have

$$(a + b)^p = a^p + \binom{p}{1} a^{p-1}b + \binom{p}{2} a^{p-2}b^2 + \cdots + \binom{p}{p-1} ab^{p-1} + b^p.$$

For any $k \in \{1, \ldots, p-1\}$ we have $\binom{p}{k} \equiv 0 \pmod{p}$ by 2.3.(iv), and this implies the assertion. $\qquad\square$

(vi) Show that for any natural number m and any $a, b \in \mathbb{Z}$ such that $a \equiv b$ $(\mod m^n)$, where $n \in \mathbb{N}$, we have $a^m = b^m \pmod{m^{n+1}}$.

SOLUTION.   By identity (5) in Chapter 1 we have

$$a^m - b^m = (a - b)(a^{m-1} + a^{m-2}b + \cdots + ab^{m-2} + b^{m-1}). \qquad (12)$$

From $n \in \mathbb{N}$ it follows that $m \mid m^n$, and so by 3.3.(vii) we have $a \equiv b \pmod{m}$. Hence all the summands in the right-hand expression in parentheses in (12) are congruent to a^{m-1} modulo m, and thus

$$a^{m-1} + a^{m-2}b + \cdots + ab^{m-2} + b^{m-1} \equiv m \cdot a^{m-1} \equiv 0 \pmod{m}.$$

Therefore, $a^{m-1} + a^{m-2} + \cdots + ab^{m-2} + b^{m-1}$ is divisible by m. From $a \equiv b \pmod{m^n}$ it follows that m^n divides $a - b$, and thus m^{n+1} divides the product, which by (12) leads to the conclusion that $a^m \equiv b^m$ $(\mod m^{n+1})$. $\qquad\square$

3.5 Exercises

Show that

- **(i)** the number $2^{60} + 7^{30}$ is divisible by 13;
- **(ii)** for any $n \in \mathbb{N}$, $72^{2n+2} - 47^{2n} + 28^{2n-1}$ is divisible by 25;
- **(iii)** for any $k, m, n \in \mathbb{N}$, $5^{5k+1} + 4^{5m+2} + 3^{5n}$ is divisible by 11;
- **(iv)** for any integers a, b, the congruence $a^2 + b^2 \equiv 0 \pmod{3}$ implies $a \equiv b \equiv 0 \pmod{3}$;
- **(v)** for any integers a, b, the congruence $a^2 + b^2 \equiv 0 \pmod{7}$ implies $a \equiv b \equiv 0 \pmod{7}$;
- **(vi)** there exist integers a, b such that $a^2 + b^2 \equiv 0 \pmod{5}$, while $a \equiv b \equiv 0 \pmod{5}$ does not hold;
- **(vii)** for any integers a, b, c, the congruence $a^3 + b^3 + c^3 \equiv 0 \pmod{9}$ implies $abc \equiv 0 \pmod{3}$;
- **(viii)** for any integers $a_1, a_2, \ldots, a_7$, the congruence $a_1^3 + a_2^3 + \cdots + a_7^3 \equiv 0$ $(\mod 9)$ implies $a_1 a_2 \cdots a_7 \equiv 0 \pmod{3}$;
- ***(ix)** no integer a satisfies $a^2 + 3a + 5 \equiv 0 \pmod{121}$.
- **(x)** Find all natural numbers n such that $n \cdot 2^n + 1$ is divisible by 3.
- **(xi)** Determine whether there exists a natural number n such that the difference $4^n - 2^{n-1}$ is the third power of an integer.

(xii) Show that one can place the numbers 1, 2, ... , 12 on the circumference of a circle in such a way that for any three neighbors a, b, and c, with b between a and c, the number $b^2 - ac$ is divisible by 13.

*(xiii) Show that $19 \cdot 8^n + 17$ is composite for any natural number n.

*(xiv) Show that $(n - 1)^2$ is a divisor of $n^n - n^2 + n - 1$ for any natural number n.

*(xv) Let $a, a_0, a_1, \ldots, a_n$ be integers ($n \in \mathbb{N}_0$). Determine whether it is true that the integer

$$(a^2 + 1)^0 \cdot a_0 + (a^2 + 1)^3 a_1 + (a^2 + 1)^6 a_2 + \cdots + (a^2 + 1)^{3k} a_k$$

is divisible by $a^2 + a + 1$ (resp. $a^2 - a + 1$) if and only if

$$a_0 - a_1 + a_2 - \cdots + (-1)^{k-1} a_{k-1} + (-1)^k a_k$$

is divisible by $a^2 + a + 1$ (resp. $a^2 - a + 1$).

3.6   Euler's φ-Function

Definition. For any natural number $m = p_1^{n_1} \cdots p_k^{n_k}$, where $p_1, \ldots, p_k$ ($k \geq 1$) are distinct primes and $n_1, \ldots, n_k \in \mathbb{N}$, we define $\varphi(m)$ as follows:

$$\varphi(m) = p_1^{n_1-1}(p_1 - 1) \cdots p_k^{n_k-1}(p_k - 1). \tag{13}$$

If, in addition, we define $\varphi(1) = 1$, we have a function $\varphi : \mathbb{N} \to \mathbb{N}$. We can easily convince ourselves that (13) can also be written as

$$\varphi(m) = m \cdot \left(1 - \frac{1}{p_1}\right) \cdots \left(1 - \frac{1}{p_k}\right)$$

and that for any relatively prime natural numbers m_1, m_2 we have $\varphi(m_1 m_2) = \varphi(m_1)\varphi(m_2)$. The reader should verify that the condition of relative primality of m_1, m_2 is indeed necessary. The clearest and most suggestive representation of Euler's φ-function is given later by the assertion in 3.10.

3.7   Examples

(i) Show that for any $n \in \mathbb{N}$ we have $\varphi(4n + 2) = \varphi(2n + 1)$.

SOLUTION.   Since $2n + 1$ is odd, the numbers 2 and $2n + 1$ are relatively prime, and thus we have $\varphi(2 \cdot (2n + 1)) = \varphi(2) \cdot \varphi(2n + 1)$. We are done if we realize that $\varphi(2) = 1$. (More generally, for each prime p we have $\varphi(p) = p - 1$.)   □

(ii) Find all $m \in \mathbb{N}$ for which $\varphi(m)$ is odd.

SOLUTION.   We have $\varphi(1) = \varphi(2) = 1$. If $m \geq 3$, then m is divisible by an odd prime p, or $m = 2^n$, $n \geq 2$ (this follows from the decomposition of m into a product of primes). In the first case, $\varphi(m)$ is divisible by the even number $p - 1$, and in the second case we have $\varphi(m) = 2^{n-1}$, so $\varphi(m)$ is always even. Therefore, the number $\varphi(m)$ is odd only when $m = 1$ or $m = 2$. $\square$

(iii) Solve the equation $\varphi(5^x) = 100$, where x is a natural number.

SOLUTION.   By (13) we have $\varphi(5^x) = 4 \cdot 5^{x-1}$; we therefore have to solve the equation $4 \cdot 5^{x-1} = 100$, i.e., $5^{x-1} = 5^2$, which means $x - 1 = 2$, or $x = 3$. $\square$

(iv) Show that for any $m, n \in \mathbb{N}$ and $d = (m, n)$ we have

$$\varphi(mn) = \varphi(m) \cdot \varphi(n) \cdot \frac{d}{\varphi(d)}.$$

SOLUTION.   We first assume that $m = p^a$ and $n = p^b$, where p is a prime and $a, b \in \mathbb{N}_0$. If $a = 0$, then $m = 1$, $d = 1$, $\varphi(d) = 1$, and the assertion clearly holds. Similarly for $b = 0$. If $a, b \in \mathbb{N}$, then

$$\varphi(mn) = \varphi(p^{a+b}) = (p - 1)p^{a+b-1},$$

and if we set $c = \min\{a, b\}$ (i.e., c is the smallest of the numbers a, b), then $d = p^c$ and

$$\varphi(m)\varphi(n) \cdot \frac{d}{\varphi(d)} = (p - 1)p^{a-1} \cdot (p - 1)p^{b-1} \cdot \frac{p^c}{p^{c-1}(p - 1)}$$
$$= (p - 1)p^{a+b-1},$$

and thus the assertion holds. Let now $m, n \in \mathbb{N}$ be arbitrary. We write $m = p_1^{a_1} \cdots p_k^{a_k}$, $n = p_1^{b_1} \cdots p_k^{b_k}$, where $a_1, \ldots, a_k, b_1, \ldots, b_k \in \mathbb{N}_0$ and $p_1, \ldots, p_k$ are distinct primes. If we set $c_i = \min\{a_i, b_i\}$ for $i = 1, \ldots, k$, then

$$\varphi(m)\varphi(n) \cdot \frac{d}{\varphi(d)} = \varphi(p_1^{a_1} \cdots p_k^{a_k})\varphi(p_1^{b_1} \cdots p_k^{b_k}) \cdot \frac{p_1^{c_1} \cdots p_k^{c_k}}{\varphi(p_1^{c_1} \cdots p_k^{c_k})}$$
$$= \varphi(p_1^{a_1}) \cdots \varphi(p_k^{a_n})\varphi(p_1^{b_1}) \cdots \varphi(p_k^{b_k}) \cdot \frac{p_1^{c_1}}{\varphi(p_1^{c_1})} \cdots \frac{p_k^{c_k}}{\varphi(p_k^{c_k})}$$
$$= \left(\varphi(p_1^{a_1})\varphi(p_1^{b_1}) \frac{p_1^{c_1}}{\varphi(p_1^{c_1})} \right) \cdots \left(\varphi(p_k^{c_k})\varphi(p_k^{b_k}) \frac{p_k^{c_k}}{\varphi(p_k^{c_k})} \right),$$

and so by the first part of the proof we have

$$\varphi(m)\varphi(n) \cdot \frac{d}{\varphi(d)} = \varphi(p_1^{a_1} p_1^{b_1}) \cdots \varphi(p_k^{a_k} p_k^{b_k})$$
$$= \varphi(p_1^{a_1} p_1^{b_1} \cdots p_k^{a_k} p_k^{b_k}) = \varphi(mn),$$

which was to be shown. $\square$

3.8   Exercises

   (i) Evaluate $\varphi(1000)$, $\varphi(635)$, $\varphi(180)$, $\varphi(360)$, $\varphi(1001)$.

  (ii) Show that for any $n \in \mathbb{N}$ we have either $\varphi(5n) = 5\varphi(n)$, or $\varphi(5n) = 4\varphi(n)$. When does each case occur?

Solve the equations (iii)–(xiv), where $x, y \in \mathbb{N}_0$, $m \in \mathbb{N}$ and p, q are primes:

   (iii) $\varphi(p^x) = p^{x-1}$.   (iv) $\varphi(3^x 5^y) = 600$.

   (v) $\varphi(p^x) = 6p^{x-2}$.   (vi) $\varphi(pq) = 120$.

  (vii) $\varphi(pm) = \varphi(m)$.   (viii) $\varphi(pm) = p\varphi(m)$.

  (ix) $\varphi(pm) = \varphi(qm)$.   (x) $\varphi(m) = 14$.

  (xi) $\varphi(m) = 16$.   (xii) $\varphi(m) = m/2$.

 (xiii) $\varphi(m) = m/3$.   (xiv) $\varphi(m) = m/4$.

  (xv) Show that $\varphi(m \cdot n) = (m, n) \cdot \varphi([m, n])$ for any $m, n \in \mathbb{N}$.

  (xvi) Evaluate the sum $\varphi(1) + \varphi(p) + \varphi(p^2) + \cdots + \varphi(p^n)$, where p is a prime and $n \in \mathbb{N}$.

*(xvii) Show that

$$\varphi(d_1) + \varphi(d_2) + \cdots + \varphi(d_s) = m,$$

where $d_1, d_2, \ldots, d_s$ are all the positive divisors of the natural number m.

3.9   Another Property of Euler's φ-Function

Theorem. *Let t and m be natural numbers. Among the integers $1, 2, \ldots,$ $tm - 1, tm$ there are exactly $t \cdot \varphi(m)$ that are relatively prime to m.*

PROOF.   We write $m = p_1^{n_1} \cdots p_k^{n_k}$, where $p_1, \ldots, p_k$ are distinct primes and $n_1, \ldots, n_k \in \mathbb{N}$. We prove the theorem by induction on k.

   If $k = 0$, then $m = 1$, $\varphi(m) = 1$, and the theorem is clearly true, since any integer is relatively prime to 1.

   We now assume that $k \geq 1$ and that the theorem is true for $m' = p_1^{n_1} \cdots p_{k-1}^{n_{k-1}}$. Then $tm = tp_k^{a_k} \cdot m'$, and by the induction hypothesis, among the integers $1, 2, \ldots, tm - 1, tm$ there are exactly $tp_k^{a_k} \cdot \varphi(m')$ numbers relatively prime to m'. Since p_k is prime, the numbers that have a factor in common with $p_k^{a_k}$ are exactly those divisible by p_k, namely $p_k, 2p_k, \ldots, (tp_k^{a_k-1}m' - 1)p_k, tp_k^{a_k-1}m'p_k$. Since $(p_k, m') = 1$, by 1.15.(iii) we have $(ap_k, m') = (a, m')$ for any $a \in \mathbb{Z}$, and thus among these integers there are exactly as many relatively prime to m' as there are among the integers $1, 2, \ldots, tp_k^{a_k-1}m' - 1, tp_k^{a_k-1}m'$; by the induction hypothesis their number is exactly $tp_k^{a_k-1} \cdot \varphi(m')$. We have thus shown that among the integers $1, 2, \ldots, tm$ there are exactly $tp_k^{a_k} \cdot \varphi(m')$ that are relatively prime to m', and $tp_k^{a_k-1}\varphi(m')$ of those have a factor in common with $p_k^{a_k}$; the remaining

$$tp_k^{a_k}\varphi(m') - tp_k^{a_k-1}\varphi(m') = t\varphi(m') \cdot (p_k^{a_k} - p_k^{a_k-1}) = t \cdot \varphi(m)$$

numbers are thus relatively prime to $p_k^{a_k}$. It remains to convince ourselves that a number is relatively prime to m if and only if it is relatively prime to m' and at the same time to $p_k^{a_k}$; this follows, for example, from 2.11.(ii).   □

3.10   An Alternative Definition

Consequence. *Exactly $\varphi(m)$ of the numbers $1, 2, \ldots, m$ are relatively prime to m.*

PROOF.   The statement follows from 3.9 with $t = 1$.   □

We remark that usually the property in 3.10 is used as definition of the number $\varphi(m)$, and then there is no need to prove it. This approach, however, requires the proof of formula (13), which is just as difficult as the proof of our Theorem 3.9.

3.11   Fermat's Theorem

Theorem. *For any prime p and any integer a we have*

$$a^p \equiv a \pmod p.$$

If furthermore $(a, p) = 1$, then $a^{p-1} \equiv 1 \pmod p$.

PROOF.   We first assume that $a \geq 0$, and we prove the assertion by induction on a.

If $a = 0$, then $a^p = 0 = a$, so certainly $a^p \equiv a \pmod p$.

We now assume that $a \geq 1$ and that $(a - 1)^p \equiv a - 1 \pmod p$, i.e., the statement of the theorem holds for $a - 1$. Then by 3.4.(v) we have

$$a^p = ((a - 1) + 1)^p \equiv (a - 1)^p + 1^p \equiv a - 1 + 1 = a \pmod p,$$

and so the statement is also true for a. It remains to show that $a^p \equiv a \pmod p$ for $a < 0$. By 1.3 there exist $q \in \mathbb{Z}$, $r \in \{0, 1, \ldots, p - 1\}$ such that $a = qp + r$. Then by 3.3.(ii) and the fact that the theorem was proved for any $r \geq 0$,

$$a^p = (qp + r)^p \equiv r^p \equiv r \equiv a \pmod p.$$

If furthermore $(a, p) = 1$, then from $a \cdot a^{p-1} \equiv a \pmod p$ we see with 3.3.(iii) that $a^{p-1} \equiv 1 \pmod p$.   □

3.12 Fermat's Theorem for Prime Powers

Consequence. *For any prime p, any $n \in \mathbb{N}$, and an arbitrary integer a relatively prime to p we have*

$$a^{p^{n-1}(p-1)} \equiv 1 \pmod{p^n}.$$

PROOF. We use induction on n. If $n = 1$, then this is the second part of Theorem 3.11. We assume now that $n > 1$ and that the statement holds for $n - 1$, i.e.,

$$a^{p^{n-2}(p-1)} \equiv 1 \pmod{p^{n-1}}.$$

By 3.4.(vi) we have

$$(a^{p^{n-2}(p-1)})^p = a^{p^{n-1}(p-1)} \equiv 1 \pmod{p^n},$$

which was to be shown.   □

3.13 Euler's Theorem

Theorem. *For any natural number m and any integer a relatively prime to m we have*

$$a^{\varphi(m)} \equiv 1 \pmod{m}.$$

PROOF. We write $m = p_1^{n_1} \cdots p_k^{n_k}$, where $p_1, \ldots, p_k$ are distinct primes and $n_1, \ldots, n_k \in \mathbb{N}$. Then for any $i \in \{1, \ldots, k\}$ we have by 3.12,

$$a^{p_i^{n_i-1}(p_i-1)} \equiv 1 \pmod{p_i^{n_i}}.$$

Raising this to the positive integer power $\varphi(m/p_i^{n_i})$, with 3.3.(ii) we obtain

$$a^{\varphi(m)} \equiv 1 \pmod{p_i^{n_i}},$$

since from the definition of Euler's function it follows that

$$\varphi\left(\frac{m}{p_i^{n_i}}\right) \cdot p_i^{n_i-1} \cdot (p_i - 1) = \varphi(m).$$

In view of the fact that $[p_1^{n_1}, \ldots, p_k^{n_k}] = m$, with 3.3.(vi) we now obtain

$$a^{\varphi(m)} \equiv 1 \pmod{m}.$$ □

3.14 Examples

(i) Show that for any integers a, b satisfying $(a, 65) = (b, 65)$, the number $a^{12} - b^{12}$ is divisible by 65.

SOLUTION. Since $65 = 5 \cdot 13$, it suffices to show that $(a, 13) = (b, 13)$ implies $13 \mid a^{12} - b^{12}$ and $(a, 5) = (b, 5)$ implies $5 \mid a^{12} - b^{12}$. Now, if $(a, 13) = (b, 13) = 1$, then by Fermat's theorem we have $a^{12} \equiv 1 \equiv b^{12}$ (mod 13), and if $(a, 13) = (b, 13) = 13$, then $13 \mid a$, $13 \mid b$. In both cases, $13 \mid a^{12} - b^{12}$. Similarly, if $(a, 5) = (b, 5) = 1$ then by Fermat's theorem we have $a^4 \equiv 1 \equiv b^4$ (mod 5), and exponentiating, we get $a^{12} \equiv b^{12}$ (mod 5); thus $5 \mid a^{12} - b^{12}$. We obtain the same relation in the case where $(a, 5) = (b, 5) = 5$, with $5 \mid a$, $5 \mid b$.   □

(ii) Show that for any two distinct primes p and q we have

$$p^{q-1} + q^{p-1} \equiv 1 \pmod{pq}.$$

SOLUTION. Since $(p, q) = 1$, by Fermat's theorem we have $p^{q-1} \equiv 1$ (mod q) and $q^{p-1} \equiv 1$ (mod p). Next, since $p \geq 2$ and $q \geq 2$, we also have $p^{q-1} \equiv 0$ (mod p) and $q^{p-1} \equiv 0$ (mod q), so the congruence $p^{q-1} + q^{p-1} \equiv 1$ holds modulo p as well as modulo q. By 3.3.(vi) it holds also modulo $[p, q] = pq$, which was to be shown.   □

(iii) Find all primes p for which $5^{p^2} + 1 \equiv 0 \pmod{p^2}$.

SOLUTION. We easily see that $p = 5$ does not satisfy the condition. For $p \neq 5$ we have $(p, 5) = 1$ and thus, by Fermat's theorem, $5^{p-1} \equiv 1$ (mod p). Raising this to the power $p+1$, we obtain $5^{p^2-1} \equiv 1$ (mod p), which means that $5^{p^2} \equiv 5$ (mod p). Now, from the condition $5^{p^2} + 1 \equiv 0$ (mod p^2) it follows that $5^{p^2} \equiv -1$ (mod p), and together $5 \equiv -1$ (mod p); thus $p \mid 6$. Therefore, $p = 2$ or $p = 3$. For $p = 2$, however, $5^4 + 1 \equiv 1^4 + 1 = 2 \not\equiv 0$ (mod 4). For $p = 3$ we get $5^9 + 1 = 5^6 \cdot 5^3 + 1 \equiv 5^3 + 1 = 126 \equiv 0$ (mod 9), where we have used $5^6 \equiv 1$ (mod 9) as a consequence of Euler's theorem. The unique prime satisfying the given condition is therefore $p = 3$.   □

(iv) Given an odd number $m > 1$, find the remainder when $2^{\varphi(m)-1}$ is divided by m.

SOLUTION. From Euler's theorem it follows that $2^{\varphi(m)} \equiv 1 \equiv 1 + m = 2r$ (mod m), where $r = (1 + m)/2$ is a natural number, $0 < r < m$. From 3.3.(iii) it follows that $2^{\varphi(m)-1} \equiv r$ (mod m), and thus the desired remainder is $r = (1 + m)/2$.   □

(v) Let m be a natural number, and a an integer relatively prime to m. Show that the smallest natural number n that satisfies the congruence $a^n \equiv 1 \pmod{m}$ divides $\varphi(m)$. Furthermore, show that any nonnegative integers r, s satisfy $a^r \equiv a^s \pmod{m}$ if and only if $r \equiv s \pmod{n}$. In particular, $a^r \equiv 1 \pmod{m}$ if and only if $n \mid r$.

SOLUTION. We will begin by proving the second assertion. Without loss of generality we may assume that $r \geq s$.

If $r \equiv s \pmod{n}$, there exists an integer t such that $r = s + nt$, and from $r \geq s$ it follows that $t \in \mathbb{N}_0$. Then we have

$$a^r = a^{s+nt} = a^s \cdot (a^n)^t \equiv a^s \pmod{m},$$

since $a^n \equiv 1 \pmod{m}$.

If $a^r \equiv a^s \pmod{m}$, in view of the fact that $(a, m) = 1$, we get $a^{r-s} \equiv 1$ $\pmod{m}$ from 3.3.(iii). By Theorem 1.3 on division with remainder, there exist $q \in \mathbb{Z}$ and $r_0 \in \{0, 1, \ldots, n - 1\}$ such that $r - s = nq + r_0$. Then

$$1 \equiv a^{r-s} = a^{nq+r_0} = (a^n)^q a^{r_0} \equiv a^{r_0} \pmod{m},$$

since $a^n \equiv 1 \pmod{m}$. Now of course, $r_0 < n$ and n was the smallest natural number such that $a^n \equiv 1 \pmod{m}$, so r_0 cannot be a natural number, and we have $r_0 = 0$. But this means that $r - s = nq$, or $r \equiv s$ $\pmod{n}$.

Now we return to the first assertion. Since by Euler's theorem we have $a^{\varphi(m)} \equiv 1 = a^0 \pmod{m}$, by what we just proved we must have $\varphi(m) \equiv 0$ $\pmod{n}$, i.e., $n \mid \varphi(m)$. $\qquad\square$

(vi) Prove the following generalization of 3.5.(iv),(v): If p is a prime, $p \equiv 3$ $\pmod 4$, then for any integers a, b the congruence $a^2 + b^2 \equiv 0 \pmod p$ implies $a \equiv b \equiv 0 \pmod p$.

SOLUTION. We assume that for $a, b \in \mathbb{Z}$ we have $a^2 + b^2 \equiv 0 \pmod p$. If $p \mid a$, then $a \equiv 0 \pmod p$; hence $b^2 \equiv 0 \pmod p$, which means that $p \mid b^2$. In view of the fact that p is a prime, we obtain therefore $p \mid b$, and thus $a \equiv b \equiv 0 \pmod p$, which was to be shown.

It remains to verify the case where a is not divisible by p. Then p does not divide b either (if $p \mid b$, we would have $p \mid a^2$). If we multiply both sides of the congruence $a^2 \equiv -b^2 \pmod p$ by b^{p-3}, we get with Fermat's theorem

$$a^2 b^{p-3} \equiv -b^{p-1} \equiv -1 \pmod p.$$

Since $p \equiv 3 \pmod 4$, $p - 3$ is an even number, and thus $\frac{p-3}{2} \in \mathbb{N}_0$. We set

$$c = ab^{(p-3)/2}.$$

Then c is not divisible by p and we have $c^2 = a^2 b^{p-3} \equiv -1 \pmod p$. Raising this last congruence to the power $\frac{p-1}{2} \in \mathbb{N}$, we obtain

$$c^{p-1} \equiv (-1)^{(p-1)/2} \pmod p.$$

Since $p \equiv 3 \pmod 4$, there exists an integer t such that $p = 3 + 4t$. Then $\frac{p-1}{2} = 1 + 2t$, which is an odd number, and thus $(-1)^{(p-1)/2} = -1$. By Fermat's theorem, on the other hand, $c^{p-1} \equiv 1 \pmod p$, which means that $1 \equiv -1 \pmod p$ and thus $p \mid 2$, a contradiction. $\qquad\square$

(vii) Show that for any $n \in \mathbb{N}$, the number $2^{2^{2n+1}} + 3$ is composite.

SOLUTION.   Since $2^2 + 3 = 7$ and $2^{2^3} + 3 = 259 = 7 \cdot 37$, we may conjecture that any number $2^{2^{2n+1}} + 3$, with $n \in \mathbb{N}_0$, is divisible by 7. Therefore, since $2 \not\equiv 1$, $2^2 \not\equiv 1$, $2^3 \equiv 1 \pmod 7$, we will be interested in the remainder of 2^{2n+1} upon division by 3:

$$2^{2n+1} = 2 \cdot 4^n \equiv 2 \pmod 3,$$

and thus by (v) we have

$$2^{2^{2n+1}} + 3 \equiv 2^2 + 3 = 7 \equiv 0 \pmod 7.$$

Since $2^{2^{2n+1}} + 3 \geq 2^{2^3} + 3 > 7$ for $n \in \mathbb{N}$, all the given numbers are composite.   □

3.15 Exercises

(i) Find the remainder when the 100th power of an integer is divided by 125.

(ii) Show that for any prime p and any $n \in \mathbb{N}_0$, $a \in \mathbb{Z}$ we have $a^{n(p-1)+1} \equiv a \pmod p$.

(iii) Show that for any integer a we have $a^{13} \equiv a \pmod{2730}$.

(iv) Show that $2^{341} \equiv 2 \pmod{341}$. Does $3^{341} \equiv 3 \pmod{341}$ also hold?

Show that for any natural number n the numbers in (v)–(x) are composite:

(v) $2^{2^{4n+1}} + 7$.

(vi) $2^{2^{6n+2}} + 13$.

(vii) $2^{2^{10n+1}} + 19$.

(viii) $2^{2^{6n+2}} + 21$.

(ix) $\frac{1}{3}(2^{2^{n+2}} + 2^{2^{n+1}} + 1)$.

(x) $(2^{2 \cdot 3^{6n+3}+4} + 1)^2 + 2^2$.

(xi) Show that among the integers $10^n + 3$ ($n \in \mathbb{N}$) there are infinitely many composite numbers.

*(xii) Show that for all odd natural numbers n we have $n \mid 2^{n!} - 1$.

*(xiii) Show that for any prime p, infinitely many numbers of the form $2^n - n$ ($n \in \mathbb{N}$) are divisible by p.

4   Congruences in One Variable

In mathematics we distinguish between the terms equality and equation. By equality we understand a true relationship between numbers, or between expressions (for instance, $6 \cdot 5 = 3 \cdot 10$, $(n+1)^2 = n^2 + 2n + 1$). On the other hand, an equation has on one side, sometimes on both sides, expressions that depend on some variables, which in this connection are also called unknowns; it is now the objective to find out for which values of these unknowns the equation becomes an equality.

In Chapter 1 we mainly dealt with algebraic equations, that is, the case where on both sides of the equation there are polynomials (mostly with real coefficients), and we were concerned with finding out for which real (or sometimes complex) values of the variables both sides of the equation attained the same values. In Section 5 of the present chapter we will solve such problems for polynomials with integer coefficients and integer values of the variables.

Similarly, given two polynomials in the variable x with integer coefficients, we can ask the question: For which integer values of x will the two polynomials have values that are congruent to each other modulo a given positive integer? We are not going to introduce new notation for this, but use the term congruence also in this situation. There is no danger of confusion; it will be clear from the context when the term "congruence" is used in this sense: e.g., "congruence in the variable $x \ldots$," "solve the congruence $\ldots$," etc. We begin by dealing with linear congruences, that is, the case where the polynomials on both sides are linear.

4.1 Linear Congruences

Let the natural number m and the integers a, b, c, d be given. Then for any $x \in \mathbb{Z}$ the congruence

$$ax + b \equiv cx + d \pmod{m},$$

is satisfied if and only if

$$(a - c)x \equiv d - b \pmod{m}.$$

Hence any linear congruence can be rewritten as a congruence of the form

$$ax \equiv b \pmod{m}, \tag{14}$$

the solutions of which we will now study.

4.2 Solutions of Linear Congruences

Theorem. *Let m be a natural number, a, b integers, and set $d = (a, m)$. Then the congruence (14) is solvable only if the number b is divisible by d; in this case x is a solution of (14) if and only if*

$$x \equiv b_1 \cdot a_1^{\varphi(m_1)-1} \pmod{m_1}, \tag{15}$$

where $a_1 = a/d$, $b_1 = b/d$, $m_1 = m/d$, and φ is Euler's function defined in 3.6.

PROOF. If some x is a solution of (14), then $d \mid ax$, so from 3.3.(viii) it follows that $d \mid b$.

Now we assume that $d \mid b$ and we set $a_1 = a/d$, $b_1 = b/d$, $m_1 = m/d$. Then by 3.3.(iv) and 3.3.(v), the congruence (14), i.e.,

$$a_1 dx \equiv b_1 d \pmod{m_1 d},$$

is satisfied for an $x \in \mathbb{Z}$ if and only if

$$a_1 x \equiv b_1 \pmod{m_1}. \tag{16}$$

By 1.14.(iii) we have $(a_1, m_1) = 1$, and by 3.3.(ii) and 3.3.(iii) the congruence (16) holds if and only if we also have

$$a_1 x \cdot a_1^{\varphi(m_1)-1} \equiv b_1 a_1^{\varphi(m_1)-1} \pmod{m_1}.$$

From Euler's theorem it follows that $a_1^{\varphi(m_1)} \equiv 1 \pmod{m_1}$, and so the last congruence is indeed the same as (15). Thus for any $x \in \mathbb{Z}$, (14) holds if and only if (15) is satisfied, which was to be shown. □

Although Theorem 4.2 enables us to solve any linear congruence, its importance is mainly of a theoretical nature, since the determination of the remainder of the number $b_1 \cdot a_1^{\varphi(m_1)-1}$ upon division by m_1 is often quite laborious. Therefore, it is often more convenient to transform the congruence (16) by multiplying or dividing by a number relatively prime to the modulus m_1, or to replace the numbers a_1, b_1 by numbers congruent modulo m_1, as will be shown in the following examples.

4.3 Examples

(i) Solve the congruence $29x \equiv 1 \pmod{17}$.

SOLUTION. Since $(17, 29) = 1$ and $\varphi(17) = 16$, the solutions of the given congruence are, by Theorem 4.2, all $x \in \mathbb{Z}$ satisfying $x \equiv 29^{15} \pmod{17}$, i.e., the numbers $x = 29^{15} + 17t$, where $t \in \mathbb{Z}$. However, determining which $x \in \{0, 1, \ldots, 16\}$ is a solution, that is, finding the remainder of 29^{15} upon division by 17, would be rather laborious, and therefore it is more appropriate to transform the given congruence as in the following steps:

$$29x \equiv 1 \pmod{17}$$

is equivalent to

$$12x \equiv 18 \pmod{17},$$

and since $(6, 17) = 1$, we obtain the equivalent condition

$$2x \equiv 3 \pmod{17},$$

which holds exactly when

$$2x \equiv 20 \quad (\text{mod } 17),$$

that is, if and only if $x \equiv 10 \ (\text{mod } 17)$, or when $x = 10 + 17t$, where $t \in \mathbb{Z}$. $\qquad\square$

(ii) Solve the congruence $21x + 5 \equiv 0 \ (\text{mod } 29)$.

SOLUTION.   For any $x \in \mathbb{Z}$ we have $21x + 5 \equiv 0 \ (\text{mod } 29)$ if and only if $50x \equiv -5 \ (\text{mod } 29)$. Since $(5, 29) = 1$, this last congruence holds if and only if $10x \equiv -1 \ (\text{mod } 29)$, i.e., exactly when $30x \equiv -3 \ (\text{mod } 29)$, since $(3, 29) = 1$. The last congruence is equivalent to $x \equiv 26 \ (\text{mod } 29)$, i.e., $x = 26 + 29t$, where $t \in \mathbb{Z}$. $\qquad\square$

(iii) Solve the congruence $(a + b)x \equiv a^2 + b^2 \ (\text{mod } ab)$, where $a, b \in \mathbb{N}$ are parameters with $(a, b) = 1$.

SOLUTION.   If we add $2ab$ to the right-hand side, we obtain

$$(a + b)x \equiv (a + b)^2 \quad (\text{mod } ab).$$

By 2.8.(i) we have $(a + b, ab) = 1$, and hence this last congruence is equivalent to $x \equiv a + b \ (\text{mod } ab)$. The solution is therefore given by all x of the form $x = a + b + tab$, where $t \in \mathbb{Z}$. $\qquad\square$

(iv) Solve the congruence $2x \equiv 1 \ (\text{mod } m)$, where $m \in \mathbb{N}$.

SOLUTION.   If m is even, then $(2, m) = 2$, and by 4.2 the congruence is not solvable. If m is odd, then $m + 1$ is even; hence there is a $k \in \mathbb{N}$ such that $m + 1 = 2k$. If we add m to the right-hand side of the given congruence, we get $2x \equiv m + 1 \ (\text{mod } m)$, i.e., $2x \equiv 2k \ (\text{mod } m)$, and since $(m, 2) = 1$, this is equivalent to $x \equiv k \ (\text{mod } m)$. The original congruence is therefore satisfied by exactly those x that are of the form $x = k + tm = (m+1)/2 + tm$, where $t \in \mathbb{Z}$. $\qquad\square$

(v) Prove *Wilson's theorem*: For each prime p we have

$$(p - 1)! \equiv -1 \quad (\text{mod } p).$$

SOLUTION.   The congruence clearly holds for $p = 2$ and $p = 3$, and hence we may assume that $p \geq 5$. For any $a \in \{2, 3, \ldots, p - 2\}$ the congruence

$$ax \equiv 1 \quad (\text{mod } p)$$

has solutions of the form $x = a^{p-2} + pt$, $t \in \mathbb{Z}$, by Theorem 4.2. There is clearly a unique $t_1 \in \mathbb{Z}$ with $0 \leq a^{p-2} + pt_1 < p$ (then $-t_1$ is the quotient in the division of a^{p-2} by p with remainder); we set $a_1 = a^{p-2} + pt_1$. Then $0 \le a_1 < p$, and

$$a \cdot a_1 \equiv 1 \pmod{p}.$$

It is easy to convince ourselves that $a_1 \notin \{0, 1, p-1\}$, and thus $a_1 \in \{2, 3, \ldots, p-2\}$. On the other hand, a is a solution of the congruence

$$a_1 x \equiv 1 \pmod{p}.$$

We also note that $a = a_1$ cannot occur. Indeed, we would then have $a^2 \equiv 1$ $\pmod{p}$, hence $(a - 1)(a + 1) \equiv 0 \pmod{p}$, and therefore $p \mid a - 1$ or $p \mid a+1$, which for $a \in \{2, 3, \ldots, p-2\}$ cannot occur. The set $\{2, 3, \ldots, p-2\}$ therefore consists of pairs of numbers whose products are congruent to 1 modulo p. Hence $(p - 2)!$, which is the product of all the numbers in $\{2, 3, \ldots, p-2\}$, is also congruent to 1 modulo p:

$$(p - 2)! \equiv 1 \pmod{p}.$$

Multiplying this congruence by $p - 1$, we obtain

$$(p - 1)! \equiv p - 1 \equiv -1 \pmod{p},$$

which was to be shown.   □

(vi) Prove the following generalization of Exercise 3.5.(vi): For any prime p with $p \not\equiv 3 \pmod 4$, there exist integers a, b such that although the congruence $a^2 + b^2 \equiv 0 \pmod{p}$ is satisfied, still $a \equiv b \equiv 0 \pmod{p}$ does not hold. Compare this with Example 3.14.(vi).

SOLUTION. If $p = 2$, we may choose $a = b = 1$. Hence we may assume that $p = 4k + 1$ for an appropriate $k \in \mathbb{N}$. If we set $a = (2k)!$, $b = 1$, then

$$a^2 = ((2k)!)^2 = 1 \cdot 2 \cdots (2k - 1) \cdot 2k \cdot (-1) \cdot (-2) \cdots (-2k + 1) \cdot (-2k).$$

Now, for any $i \in \{1, 2, \ldots, 2k\}$ we have $-i \equiv p - i \pmod{p}$, and thus

$$a^2 \equiv 1 \cdot 2 \cdots (2k - 1) \cdot 2k \cdot 4k \cdot (4k - 1) \cdots (2k + 2) \cdot (2k + 1) \pmod{p}.$$

On the right we have $(4k)! = (p - 1)!$, which by Wilson's theorem gives

$$a^2 \equiv (p - 1)! \equiv -1 \pmod{p},$$

and thus

$$a^2 + b^2 \equiv -1 + 1 = 0 \pmod{p}.$$

Since $b = 1$, certainly $b \equiv 0 \pmod{p}$ does not hold. Comparing this result with 3.14.(vi), we see that the primes with the property that for any $a, b \in \mathbb{Z}$ the congruence $a^2 + b^2 \equiv 0 \pmod{p}$ implies $a \equiv b \equiv 0 \pmod{p}$ are exactly the primes p of the form $4k + 3$ ($k \in \mathbb{N}_0$).   □

(vii) Let a, b, c, d be integers satisfying $(a, b, c, d) = 1$. Show that every prime dividing $ad - bc$ also divides a and c if and only if for any integer n the numbers $an + b$ and $cn + d$ are relatively prime to each other.

SOLUTION.   We assume that every prime dividing $ad - bc$ divides also a, c, but that for an appropriate $n \in \mathbb{Z}$ there exists a prime p such that $p \mid an + b$ and $p \mid cn + d$. But then p also divides

$$ad - bc = a(cn + d) - c(an + b),$$

and thus $p \mid a$, $p \mid c$. Therefore, p divides $b = (an + b) - an$ and $d = (cn + d) - cn$ as well; but this means that $p \mid (a, b, c, d)$, which is a contradiction.

On the other hand, we assume that $(an + b, cn + d) = 1$ for all $n \in \mathbb{N}$ and that there exists a prime p such that $ad - bc \equiv 0 \pmod{p}$, $a \not\equiv 0 \pmod{p}$ (the case $c \not\equiv 0 \pmod{p}$ can be considered analogously). By Theorem 4.2 the linear congruence $ax \equiv -b \pmod{p}$ has a solution, which we denote by n. Then $an + b \equiv 0 \pmod{p}$; therefore,

$$a(cn + d) = c(an + b) + (ad - bc) \equiv 0 \pmod{p},$$

and since $(p, a) = 1$, we get $cn + d \equiv 0 \pmod{p}$. Thus, $p \mid (an + b, cn + d)$, which is a contradiction. $\square$

4.4   Exercises

Solve the congruences (i)–(vi):

(i) $7x \equiv 15 \pmod{9}$.   **(ii)** $7x \equiv 9 \pmod{10}$.

(iii) $14x \equiv 23 \pmod{31}$.   **(iv)** $72x \equiv 2 \pmod{10}$.

(v) $8x \equiv 20 \pmod{12}$.   **(vi)** $6x \equiv 27 \pmod{12}$.

*(vii) Let a, b be relatively prime positive integers. Show that for any natural number n there exist $k_1, k_2, \ldots, k_n \in \mathbb{N}$ such that $k_1 < k_2 < \cdots < k_n$ and that the numbers $ak_1 + b, ak_2 + b, \ldots, ak_n + b$ are pairwise relatively prime.

4.5   Systems of Linear Congruences in One Variable

If we have a system of linear congruences in the same variable, we can use Theorem 4.2 to decide whether each one of the congruences is solvable. If at least one of them does not have a solution, then the entire system has no solution. On the other hand, if all congruences have solutions, we can write them in the form $x \equiv c_i \pmod{m_i}$. We thus obtain the system of congruences

$$x \equiv c_1 \pmod{m_1},$$

$$\vdots \tag{17}$$

$$x \equiv c_k \pmod{m_k}.$$

We first examine the case $k = 2$, which—as we will see later—is crucial in the solution of the system (17) for $k > 2$.

4.6   Existence of Solutions

Theorem. *Let c_1, c_2 be integers, m_1, m_2 natural numbers, and set $d = (m_1, m_2)$. Then the system of two congruences*

$$x \equiv c_1 \pmod{m_1},$$
$$x \equiv c_2 \pmod{m_2} \tag{18}$$

has no solutions when $c_1 \not\equiv c_2 \pmod{d}$. However, if $c_1 \equiv c_2 \pmod{d}$, then there is an integer c such that $x \in \mathbb{Z}$ satisfies the system (18) if and only if it satisfies

$$x \equiv c \pmod{[m_1, m_2]}.$$

PROOF.   If the system (18) has a solution $x \in \mathbb{Z}$, then by 3.3.(vii) we have $x \equiv c_1 \pmod{d}$ and $x \equiv c_2 \pmod{d}$, and thus $c_1 \equiv c_2 \pmod{d}$ as well. This means that in the case $c_1 \not\equiv c_2 \pmod{d}$, (18) cannot have a solution.

Let us now assume that $c_1 \equiv c_2 \pmod{d}$. The first congruence in (18) is satisfied by all integers x of the form $x = c_1 + tm_1$, for arbitrary $t \in \mathbb{Z}$. This x will also satisfy the second congruence in (18) if and only if $c_1 + tm_1 \equiv c_2 \pmod{m_2}$, that is,

$$tm_1 \equiv c_2 - c_1 \pmod{m_2}.$$

By 4.2, this congruence has a solution (in terms of t), since $d = (m_1, m_2)$ divides $c_2 - c_1$, and $t \in \mathbb{Z}$ satisfies the congruence if and only if

$$t \equiv \frac{c_2 - c_1}{d} \cdot \left(\frac{m_1}{d}\right)^{\varphi(m_2/d)-1} \left(\bmod \frac{m_2}{d}\right),$$

i.e., if and only if

$$t = \frac{c_2 - c_1}{d} \cdot \left(\frac{m_1}{d}\right)^{\varphi(m_2/d)-1} + r \cdot \frac{m_2}{d},$$

with arbitrary $r \in \mathbb{Z}$. By substituting, we obtain

$$x = c_1 + tm_1 = c_1 + (c_2 - c_1) \cdot \left(\frac{m_1}{d}\right)^{\varphi(m_2/d)} + r\frac{m_1 m_2}{d} = c + r \cdot [m_1, m_2],$$

where $c = c_1 + (c_2 - c_1) \cdot (m_1/d)^{\varphi(m_2/d)}$, since by Theorem 1.9 we have $m_1 m_2 = d \cdot [m_1, m_2]$. We have thus found the desired $c \in \mathbb{Z}$ with the property that an $x \in \mathbb{Z}$ satisfies the system (18) if and only if

$$x \equiv c \pmod{[m_1, m_2]},$$

and this completes the proof.   □

We note that this proof is constructive, i.e., it gives a formula for finding the number c. Theorem 4.6 thus provides us with a method for reducing the system (18) to a single congruence. It is now important to realize that this new congruence is of the same form as the two original congruences. We can therefore also apply this method to the system (17): To begin, we combine the first and the second congruences to form the new single congruence as described above; then we combine this new congruence with the third one, etc. With each step the number of congruences in the system decreases by 1, after $k - 1$ steps we are therefore left with a unique congruence that describes all solutions of the system (17).

We further remark that the number c in Theorem 4.6 does not have to be determined by way of the formula given above. We may take any $t \in \mathbb{Z}$ satisfying the congruence

$$t \cdot \frac{m_1}{d} \equiv \frac{c_2 - c_1}{d} \quad \left(\mathrm{mod}\,\frac{m_2}{d}\right)$$

and set $c = c_1 + tm_1$.

4.7 Examples

In (i)–(iii), solve the given systems of congruences:

(i) $x \equiv -3 \pmod{49}$,

$\qquad x \equiv 2 \pmod{11}$.

SOLUTION.   The first congruence is satisfied exactly by the numbers x of the form $x = -3 + 49t$, where $t \in \mathbb{Z}$. Substituting this into the second congruence, we obtain

$$-3 + 49t \equiv 2 \pmod{11},$$

which means

$$5t \equiv 5 \pmod{11},$$

and since $(5, 11) = 1$, dividing by 5 gives

$$t \equiv 1 \pmod{11},$$

or $t = 1 + 11s$ with $s \in \mathbb{Z}$. Hence $x = -3 + 49(1 + 11s) = 46 + 539s$, where $s \in \mathbb{Z}$, which can also be written as $x \equiv 46 \pmod{539}$.   □

(ii) $x \equiv 1 \pmod{10}$,

$\qquad x \equiv 5 \pmod{18}$,

$\qquad x \equiv -4 \pmod{25}$.

SOLUTION.   From the first congruence we get $x = 1 + 10t$ for $t \in \mathbb{Z}$. Substituting this into the second congruence gives

$$1 + 10t \equiv 5 \pmod{18},$$

hence $10t \equiv 4 \pmod{18}$. Since $(10, 18) = 2$ divides 4, by Theorem 4.2 this congruence has the solution $t \equiv 2 \cdot 5^5 \pmod 9$; now $2 \cdot 5^5 = 10 \cdot 25^2 \equiv 1 \cdot (-2)^2 = 4 \pmod 9$, and thus $t = 4 + 9s$, with $s \in \mathbb{Z}$. By substituting, we get $x = 1 + 10(4 + 9s) = 41 + 90s$, and with the third congruence,

$$41 + 90s \equiv -4 \pmod{25},$$

hence $90s \equiv -45 \pmod{25}$. Dividing by 5 (including the modulus, since $5 \mid 25$), we get

$$18s \equiv -9 \pmod 5,$$

which means $-2s \equiv 1 \pmod 5$, so $2s \equiv 4 \pmod 5$, $s \equiv 2 \pmod 5$, and therefore $s = 2 + 5r$, where $r \in \mathbb{Z}$. Substituting, $x = 41 + 90(2 + 5r) = 221 + 450r$, so finally $x \equiv 221 \pmod{450}$. $\qquad\qquad\square$

(iii) $x \equiv 18 \pmod{25}$,

$\qquad x \equiv 21 \pmod{45}$,

$\qquad x \equiv 25 \pmod{73}$.

SOLUTION.   From the first congruence we get $x = 18 + 25t$, with $t \in \mathbb{Z}$, and substituting this into the second one,

$$18 + 25t \equiv 21 \pmod{45},$$

hence

$$25t \equiv 3 \pmod{45}.$$

But by Theorem 4.2 this congruence has no solution, since $(25, 45) = 5$ does not divide 3. Hence the whole system has no solution. We could also have obtained this result directly from Theorem 4.6, since $18 \not\equiv 21 \pmod{(25, 45)}$. $\qquad\qquad\square$

(iv) Solve the congruence $23\,941x \equiv 915 \pmod{3564}$.

SOLUTION.   We factor $3564 = 2^2 \cdot 3^4 \cdot 11$. Since neither 2, nor 3, nor 11 divides $23\,941$, we have $(23\,941, 3564) = 1$, and thus by Theorem 4.2 the congruence has a solution. Since $\varphi(3564) = 2 \cdot (3^3 \cdot 2) \cdot 10 = 1080$, by 4.2 the solution is of the form

$$x \equiv 915 \cdot 23\,941^{1079} \pmod{3564}.$$

However, determining the remainder (modulo 3564) of the number on the right would require considerable effort. We will therefore solve the congruence quite differently. By 3.3.(vi), $x \in \mathbb{Z}$ is a solution of the given congruence if and only if it is a solution of the system

$$\begin{aligned} 23941x &\equiv 915 \pmod{2^2}, \\ 23941x &\equiv 915 \pmod{3^4}, \\ 23941x &\equiv 915 \pmod{11}. \end{aligned} \tag{19}$$

We now solve each congruence of the system (19) separately. The first one holds if and only if

$$x \equiv 3 \pmod{4},$$

and the second one exactly when

$$46x \equiv 24 \pmod{81},$$

which, upon multiplying by 2, gives $92x \equiv 48 \pmod{81}$, i.e., $11x \equiv -33 \pmod{81}$, and after dividing by 11,

$$x \equiv -3 \pmod{81}.$$

The third congruence holds if and only if

$$5x \equiv 2 \pmod{11},$$

which, upon multiplying by -2, gives $-10x \equiv -4 \pmod{11}$, and thus

$$x \equiv -4 \pmod{11}.$$

An $x \in \mathbb{Z}$ is therefore a solution of the system (19) if and only if it is a solution of the system

$$
\begin{aligned}
x &\equiv 3 \pmod{4}, \\
x &\equiv -3 \pmod{81}, \\
x &\equiv -4 \pmod{11}.
\end{aligned}
\tag{20}
$$

The second congruence gives $x = -3 + 81t$, where $t \in \mathbb{Z}$, and substituting this into the third one,

$$-3 + 81t \equiv -4 \pmod{11};$$

thus $81t \equiv -1 \pmod{11}$, or $4t \equiv 32 \pmod{11}$; then $t \equiv 8 \pmod{11}$, and therefore $t = -3 + 11s$, with $s \in \mathbb{Z}$. Substituting $x = -3 + 81(-3 + 11s) = -3 - 3 \cdot 81 + 11 \cdot 81s$ into the first congruence of (20), we obtain

$$-3 - 3 \cdot 81 + 11 \cdot 81s \equiv 3 \pmod{4},$$

hence

$$1 + 1 \cdot 1 + (-1) \cdot 1s \equiv 3 \pmod{4},$$

i.e., $-s \equiv 1 \pmod{4}$ and thus $s = -1 + 4r$, where $r \in \mathbb{Z}$. Finally,

$$
\begin{aligned}
x &= -3 - 3 \cdot 81 + 11 \cdot 81(-1 + 4r) \\
&= -3 - 14 \cdot 81 + 4 \cdot 11 \cdot 81r = -1137 + 3564r,
\end{aligned}
$$

or $x \equiv -1137 \pmod{3564}$, which is also the solution of the original congruence.   □

4.8   Exercises

Solve the systems of congruences (i)–(viii), where $a \in \mathbb{Z}$ is a parameter:

(i) $x \equiv 2 \pmod 5$,
   $x \equiv 8 \pmod{11}$.

(ii) $x \equiv 7 \pmod{33}$,
   $x \equiv 3 \pmod{63}$.

(iii) $4x \equiv 3 \pmod 7$,
   $5x \equiv 4 \pmod 6$.

(iv) $17x \equiv 5 \pmod 6$,
   $11x \equiv 35 \pmod{36}$.

(v) $17x \equiv 7 \pmod 2$,
   $2x \equiv 1 \pmod 3$,
   $2x \equiv 7 \pmod 5$.

(vi) $3x \equiv 5 \pmod 7$,
   $2x \equiv 3 \pmod 5$,
   $3x \equiv 3 \pmod 9$.

(vii) $x \equiv a \pmod 6$,
   $x \equiv 1 \pmod 8$.

(viii) $2x \equiv a \pmod 4$,
   $3x \equiv 4 \pmod{10}$.

(ix) Show that if the number c satisfies all congruences of the system (17), then all solutions of this system are given by the congruence $x \equiv c \pmod m$, where $m = [m_1, \ldots, m_k]$.

(x) Prove the so-called *Chinese remainder theorem*: If the numbers $m_1, m_2, \ldots, m_k$ are pairwise relatively prime, then the system (17) is solvable.

(xi) Let $k, n \in \mathbb{N}$. Show that there exist k consecutive integers, each of which is of the form ab^n, where $a, b \in \mathbb{Z}$ and $b > 1$.

*(xii) Show that the system (17) is solvable if and only if for all $i, j \in \mathbb{N}$, $i < j \le k$, we have $c_i \equiv c_j \pmod{(m_i, m_j)}$.

4.9   Congruences of Higher Degree

We return now to the more general case, where on both sides of the congruence we have a polynomial of the same variable x with integer coefficients. By subtracting we can easily change this to

$$F(x) \equiv 0 \pmod m, \qquad (21)$$

where $F(x)$ is a polynomial with integer coefficients, and $m \in \mathbb{N}$. Based on the following theorem, we will describe a method for solving such congruences; it may be laborious, but it is universally applicable.

4.10   A Substitution Result

Theorem. *For any polynomial $F(x)$ with integer coefficients, and for a natural number m and integers a, b such that $a \equiv b \pmod m$, we have $F(a) \equiv F(b) \pmod m$.*

PROOF. Let $F(x) = c_n x^n + c_{n-1} x^{n-1} + \cdots + c_1 x + c_0$, where $c_0, c_1, \ldots, c_n \in \mathbb{Z}$. Since $a \equiv b \pmod m$, by 3.3.(ii) we have for all $i = 1, 2, \ldots, n$,

$$c_i a^i \equiv c_i b^i \pmod m,$$

and thus, adding these congruences for $i = 1, 2, \ldots, n$ and the congruence $c_0 \equiv c_0 \pmod{m}$, we get

$$c_n a^n + a_{n-1} a^{n-1} + \cdots + c_1 a + c_0 \equiv c_n b^n + c_{n-1} b^{n-1} + \cdots + c_1 b + c_0 \quad \pmod{m},$$

i.e., $F(a) \equiv F(b) \pmod{m}$. $\square$

4.11 Existence of Solutions

Consequence. *If no integer a, $0 \le a < m$, satisfies $F(a) \equiv 0 \pmod{m}$, then the congruence (21) has no solution. If $a_1, a_2, \ldots, a_s$ are all integers such that $0 \le a_i < m$ and $F(a_i) \equiv 0 \pmod{m}$, then the solutions of the congruence (21) are exactly all integers x that satisfy one of the congruences*

$$x \equiv a_1 \pmod{m}, \quad x \equiv a_2 \pmod{m}, \quad \ldots, \quad x \equiv a_s \pmod{m}.$$

PROOF. If b is any integer, then from the division theorem 1.3 it follows that there exists a unique $a \in \mathbb{Z}$ such that $0 \le a < m$ and $b \equiv a \pmod{m}$. By Theorem 4.10, b is a solution of (21) if and only if a is a solution. $\square$

To solve the congruence (21) it therefore suffices to find out for which integers a, $0 \le a < m$, we have $F(a) \equiv 0 \pmod{m}$. A disadvantage of this method is the amount of work involved, which increases with the size of m. If m is composite, $m = p_1^{n_1} \cdots p_k^{n_k}$, where $p_1, \ldots, p_k$ are distinct primes, and if furthermore $k > 1$, we can proceed as in 4.7.(iv) and replace the congruence (21) by the system

$$F(x) \equiv 0 \quad \pmod{p_1^{n_1}},$$
$$\vdots \tag{22}$$
$$F(x) \equiv 0 \quad \pmod{p_k^{n_k}}$$

(which has the same set of solutions), and then solve each congruence of this system separately. Doing this, we obtain in general several systems of congruences of the type (17), which we know how to solve. The advantage of this method lies in the fact that the moduli of the congruences in (22) are smaller than the modulus of the original congruence (21).

4.12 Examples

(i) Solve the congruence $x^5 + 1 \equiv 0 \pmod{11}$.

SOLUTION. We set $F(x) = x^5 + 1$. Then we have $F(0) = 1$, $F(1) = 2$, and

$$F(2) = 33 \equiv 0 \quad \pmod{11},$$
$$F(3) = 3^5 + 1 = 9 \cdot 9 \cdot 3 + 1 \equiv (-2)^2 \cdot 3 + 1 = 12 + 1 \equiv 2 \quad \pmod{11},$$
$$F(4) = 4^5 + 1 = 2^{10} + 1 \equiv 1 + 1 = 2 \quad \pmod{11},$$

where we have used Fermat's theorem to obtain $2^{10} \equiv 1 \pmod{11}$. Similarly,

$$F(5) = 5^5 + 1 \equiv 16^5 + 1 = 4^{10} + 1 \equiv 1 + 1 = 2 \pmod{11},$$
$$F(6) = 6^5 + 1 \equiv (-5)^5 + 1 \equiv -16^5 + 1 = -4^{10} + 1 \equiv 0 \pmod{11},$$
$$F(7) = 7^5 + 1 \equiv (-4)^5 + 1 = -2^{10} + 1 \equiv -1 + 1 = 0 \pmod{11},$$
$$F(8) = 8^5 + 1 \equiv 2^5 \cdot 2^{10} + 1 \equiv 32 + 1 \equiv 0 \pmod{11},$$
$$F(9) = 9^5 + 1 \equiv 3^{10} + 1 \equiv 1 + 1 = 2 \pmod{11},$$
$$F(10) = 10^5 + 1 \equiv (-1)^5 + 1 = 0 \pmod{11},$$

and thus the congruence $x^5 + 1 \equiv 0 \pmod{11}$ is solved exactly by those x that satisfy one of the congruences $x \equiv 2 \pmod{11}$, $x \equiv 6 \pmod{11}$, $x \equiv 7 \pmod{11}$, $x \equiv 8 \pmod{11}$, $x \equiv 10 \pmod{11}$.   □

(ii) Solve the congruence $x^8 - 2 \equiv 0 \pmod 7$.

SOLUTION.   Although we could proceed exactly as in example (i), we will use a different approach here. We use the fact that, by Fermat's theorem 3.11, for all integer x we have

$$x^7 \equiv x \pmod 7,$$

and so the given congruence can be rewritten as

$$x^8 - 2 = x \cdot x^7 - 2 \equiv x \cdot x - 2 \equiv x^2 - 9 = (x-3)(x+3) \pmod 7.$$

Our congruence is therefore satisfied if and only if $7 \mid (x-3)(x+3)$, which is the case if and only if $7 \mid x - 3$ or $7 \mid x + 3$. As solutions we therefore have exactly those x which satisfy one of the congruences $x \equiv 3 \pmod 7$, $x \equiv -3 \pmod 7$.   □

(iii) Solve the congruence $x^3 - 3x + 5 \equiv 0 \pmod{105}$.

SOLUTION.   To proceed as in 4.11 for $m = 105$, we would have to evaluate $F(x) = x^3 - 3x + 5$ for the 105 values $x = 0, 1, \ldots, 104$. For this reason it is better to factor $105 = 3 \cdot 5 \cdot 7$ and to successively solve the congruence $F(x) \equiv 0$ for the moduli 3, 5, 7. We have $F(0) = 5$, $F(1) = 3$, $F(2) = 7$, $F(3) = 23$, $F(-1) = 7$, $F(-2) = 3$, $F(-3) = -13$ (note that for easier calculation we have evaluated, for example, $F(-1)$ instead of $F(6)$, using the fact that by 4.10 we have $F(6) \equiv F(-1) \pmod 7$, etc.). The congruence $F(x) \equiv 0$ $\pmod 3$ has therefore the solution $x \equiv 1 \pmod 3$; the congruence $F(x) \equiv 0$ $\pmod 5$ has the solution $x \equiv 0 \pmod 5$; and $F(x) \equiv 0 \pmod 7$ is solved by $x \in \mathbb{Z}$ satisfying $x \equiv 2 \pmod 7$ or $x \equiv -1 \pmod 7$. It therefore remains to solve the following two systems of congruences:

$$x \equiv 1 \pmod 3, \qquad\qquad x \equiv 1 \pmod 3,$$
$$x \equiv 0 \pmod 5, \qquad \text{and} \qquad x \equiv 0 \pmod 5,$$
$$x \equiv 2 \pmod 7, \qquad\qquad x \equiv -1 \pmod 7.$$

Since the first two congruences are the same in both systems, we will deal with them first. From the second congruence we obtain $x = 5t$ for $t \in \mathbf{Z}$, and substituting this into the first one, we get

$$5t \equiv 1 \pmod 3,$$

hence $-t \equiv 1 \pmod 3$, i.e., $t = -1 + 3s$ for $s \in \mathbf{Z}$, and thus $x = -5 + 15s$. First we substitute this into the third congruence of the first system:

$$-5 + 15s \equiv 2 \pmod 7,$$

so $s \equiv 0 \pmod 7$, i.e., $s = 7r$ for $r \in \mathbf{Z}$ and thus $x = -5 + 105r$. If we substitute $x = -5 + 15s$ into the third congruence of the second system, we get

$$-5 + 15s \equiv -1 \pmod 7,$$

hence $s \equiv 4 \pmod 7$, i.e., $s = 4 + 7r$ for $r \in \mathbf{Z}$, and thus $x = -5 + 15(4 + 7r) = 55 + 105r$. Together, the solution of the given congruence $F(x) \equiv 0 \pmod{105}$ consists of all integers x satisfying $x \equiv -5 \pmod{105}$ or $x \equiv 55 \pmod{105}$. $\qquad\qquad\square$

(iv) Solve the congruence $x^2 - 3x - 10 \equiv 0 \pmod{121}$.

SOLUTION. Since $121 = 11^2$, if the integer x is a solution of the given congruence, then by 3.3.(vii) it also has to be a solution of the same congruence modulo 11 (but not conversely!). Therefore, we will first solve the congruence $x^2 - 3x - 10 \equiv 0 \pmod{11}$. For each $x \in \mathbf{Z}$ we have

$$x^2 - 3x - 10 \equiv x^2 + 8x - 10 = (x+4)^2 - 16 - 10$$
$$\equiv (x+4)^2 - 4 = (x+4-2)(x+4+2) \pmod{11},$$

so $x \in \mathbf{Z}$ is a solution of this congruence if and only if $11 \mid x + 2$ or $11 \mid x + 6$ (here we have used the fact that 11 is prime). Now we substitute $x = -2 + 11r$, resp., $x = -6 + 11r$, where $r \in \mathbf{Z}$, into the polynomial $F(x) = x^2 - 3x - 10$, to obtain

$$F(-2 + 11r) = 4 - 44r + 121r^2 + 6 - 33r - 10 \equiv -77r \pmod{121},$$
$$F(-6 + 11r) = 36 - 132r + 121r^2 + 18 - 33r - 10$$
$$\equiv -44r + 44 \pmod{121},$$

which means that in the first case x is a solution if and only if

$$-77r \equiv 0 \pmod{121},$$

i.e., $-7r \equiv 0 \pmod{11}$, so $r = 11s$, where $s \in \mathbb{Z}$, and therefore $x = -2 + 121s$. In the second case x is a solution if and only if

$$-44r + 44 \equiv 0 \pmod{121},$$

i.e., $-4r \equiv -4 \pmod{11}$, so $r = 1 + 11s$, where $s \in \mathbb{Z}$, and thus $x = -6 + 11(1 + 11s) = 5 + 121s$. The solutions of the original congruence are therefore given by all integers x that satisfy the congruences $x \equiv -2 \pmod{121}$ or $x \equiv 5 \pmod{121}$. $\square$

(v) Solve the congruence $x^2 - 3x - 10 \equiv 0 \pmod{49}$.

SOLUTION. We proceed as in (iv) and first solve the given congruence modulo 7. From

$$x^2 - 3x - 10 \equiv x^2 + 4x - 10 = (x+2)^2 - 4 - 10 \equiv (x+2)^2 \pmod{7}$$

we see that $x^2 - 3x - 10 \equiv 0 \pmod{7}$ if and only if $x = -2 + 7r$, where $r \in \mathbb{Z}$. Substituting this into $F(x) = x^2 - 3x - 10$, we obtain

$$F(-2 + 7r) = 4 - 28r + 49r^2 + 6 - 21r - 10 = 49r^2 - 49r \equiv 0 \pmod{49},$$

and so the solution of the original congruence is given by all integers x satisfying $x \equiv -2 \pmod{7}$. $\square$

4.13   Exercises

Solve the congruences (i)–(xii):

(i) $x^2 + 5x + 1 \equiv 0 \pmod{7}$.

(ii) $27x^2 - 13x + 11 \equiv 0 \pmod{5}$.

(iii) $x^3 + x + 4 \equiv 0 \pmod{5}$.

(iv) $5x^2 + x + 4 \equiv 0 \pmod{10}$.

(v) $x^2 - 4x + 3 \equiv 0 \pmod{6}$.

(vi) $x^6 + x^5 - 2x^2 - x \equiv 0 \pmod{5}$.

(vii) $x^{12} + x^{11} - x^2 - 1 \equiv 0 \pmod{11}$.

(viii) $x^7 - 3x^6 + x^5 - x^3 + 4x^2 - 4x + 2 \equiv 0 \pmod{5}$.

(ix) $x^5 + 2 \equiv 0 \pmod{35}$.

(x) $5x^4 + 3x^2 + x + 3 \equiv 0 \pmod{45}$.

(xi) $x^2 + 10x + 3 \equiv 0 \pmod{121}$.

(xii) $x^6 - 1 \equiv 0 \pmod{35}$.

*(xiii) Find all integers $m > 1$ satisfying the following condition: There is a polynomial $F(x)$ with integer coefficients such that for any integer a either $F(a) \equiv 0 \pmod{m}$ or $F(a) \equiv 1 \pmod{m}$ and each of these two cases is satisfied for at least one integer a.

The theory of congruences of higher degree is a well-developed topic in number theory; here we have presented only part of it. Those interested in further studying this topic should consult, for instance, the book by I. M. Vinogradov [12], which treats the theory of higher-degree congruences, especially quadratic congruences, in a very accessible way.

In conclusion, we state without proof a criterion for the solvability of the congruence

$$x^2 \equiv t \pmod{p}, \tag{23}$$

where p is an odd prime and t an integer not divisible by p. We note that any quadratic congruence modulo p can easily be reduced to this congruence. Indeed, if we multiply the congruence

$$ax^2 + bx + c \equiv 0 \pmod{p},$$

where $a \not\equiv 0 \pmod{p}$, by the number t obtained as a solution of the congruence $at \equiv 1 \pmod{p}$, then we get

$$x^2 + btx + ct \equiv 0 \pmod{p}.$$

This last congruence can be further transformed by completing the square: We set $d = bt/2$ if bt is even, and $d = (bt + p)/2$ if bt is odd. Then d is always an integer, and $2d \equiv bt \pmod{p}$. Substituting, we obtain

$$x^2 + 2dx + ct \equiv 0 \pmod{p};$$

thus

$$(x + d)^2 \equiv d^2 - ct \pmod{p},$$

and this is a congruence of the form (23), if we set $y = x + d$.

4.14 Euler's Criterion

Theorem. *For any prime $p > 2$ and integer t, not divisible by p, the congruence (23) has a solution if and only if*

$$t^{(p-1)/2} \equiv 1 \pmod{p}.$$

A proof of this result can be found, e.g., in [12].

4.15 Exercises

Let p be an odd prime.

(i) Prove the following part of Theorem 4.14: If the congruence $x^2 \equiv t$ $\pmod{p}$ has a solution, then $t^{(p-1)/2} \equiv 1 \pmod{p}$.

(ii) Let $p = 4k + 3$, where $k \in \mathbb{N}_0$, and suppose that $t^{2k+1} \equiv 1 \pmod{p}$. Show that $x \equiv \pm t^{k+1} \pmod{p}$ is a solution of the congruence $x^2 \equiv t$ $\pmod{p}$.

5   Diophantine Equations

As early as in the third century of our era the Greek mathematician Diophantus considered equations for whose solutions only integers were admissible. This is not so surprising: There are many practical problems that are solvable by equations where nonintegral solutions have no reasonable interpretation. (Consider, for instance, the problem of how to measure 8 litres of water using two containers of 5 and 7 litres each; this gives the equation $5x+7y = 8$.) In honour of Diophantus, equations for which we consider only integer solutions are called *Diophantine equations*. Unfortunately, there is no universal method for solving such equations. Nevertheless, we will introduce several very useful methods that turn out to be successful in many specific examples.

5.1   Linear Diophantine Equations

We consider equations of the form

$$a_1 x_1 + a_2 x_2 + \cdots + a_n x_n = b, \tag{24}$$

where $x_1, \ldots, x_n$ are variables and $a_1, \ldots, a_n, b$ are fixed integers. We assume that $a_i \neq 0$ for all $i = 1, \ldots, n$ (if $a_i = 0$, then the variable x_i "disappears" from the equation). Congruences can be used for the solution of such equations.

First we try to decide when the equation (24) can, or cannot, have solutions. If b is not divisible by $d = (a_1, \ldots, a_n)$, then (24) cannot have any solution, since for any integers $x_1, \ldots, x_n$ the left-hand side of (24) is divisible by d. If, on the other hand, $d \mid b$, then we can divide the whole equation (24) by d. We then obtain the equivalent equation

$$a_1' x_1 + a_2' x_2 + \cdots + a_n' x_n = b',$$

where $a_i' = a_i/d$ for $i = 1, \ldots, n$ and $b' = b/n$. By 1.14.(i), and using (6), we have

$$d \cdot (a_1', \ldots, a_n') = (da_1', \ldots, da_n') = (a_1, \ldots, a_n) = d,$$

and thus $(a_1', \ldots, a_n') = 1$. In the following theorem we will show that such an equation always has solutions; we can therefore summarize this discussion as follows: The equation (24) has integer solutions if and only if the number b is divisible by the greatest common divisor of $a_1, a_2, \ldots, a_n$.

5.2   Existence of Solutions

Theorem. *Let $n \geq 2$. The equation*

$$a_1 x_1 + a_2 x_2 + \cdots + a_n x_n = b, \tag{25}$$

where $a_1, a_2, \ldots, a_n, b$ are integers such that $(a_1, \ldots, a_n) = 1$, always has integer solutions. All integer solutions of this equation can be written in terms of $n - 1$ integer parameters.

PROOF. We use induction on the number n of the unknowns x_i in (25). It is convenient to formally begin with the case $n = 1$, where the condition $(a_1) = 1$ means that $a_1 = \pm 1$. Then (25) is of the form $x_1 = b$, or $-x_1 = b$, and thus the unique solutions clearly do not depend on any parameter, which corresponds to the fact that $n - 1 = 0$.

We now assume that $n \geq 2$ and that the theorem holds for equations in $n - 1$ variables; our aim is to prove (25) for n variables. We set $d = (a_1, \ldots, a_{n-1})$. Then any solution $x_1, \ldots, x_n$ of (25) trivially satisfies the congruence

$$a_1 x_1 + a_2 x_2 + \cdots + a_n x_n \equiv b \pmod{d}.$$

Since d is a common divisor of $a_1, \ldots, a_{n-1}$, this congruence is of the form

$$a_n x_n \equiv b \pmod{d},$$

and since by (6) we have $(d, a_n) = (a_1, \ldots, a_{n-1}, a_n) = 1$, Theorem 4.2 shows that it has solutions

$$x_n \equiv c \pmod{d},$$

where c is an appropriate integer, that is, $x_n = c + d \cdot t$, with arbitrary $t \in \mathbb{Z}$. Substituting this into (25) and rearranging, we get

$$a_1 x_1 + \cdots + a_{n-1} x_{n-1} = b - a_n c - a_n d t.$$

Since $a_n c \equiv b \pmod{d}$, the number $(b - a_n c)/d$ is an integer. We can therefore divide this last equation by d, and obtain

$$a_1' x_1 + \cdots + a_{n-1}' x_{n-1} = b',$$

where $a_i' = a_i/d$ for $i = 1, \ldots, n-1$ and $b' = (b - a_n c)/d - a_n t$. Since

$$(a_1', \ldots, a_{n-1}') = (d a_1', \ldots, d a_{n-1}') \cdot \frac{1}{d} = (a_1, \ldots, a_{n-1}) \cdot \frac{1}{d} = 1,$$

by the induction hypothesis this last equation has, for each $t \in \mathbb{Z}$, solutions that can be written in terms of $n-2$ integer parameters. If we add to these solutions the condition $x_n = c + dt$, we obtain solutions of (25) written in terms of the $n-2$ original parameters and the new parameter t. The proof by induction is now complete. □

5.3 Examples

We use the method of proof of Theorem 5.2 to solve the following Diophantine equations. For clarity we will denote the variables by $x, y, z, \ldots$ instead of $x_1, x_2, x_3, \ldots$.

(i) $5x + 7y = 8$.

SOLUTION.   Any solution of this equation has to satisfy the congruence

$$5x + 7y \equiv 8 \pmod 5,$$

hence $2y \equiv -2 \pmod 5$, which means that $y \equiv -1 \pmod 5$, i.e., $y = -1 + 5t$ for $t \in \mathbb{Z}$. Substituting this into the given equation, we get

$$5x + 7(-1 + 5t) = 8,$$

which simplifies to $x = 3 - 7t$. The solutions to our equation are therefore

$$x = 3 - 7t, \quad y = -1 + 5t,$$

where t is an arbitrary integer.   □

(ii) $91x - 28y = 35$.

SOLUTION.   Since $(91, 28) = 7$ and $7 \mid 35$, the equation does have solutions. Dividing by 7, we get

$$13x - 4y = 5.$$

Any solution of this equation will satisfy the congruence

$$13x - 4y \equiv 5 \pmod{13},$$

i.e., $-4y \equiv -8 \pmod{13}$, which means that $y \equiv 2 \pmod{13}$ and $y = 2 + 13t$ for $t \in \mathbb{Z}$. By substituting, we obtain

$$13x - 4(2 + 13t) = 5,$$

which simplifies to $x = 1 + 4t$. Hence the solutions are $x = 1 + 4t$, $y = 2 + 13t$, where t is an arbitrary integer.   □

Of course, we would have obtained the same result by considering the congruence modulo 4 instead of 13. Since solving congruences with smaller moduli is easier, it is well worth remembering this fact so as to avoid congruences with large moduli.

(iii) $18x + 20y + 15z = 1$.

SOLUTION.   Since $(18, 20, 15) = 1$, the equation has integer solutions. Any solution will satisfy the congruence (as modulus we choose $(18,20)$)

$$18x + 20y + 15z \equiv 1 \pmod 2;$$

thus $z \equiv 1 \pmod 2$, and so $z = 1 + 2s$, where $s \in \mathbb{Z}$. Substituting,

$$18x + 20y + 15(1 + 2s) = 1,$$

and after dividing by 2 and rearranging, we get

$$9x + 10y = -7 - 15s,$$

which we can solve for any $s \in \mathbb{Z}$. If this equation is satisfied, we will have

$$9x + 10y \equiv -7 - 15s \pmod{9},$$

hence $y \equiv 2 + 3s \pmod{9}$, and therefore $y = 2 + 3s + 9t$, where $t \in \mathbb{Z}$. Substituting, we get

$$9x + 10(2 + 3s + 9t) = -7 - 15s,$$

and after simplifying, $x = -3 - 5s - 10t$. The solutions of the given equation are therefore the triples

$$x = -3 - 5s - 10t,$$
$$y = 2 + 3s + 9t,$$
$$z = 1 + 2s,$$

where s, t are arbitrary integers. □

(iv) $15x - 12y + 48z - 51u = 1.$

SOLUTION. Since $(15, 12, 48, 51) = 3$ does not divide 1, the equation has no integer solution. □

5.4 Exercises

Solve the following equations in integers.

(i) $2x - 3y = 1.$ (ii) $17x + 15y = 2.$

(iii) $16x + 48y = 40.$ (iv) $10x + 4y = 16.$

(v) $17x + 13y = 1.$ (vi) $60x - 77y = 1.$

(vii) $12x + 10y + 15z = 1.$ (viii) $63x + 70y + 75z = 91.$

(ix) $105x + 119y + 161z = 83.$ (x) $10x + 11y + 12z + 13u = 14.$

(xi) $36x + 45y + 105z + 84u = 39.$ (xii) $2x + 4y + 8z + 16u = 32.$

5.5 Diophantine Equations Linear in Some Variable

We will now consider equations that can be written in the form

$$mx_n = F(x_1, \ldots, x_{n-1}), \tag{26}$$

where m is a natural number and $F(x_1,\ldots,x_{n-1})$ is a polynomial with integer coefficients. It is clear that for $x_1, x_2, \ldots, x_n$ to be an integer solution of (26), we require

$$F(x_1,\ldots,x_{n-1}) \equiv 0 \pmod{m}. \tag{27}$$

On the other hand, if $x_1,\ldots,x_{n-1}$ is a solution of (27), then with $x_n = F(x_1,\ldots,x_{n-1})/m$ we obtain an integer solution $x_1,\ldots,x_{n-1},x_n$ of (26). Hence, to solve the equation (26), it suffices to solve the congruence (27). In the case $n = 2$, i.e., the case where $F(x_1)$ is a polynomial in one variable, this reduces to the problem we have dealt with in 4.9. The case $n > 2$ can be solved analogously, with the help of the following theorem.

5.6   A Substitution Result

Theorem. *For any polynomial $F(x_1,\ldots,x_s)$ with integer coefficients, and for any natural number m and integers $a_1,\ldots,a_s$, $b_1,\ldots,b_s$ such that $a_1 \equiv b_1 \pmod{m}$, $\ldots$, $a_s \equiv b_s \pmod{m}$, we have $F(a_1,\ldots,a_s) \equiv F(b_1,\ldots,b_s) \pmod{m}$.*

PROOF.   Similar to the proof of Theorem 4.10.   □

To find all solutions of the congruence (27), it therefore suffices to substitute the variables $x_1,\ldots,x_{n-1}$ in $F(x_1,\ldots,x_{n-1})$ independently of each other by the numbers $0,1,2,\ldots,m-1$ (i.e., m^{n-1} choices altogether). Exactly in the case where the numbers $a_1,\ldots,a_{n-1}$ satisfy the condition $F(a_1,\ldots,a_{n-1}) \equiv 0 \pmod{m}$, we obtain solutions of the congruence (27) in the form

$$x_1 = a_1 + mt_1, \quad\ldots,\quad x_{n-1} = a_{n-1} + mt_{n-1},$$

where the $t_1,\ldots,t_{n-1}$ can take on any integer values. Finally, this gives us solutions to the original equation (26):

$$x_1 = a_1 + mt_1,$$
$$\vdots$$
$$x_{n-1} = a_{n-1} + mt_{n-1},$$
$$x_n = \frac{1}{m}F(a_1 + mt_1,\ldots,a_{n-1} + mt_{n-1}).$$

5.7   Examples

Solve the following Diophantine equations:

(i) $7x^2 + 5y + 13 = 0$.

SOLUTION.   We rewrite the equation as $5y = -7x^2 - 13$, and solve the congruence

$$-7x^2 - 13 \equiv 0 \pmod{5},$$

i.e., $3x^2 \equiv 3 \pmod 5$, which means that $x^2 \equiv 1 \pmod 5$. If we replace x by 0, 1, 2, 3, 4, we find that the congruence is satisfied by the numbers 1 and 4. By 4.11, the solutions of this congruence are therefore exactly the numbers

$$x = 1 + 5t \quad \text{or} \quad x = 4 + 5t,$$

where $t \in \mathbb{Z}$. By substituting we get in the first case

$$5y = -7(1 + 5t)^2 - 13 = -7 - 70t - 175t^2 - 13$$

and thus

$$y = -4 - 14t - 35t^2,$$

and in the second case,

$$5y = -7(4 + 5t)^2 - 13 = -112 - 280t - 175t^2 - 13,$$

which means that

$$y = -25 - 56t - 35t^2.$$

The solutions of the given equation are therefore exactly all the pairs x, y of the form

$$x = 1 + 5t, y = -4 - 14t - 35t^2 \quad \text{or} \quad x = 4 + 5t, y = -25 - 56t - 35t^2,$$

where t is an arbitrary integer.   □

(ii) $x(x + 3) = 4y - 1.$

SOLUTION.   We write the equation in the form $4y = x^2 + 3x + 1$, and solve the congruence

$$x^2 + 3x + 1 \equiv 0 \pmod 4.$$

By substituting we find that none of the numbers 0, 1, 2, 3 satisfies this congruence, and thus by 4.11 it has no solution. Therefore, the given equation has no solution either.   □

(iii) $x^2 + 4z^2 + 6x + 7y + 8z = 1.$

SOLUTION.   We rewrite the equation as

$$7y = -x^2 - 6x - 4z^2 - 8z + 1,$$

and completing squares, we get

$$7y = -(x + 3)^2 - (2z + 2)^2 + 14.$$

We will therefore solve the congruence

$$(x+3)^2 + (2z+2)^2 \equiv 0 \pmod{7}. \tag{28}$$

Now, we could substitute the ordered pair (x, z) consecutively by the ordered pairs $(0,0)$, $(0,1)$, ..., $(0,6)$, $(1,0)$, $(1,1)$, ..., $(6,5)$, $(6,6)$, and evaluate the left-hand side of (28) for all 49 pairs. However, it will be more convenient to exploit the particular form of the congruence (28) and refer to 3.14.(vi), which shows with $p = 7$, $a = x + 3$, $b = 2z + 2$ that (28) implies

$$x + 3 \equiv 2z + 2 \equiv 0 \pmod{7},$$

and therefore all the solutions of (28) are of the form

$$x = -3 + 7t, \qquad z = -1 + 7s,$$

where t, s are integers. Substituting this into the equation, we obtain

$$7y = -(x+3)^2 - (2z+2)^2 + 14 = -49t^2 - 196s^2 + 14,$$

and thus

$$y = -7t^2 - 28s^2 + 2.$$

The solutions of the given equation are therefore exactly all the triples x, y, z of the form

$$x = -3 + 7t, \quad y = -7t^2 - 28s^2 + 2, \quad z = -1 + 7s,$$

where s, t are arbitrary integers. $\qquad \square$

(iv) $(x - y)^2 = x + y.$

SOLUTION.   Although this equation is not of the form (26), we can bring it into this form by introducing the variable $z = x - y$. Then $x + y = z + 2y$, and we get the new equation

$$z^2 = z + 2y,$$

in the variables z and y, or

$$2y = z(z - 1).$$

We will therefore solve the congruence

$$z(z - 1) \equiv 0 \pmod{2}.$$

By substituting we see that both numbers 0, 1 satisfy this congruence, and so by 4.11 it is satisfied by all integers z (if we wanted to strictly "follow the instructions," we would now have to distinguish between the cases $z = 2t$, $t \in \mathbb{Z}$, and $z = 2t + 1$, $t \in \mathbb{Z}$, but this is not necessary here). Therefore,

$$y = \frac{z(z-1)}{2}$$

is an integer for any $z \in \mathbb{Z}$. The original variable x is then of the form

$$x = z + y = z + \frac{z(z-1)}{2} = \frac{z(z+1)}{2}.$$

The solutions of the given equation are therefore all pairs x, y of the form

$$x = \frac{1}{2}z(z+1), \qquad y = \frac{1}{2}z(z-1),$$

where z is an arbitrary integer. $\square$

5.8 Exercises

Solve the Diophantine equations (i)–(xii):

(i) $x^2 - 11y - 4 = 0$.   (ii) $x^2 - 11y - 6 = 0$.

(iii) $x^3 + 2x^2 + 5 = 21y$.   *(iv) $(x^2 + 2)x = 125y - 2$.

(v) $x^3 - 11x^2 - 11y - 2 = 0$.   (vi) $x^3 - 13x^2 - 13y - 2 = 0$.

(vii) $x^2 - 39 = 30y$.   (viii) $(x + y)^2 + 8z = 1$.

(ix) $x^2 + xy + y^2 = 3(x + y + z)$.

*(x) $x^2 + 4y^2 + 9z^2 = 8u - 1$.

*(xi) $x^2 + 4xy + 4y^2 - 3x - 4y - 600 = 0$.

*(xii) $x^4 + 2x^2y + y^2 + x^2 + y + 7x + 1 = 0$.

The method used for solving the examples in 5.7 can also be described as follows: "Express one of the variables in terms of the others and find out when it is integral." Indeed, if we solve the equation (26) for x_n, we get

$$x_n = \frac{F(x_1, \ldots, x_{n-1})}{m},$$

which is an integer if and only if $m \mid F(x_1, \ldots, x_{n-1})$, i.e., if and only if the congruence (27) is satisfied. We will see in the following examples that this approach is also useful for equations that are not of the form (26). In Example 5.9.(iv) we will see that instead of discussing the integrality of some variable it may be more appropriate to study the integrality of some other suitable expression and find out under which circumstances it will be integral.

5.9 Examples

Solve the following Diophantine equations.

(i) $3^x = 4y + 5$.

SOLUTION.   We solve this equation for y:

$$y = \frac{1}{4}(3^x - 5).$$

If $x < 0$, then $0 < 3^x < 1$, and thus $\frac{1}{4}(3^x - 5) \notin \mathbb{Z}$. For $x \geq 0$ we have

$$3^x - 5 \equiv (-1)^x - 1 \pmod 4.$$

The number $(-1)^x - 1$ is congruent to 0 modulo 4 if and only if x is even, i.e., $x = 2k$, where $k \in \mathbb{N}_0$. The solutions of the given equation are therefore all pairs

$$x = 2k, \qquad y = \frac{9^k - 5}{4},$$

where $k \in \mathbb{N}_0$ is arbitrary.   □

(ii) $x(y + 1)^2 = 243y$.

SOLUTION.   We solve for x:

$$x = \frac{243y}{(y + 1)^2}.$$

In order for x to be an integer, $(y+1)^2$ must divide $243y$. Since y and $y+1$ are relatively prime for any $y \in \mathbb{Z}$, then $(y+1)^2$ has to divide $243 = 3^5$. But this number has only three divisors that are squares of integers, namely $1, 9$, and 81, and this leads to the following possibilities: $y+1 = \pm 1$, $y+1 = \pm 3$, or $y + 1 = \pm 9$. Hence we obtain the following six solutions of the given equation:

$y = 0,$	$x = 0,$
$y = -2,$	$x = -2 \cdot 243 = -486,$
$y = 2,$	$x = 2 \cdot 27 = 54,$
$y = -4,$	$x = -4 \cdot 27 = -108,$
$y = 8,$	$x = 8 \cdot 3 = 24,$
$y = -10,$	$x = -10 \cdot 3 = -30.$

The equation has no other solutions.   □

(iii) $\sqrt{x} + \sqrt{y} = \sqrt{1988}$.

SOLUTION. If we subtract $\sqrt{y}$ from both sides and then square, we obtain

$$x = 1988 - 4\sqrt{7 \cdot 71 \cdot y} + y.$$

If x, y are integers, then $4\sqrt{7 \cdot 71y}$ is also an integer, and thus $\sqrt{7 \cdot 71y}$ is a rational number. By 2.11.(iii), $\sqrt{7 \cdot 71y} = k$ is a nonnegative integer. Hence $7 \cdot 71y = k^2$, which means that k^2 and thus also k are divisible by the primes 7 and 71. Therefore, $k = 7 \cdot 71t$ for an appropriate $t \in \mathbb{N}_0$, and thus

$$y = \frac{k^2}{7 \cdot 71} = 497t^2.$$

In complete analogy we can find that there exists an $s \in \mathbb{N}_0$ such that

$$x = 497s^2.$$

Substituting this into the original equation, we get

$$\sqrt{497}s + \sqrt{497}t = \sqrt{1988},$$

which, after dividing, gives $s + t = 2$. Hence there are three possibilities: $s = 0, t = 2$, or $s = t = 1$, or $s = 2, t = 0$, and so the given Diophantine equation has the three solutions

$$x = 0, \quad y = 1988; \qquad x = y = 497; \qquad x = 1988, \quad y = 0. \qquad \square$$

(iv) $x^2 + y^2 = (x - y)^3$.

SOLUTION. The equation is certainly satisfied by $x = y = 0$, so this is one solution. For what follows we assume that at least one of the numbers x, y is nonzero. We rewrite the equation as

$$x^2 - 2xy + y^2 = (x - y)^3 - 2xy,$$
$$(x - y)^2(x - y - 1) = 2xy.$$

We denote the greatest common divisor of x, y by k, and set $x_1 = x/k$, $y_1 = y/k$. By 1.14.(iii) the integers x_1, y_1 are relatively prime. By substituting into the last equation, we obtain

$$k^2(x_1 - y_1)^2(kx_1 - ky_1 - 1) = 2k^2x_1y_1.$$

If $x_1 - y_1 = 0$, then $x = y$ and the original equation gives $x^2 + y^2 = 0$; this implies $x = y = 0$, which contradicts our assumption. Hence $x_1 - y_1 \neq 0$; moreover, since the original equation implies $x \geq y$, we have $x_1 > y_1$. If we divide the last equation by $k^2(x_1 - y_1)^2$, we obtain

$$kx_1 - ky_1 - 1 = \frac{2x_1y_1}{(x_1 - y_1)^2}.$$

Hence $(x_1 - y_1)^2$ is a divisor of $2x_1y_1$, but this can occur only if $(x_1 - y_1)^2 = 1$. Indeed, if we assume that $(x_1 - y_1)^2 > 1$, then by 2.5 there exists a prime p that divides $(x_1 - y_1)^2$. But then, by 2.2, $p \mid x_1 - y_1$, and thus from $p \mid x_1$ it would follow that $p \mid y_1$, and vice versa. Since $(x_1, y_1) = 1$, p divides neither x_1 nor y_1. However, in this case the condition $p^2 \mid (x_1 - y_1)^2 \mid 2x_1y_1$ implies $p^2 \mid 2$, which is a contradiction. Hence $(x_1 - y_1)^2 = 1$, and from $x_1 > y_1$ it follows that $x_1 - y_1 = 1$, i.e., $x_1 = 1 + y_1$. Substituting this into the last equation, we obtain

$$k - 1 = 2(1 + y_1)y_1,$$

and thus

$$x = kx_1 = (2(1 + y_1)y_1 + 1)(1 + y_1),$$
$$y = ky_1 = (2(1 + y_1)y_1 + 1)y_1.$$

If we set $y_1 = t$, then all the solutions of the original equation are given by the pair $x = 0$, $y = 0$, and all pairs

$$x = (2t^2 + 2t + 1)(t + 1), \qquad y = (2t^2 + 2t + 1) \cdot t,$$

where t is an arbitrary integer.   □

5.10 Exercises

Solve the following Diophantine equations:

 (i) $5^x + 3^y = 8z - 2$.
 (ii) $2^x = 3 + 13y$.
 (iii) $x^2 - xy + 2x - 3y = 11$.
 (iv) $2x^3 - x^2y + x^2 + 14x - 7y + 40 = 0$.
 (v) $(x - 2)^4 = (y + 3)^5$.
 (vi) $\sqrt{x} = 1 + \sqrt{y}$.
 (vii) $2xy + 3y^2 = 24$.
 (viii) $x^2y^2 + x^2y + xy^2 + xy = xyz - 1$.

5.11 Solving Diophantine Equations Using Inequalities

This method is based on the fact that for any real numbers a, b there exist only finitely many integers x such that $a < x < b$. Therefore, in order to solve a given equation, we try to find numbers a, b such that the inequalities $a < x < b$, for some variables x, follow from this equation. The finitely many integers lying between a and b are then substituted for x in the given equation, which in this way becomes simpler. Let us consider the following examples.

5.12 Examples

Solve the Diophantine equations (i)–(iii):

(i) $6x^2 + 5y^2 = 74$.

SOLUTION. Since all $y \in \mathbb{Z}$ satisfy $5y^2 \geq 0$, any solution x, y of the given equation must satisfy

$$74 = 6x^2 + 5y^2 \geq 6x^2,$$

which means $x^2 < 37/3$, and thus $-3 \leq x \leq 3$, so x^2 is among the numbers $0, 1, 4, 9$. Substituting these in succession, we obtain $5y^2 = 74$, $5y^2 = 68$, $5y^2 = 50$, $5y^2 = 20$. The first three contradict $y \in \mathbb{Z}$, while from the last one we get $y^2 = 4$, i.e., $y = \pm 2$.

Our equation therefore has the following four solutions: $x = 3$, $y = 2$; $x = 3$, $y = -2$; $x = -3$, $y = 2$; $x = -3$, $y = -2$.   □

(ii) $x^2 + xy + y^2 = x^2 y^2$.

SOLUTION. Since this equation is symmetric in the variables x, y, we may assume that $x^2 \leq y^2$, which implies $xy \leq y^2$, and thus

$$x^2 y^2 = x^2 + xy + y^2 \leq y^2 + y^2 + y^2 = 3y^2.$$

We have therefore $y = 0$ or $x^2 \leq 3$. Substituting these into the equation, we obtain $x = 0$ in the first case, and in the second case for $x = 0$ again $y = 0$, while for $x = 1$ we get $y = -1$ and for $x = -1$ it is $y = 1$. Our equation has therefore the following three solutions:

$$x = 0, \quad y = 0; \qquad x = 1, \quad y = -1; \qquad x = -1, \quad y = 1. \qquad \square$$

(iii) $2^x = 1 + 3^y$.

SOLUTION. If $y < 0$, then $1 < 1 + 3^y < 2$, which means that $0 < x < 1$, but this is a contradiction. Hence $y \geq 0$ and therefore $2^x = 1 + 3^y \geq 2$, which means that $x \geq 1$. We will now show that also $x \leq 2$. Indeed, if we had $x \geq 3$, then

$$1 + 3^y = 2^x \equiv 0 \pmod 8,$$

which would imply

$$3^y \equiv -1 \pmod 8;$$

but this is impossible, since for even numbers y we have $3^y \equiv 1 \pmod 8$, and for y odd, $3^y \equiv 3 \pmod 8$. It therefore remains to consider the case $1 \leq x \leq 2$. For $x = 1$ we get

$$3^y = 2^1 - 1 = 1,$$

and thus $y = 0$. The case $x = 2$ gives

$$3^y = 2^2 - 1 = 3;$$

hence $y = 1$. The given equation has therefore the two solutions $x = 1$, $y = 0$ and $x = 2$, $y = 1$. □

(iv) Solve the equation $x + y + z = xyz$ in natural numbers.

SOLUTION. Since the variables in this equation are symmetric, we may assume that $x \le y \le z$. But then

$$xyz = x + y + z \le z + z + z = 3z,$$

which means that $xy \le 3$. Hence $xy = 1$, or $xy = 2$, or $xy = 3$.

If $xy = 1$, then $x = 1$, $y = 1$, and by substituting into the given equation we get $2 + z = z$, which is impossible.

If $xy = 2$, then $x = 1$, $y = 2$ (recall that we have $x \le y$), which means that $3 + z = 2z$, and thus $z = 3$.

If $xy = 3$, then $x = 1$, $y = 3$; hence $4 + z = 3z$, and thus $z = 2$. But this is a contradiction to the assumption $y \le z$.

Our equation has therefore the unique solution $x = 1$, $y = 2$, $z = 3$ under the assumption $x \le y \le z$. The complete set of solutions in natural numbers is therefore given by all permutations of the numbers 1, 2, 3:

$$(x, y, z) \in \{(1,2,3), (1,3,2), (2,1,3), (2,3,1), (3,1,2), (3,2,1)\}. \quad □$$

5.13 Exercises

Solve the equations (i)–(iv) in natural numbers, and the equations (v) and (vi) in integers:

(i) $6(x - y) = xy$.

(ii) $\frac{1}{x} + \frac{1}{y} + \frac{1}{z} = 1$.

(iii) $x^x + y^y = 2x + y$.

(iv) $x^3 + y^3 + z^3 = 99$.

(v) $3^y = 1 + 2^x$.

(vi) $x^2 - xy + y^2 = x + y$.

It is often advantageous to use contradictions to show that the set of values of the variable x is finite and bounded by the inequalities $a < x < b$; this is done by deriving a false statement from the assumption $x \le a$ (resp. $x \ge b$). In the following examples, such a false statement will be the pair of inequalities

$$c^n < d^n < (c + 1)^n,$$

where c, d are integers and n is a natural number.

5.14 Examples

(i) Solve the Diophantine equation $x(x + 1)(x + 7)(x + 8) = y^2$.

SOLUTION. Rewriting, we get

$$y^2 = (x^2 + 8x)(x^2 + 8x + 7).$$

If we set $x^2 + 8x = z$, then our equation becomes

$$y^2 = z^2 + 7z.$$

We will show that $z \leq 9$. We assume to the contrary that $z > 9$. Then

$$(z + 3)^2 = z^2 + 6z + 9 < z^2 + 7z = y^2 < z^2 + 8z + 16 = (z + 4)^2,$$

which is a contradiction, since $z + 3$, y, $z + 4$ are integers, and these inequalities would imply

$$|z + 3| < |y| < |z + 4|.$$

Hence $z \leq 9$, i.e., $x^2 + 8x \leq 9$, which means that

$$(x + 4)^2 = x^2 + 8x + 16 \leq 25,$$

and thus $-5 \leq x + 4 \leq 5$, or $-9 \leq x \leq 1$. Substituting these values into the original equation, we obtain all solutions: $(x, y) \in \{(-9, 12), (-9, -12),$ $(-8, 0), (-7, 0), (-4, 12), (-4, -12), (-1, 0), (0, 0), (1, 12), (1, -12)\}$.   □

(ii) Solve the Diophantine equation $(x + 2)^4 - x^4 = y^3$.

SOLUTION.   Rewriting, we obtain

$$y^3 = 8x^3 + 24x^2 + 32x + 16 = 8(x^3 + 3x^2 + 4x + 2),$$

which means that y is even, and we can set $y = 2z$, $z \in \mathbb{Z}$. Then

$$z^3 = x^3 + 3x^2 + 4x + 2.$$

If $x \geq 0$, we have

$$\begin{aligned}(x + 1)^3 = x^3 + 3x^2 + 3x + 1 &< x^3 + 3x^2 + 4x + 2 \\ &= z^3 < x^3 + 6x^2 + 12x + 8 = (x + 2)^3,\end{aligned}$$

hence $x + 1 < z < x + 2$, which is impossible. The given equation therefore has no solutions $x, y \in \mathbb{Z}$ such that $x \geq 0$. Let us now assume that it has a solution $x_1, y_1 \in \mathbb{Z}$ such that $x_1 \leq -2$. Then

$$(x_1 + 2)^4 - x_1^4 = y_1^3,$$

and if we set $x_2 = -x_1 - 2$, $y_2 = -y_1$, we obtain

$$x_2^4 - (x_2 + 2)^4 = -y_2^3,$$

and so x_2, y_2 is also a solution of the given equation. But $x_2 = -x_1 - 2 \geq 0$, and from the above it follows that this case cannot occur. In summary, then, $-2 < x < 0$, i.e., $x = -1$. The original equation then gives $y = 0$; the pair $x = -1$, $y = 0$ is therefore the unique solution.   □

5.15 Exercises

Solve the Diophantine equations (i)–(iv):

(i) $x^2 = 1 + y + y^2 + y^3 + y^4$ in natural numbers.

*(ii) $x^2 + x = y^4 + y^3 + y^2 + y$ in integers.

*(iii) $x^6 + 3x^3 + 1 = y^4$ in integers.

(iv) $x^8 + 2x^6 + 2x^4 + 2x^2 + 1 = y^2$ in integers.

*(v) For any natural number n and all natural numbers d dividing $2n^2$ show that $n^2 + d$ is not the square of a natural number.

Some other, more complicated, approaches based on inequalities introduced in Chapter 2 are sometimes also useful for solving Diophantine equations. We will now consider several such examples.

5.16 Examples

(i) Find solutions in natural numbers of the equation

$$\frac{1}{x^2} + \frac{1}{y^2} + \frac{1}{z^2} + \frac{1}{u^2} = 1.$$

SOLUTION.   Since the equation is symmetric in the variables x, y, z, u, we may assume that $x \leq y \leq z \leq u$. It is clear that $x > 1$. If $u \geq 3$, we would have

$$\frac{1}{x^2} + \frac{1}{y^2} + \frac{1}{z^2} + \frac{1}{u^2} \leq \frac{1}{4} + \frac{1}{4} + \frac{1}{4} + \frac{1}{9} < 1;$$

hence $u \leq 2$, and consequently $2 \leq x \leq y \leq z \leq u \leq 2$. By substituting into the given equation we verify that $x = y = z = u = 2$ is indeed the unique solution. $\square$

(ii) Given any $s \geq 2$, show that the equation

$$\frac{1}{x_1^2} + \frac{1}{x_2^2} + \cdots + \frac{1}{x_s^2} = 1$$

has no solutions in natural numbers $x_1 < x_2 < \cdots < x_s$.

SOLUTION.   Since $s \geq 2$, we have $x_1 > 1$; hence from $x_1 < x_2 < \cdots < x_s$ and $x_1, x_2, \ldots, x_s \in \mathbb{N}$ it follows that $x_k \geq k+1$ for all $k = 1, 2, \ldots, s$. Since for any $k = 1, 2, \ldots, s$,

$$\frac{1}{x_k^2} \leq \frac{1}{(k+1)^2} < \frac{1}{k(k+1)} = \frac{1}{k} - \frac{1}{k+1},$$

we obtain

$$\frac{1}{x_1^2} + \frac{1}{x_2^2} + \cdots + \frac{1}{x_s^2} \leq \frac{1}{2^2} + \frac{1}{3^2} + \cdots + \frac{1}{(s+1)^2}$$
$$< \left(1 - \frac{1}{2}\right) + \left(\frac{1}{2} - \frac{1}{3}\right) + \cdots + \left(\frac{1}{s} - \frac{1}{s+1}\right)$$
$$= 1 - \frac{1}{s+1} < 1;$$

hence the given equation has no solution in integers with the required properties.   □

(iii) Find solutions in natural numbers for the equation

$$\frac{x}{y} + \frac{y}{z} + \frac{z}{x} = 3.$$

SOLUTION.   The quotient of natural numbers is a positive number, so we may apply the arithmetic–geometric mean inequality (see Section 8.1 in Chapter 2) to the numbers x/y, y/z and z/x. The geometric mean of these three numbers is 1; hence their arithmetic mean satisfies

$$\frac{1}{3}\left(\frac{x}{y} + \frac{y}{z} + \frac{z}{x}\right) \geq 1,$$

with equality if and only if

$$\frac{x}{y} = \frac{y}{z} = \frac{z}{x} = 1.$$

If we compare the above inequality with the given equation, we see that there are infinitely many solutions of the form $x = y = z$, where z is an arbitrary natural number, but no other solutions.   □

(iv) Show that, given any integer $n > 2$, the equation

$$x^n + (x+1)^n = (x+2)^n$$

has no solutions in natural numbers.

SOLUTION.   Let us assume to the contrary that for some natural numbers x, n, with $n > 2$, the given equation is satisfied, and set $y = x + 1 \geq 2$. Then

$$(y-1)^n + y^n = (y+1)^n, \tag{29}$$

and we get

$$0 = (y+1)^n - y^n - (y-1)^n \equiv 1 - (-1)^n \pmod{y}.$$

If n is odd, then $0 \equiv 2 \pmod{y}$, and thus $y = 2$ and

$$0 = 3^n - 2^n - 1,$$

which holds only for $n = 1$. Hence n is even, and by the binomial theorem we have

$$(y + 1)^n \equiv \binom{n}{2} y^2 + \binom{n}{1} y + 1 \pmod{y^3},$$

$$(y - 1)^n \equiv \binom{n}{2} y^2 - \binom{n}{1} y + 1 \pmod{y^3},$$

which implies

$$0 = (y + 1)^n - y^n - (y - 1)^n \equiv 2ny \pmod{y^3};$$

hence $0 \equiv 2n \pmod{y^2}$ by 3.3.(v), and therefore $2n \geq y^2$. If we divide (29) by y^n, we get

$$\left(1 + \frac{1}{y}\right)^n = 1 + \left(1 - \frac{1}{y}\right)^n < 2.$$

On the other hand, Bernoulli's inequality (see 3.3.(iii), Chapter 2) gives

$$\left(1 + \frac{1}{y}\right)^n > 1 + \frac{n}{y} = 1 + \frac{2n}{2y} \geq 1 + \frac{y^2}{2y} = 1 + \frac{y}{2} \geq 2.$$

Putting everything together, it follows that the given equation has no solution in natural numbers for any integer $n > 2$. $\square$

5.17 Exercises

Solve the following equations in natural numbers.

(i) $x^3 + y^3 + z^3 = 3xyz$.

(ii) $2\left(\dfrac{x}{y} + \dfrac{y}{z} + \dfrac{z}{u} + \dfrac{u}{x}\right) = 7$.

(iii) $(x + 2y + 3z)^2 = 14(x^2 + y^2 + z^2)$.

(iv) $\dfrac{1}{x^2} + \dfrac{1}{y^2} + \dfrac{1}{z^2} + \dfrac{1}{u^2} + \dfrac{1}{v^2} = 1$.

5.18 Solving Diophantine Equations by Decomposition

This method is based on writing a given equation in the form

$$A_1 \cdot A_2 \cdots A_n = B, \tag{30}$$

where $A_1, \ldots, A_n$ are expressions containing variables such that they take on integer values when the variables are replaced by integers, and B is a number (possibly an expression) whose prime decomposition we know. In this case there exist only finitely many decompositions of the number B into n integer factors $a_1, \ldots, a_n$. If for each such decomposition we then consider the system of equations

$$A_1 = a_1, \quad A_2 = a_2, \quad \ldots, \quad A_n = a_n,$$

we obtain all solutions of the equation (30). The following examples will illustrate this.

5.19 Examples

(i) Solve the Diophantine equation $y^3 - x^3 = 91$.

SOLUTION.   Factoring the left-hand side of the equation, we get

$$(y - x)(y^2 + xy + x^2) = 91.$$

Since

$$y^2 + xy + x^2 = \left(y + \frac{x}{2}\right)^2 + \frac{3}{4}x^2 \geq 0,$$

we will also have $y - x > 0$. The number 91 can be written as a product of two natural numbers in four different ways: $91 = 1 \cdot 91 = 7 \cdot 13 = 13 \cdot 7 = 91 \cdot 1$. Therefore, we will separately solve the following four systems of equations:

(a) $y - x = 1$, $y^2 + xy + x^2 = 91$. Substituting $y = x + 1$ into the second equation, we obtain $x^2 + x - 30 = 0$, which gives $x = 5$ or $x = -6$. The corresponding values of the second variable are then $y = 6$, $y = -5$.

(b) $y - x = 7$, $y^2 + xy + x^2 = 13$. Then $x^2 + 7x + 12 = 0$, so $x = -3$ and $y = 4$, or $x = -4$ and $y = 3$.

(c) $y - x = 13$, $y^2 + xy + x^2 = 7$. Here we have $x^2 + 13x + 54 = 0$. But this equation has no real roots, and thus no integer roots.

(d) $y - x = 91$, $y^2 + xy + x^2 = 1$. In this case $x^2 + 91x + 2760 = 0$, and this equation has no real roots either.

The given equation has therefore the following four solutions: $(x, y) \in \{(5, 6), (-6, -5), (-3, 4), (-4, 3)\}$.   □

(ii) Solve the Diophantine equation $x^4 + 2x^7 y - x^{14} - y^2 = 7$.

SOLUTION.   We rewrite the left-hand side,

$$x^4 + 2x^7 y - x^{14} - y^2 = x^4 - (x^7 - y)^2 = (x^2 - x^7 + y)(x^2 + x^7 - y),$$

and note that the number 7 can be written in four different ways as a product of two integers: $7 = 1 \cdot 7 = 7 \cdot 1 = (-1) \cdot (-7) = (-7) \cdot (-1)$. We will therefore solve the following four systems of equations:

(a) $x^2 - x^7 + y = 1$, $x^2 + x^7 - y = 7$. Adding these two equations, we obtain $x^2 = 4$, thus $x = 2$ and $y = 125$, or $x = -2$ and $y = -131$.

(b) $x^2 - x^7 + y = 7$, $x^2 + x^7 - y = 1$. Again $x^2 = 4$, and thus $x = 2$, $y = 131$ or $x = -2$, $y = -125$.

(c) $x^2 - x^7 + y = -1$, $x^2 + x^7 - y = -7$. By adding we now get $x^2 = -4$, which is a contradiction.

(d) $x^2 - x^7 + y = -7$, $x^2 + x^7 - y = -1$. Again $x^2 = -4$, which is impossible.

The equation therefore has the four solutions

$$(x, y) \in \{(-2, -131), (-2, -125), (2, 125), (2, 131)\}. \qquad \square$$

(iii) Solve the Diophantine equation

$$\frac{1}{x} + \frac{1}{y} = \frac{1}{p},$$

where p is an arbitrary prime.

SOLUTION.   Multiplying by xyp and rearranging, we get

$$xy - px - py = 0.$$

Bringing this into the form (30) now requires a trick. Add p^2 to both sides of the equation, so that the left-hand side can be written as a product:

$$(x - p)(y - p) = p^2.$$

Since p is a prime, p^2 can be written as a product of two integers in no more than the following six ways: $p^2 = 1 \cdot p^2 = p \cdot p = p^2 \cdot 1 = (-1) \cdot (-p^2) = (-p) \cdot (-p) = (-p^2) \cdot (-1)$. This leads to the following six systems of equations:

(a) $x - p = 1$, $y - p = p^2$, hence $x = p + 1$, $y = p^2 + p$;
(b) $x - p = p$, $y - p = p$, hence $x = 2p$, $y = 2p$;
(c) $x - p = p^2$, $y - p = 1$, hence $x = p^2 + p$, $y = p + 1$;
(d) $x - p = -1$, $y - p = -p^2$, hence $x = p - 1$, $y = p - p^2$;
(e) $x - p = -p$, $y - p = -p$, hence $x = y = 0$, which is impossible;
(f) $x - p = -p^2$, $y - p = -1$, hence $x = p - p^2$, $y = p - 1$.

The given equation has therefore five solutions, as described in the cases (a)–(d) and (f). $\qquad \square$

(iv) Solve the Diophantine equation

$$1 + x + x^2 + x^3 = 2^y.$$

SOLUTION.   Again, we rewrite the left-hand side as a product:

$$1 + x + x^2 + x^3 = 1 + x + x^2(1 + x) = (1 + x^2)(1 + x).$$

Since the left-hand side of the given equation is integer-valued for $x \in \mathbb{Z}$, the right-hand side must be an integer as well, hence $y \geq 0$. Now, the right-hand side is a power of 2, and so each factor of the left-hand side must be a power of 2 as well; hence there exists an integer m, $0 \leq m \leq y$, such that

$$1 + x = 2^m, \qquad 1 + x^2 = 2^{y-m},$$

or

$$x = 2^m - 1, \qquad x^2 = 2^{y-m} - 1.$$

Squaring the first equation and comparing with the second one, we obtain

$$2^{2m} - 2^{m+1} + 1 = 2^{y-m} - 1,$$

or

$$2^{y-m} = 2^{2m} - 2^{m+1} + 2. \tag{31}$$

If $m = 0$, then $2^y = 1$, which means that $y = 0$ and $x = 2^m - 1 = 0$. If $m \geq 1$, then the odd number $2^{2m-1} - 2^m + 1$ is, by (31), a power of 2, which is possible only when $2^{2m-1} - 2^m + 1 = 1$, which means that $2m - 1 = m$, i.e., $m = 1$, and thus $x = 1$, $y = 2$. By substituting into the original equation, we convince ourselves that $x = y = 0$ and $x = 1$, $y = 2$ are indeed solutions. $\qquad \square$

(v) Solve the equation
$$x^2 + y^2 = z^2$$

in natural numbers.

SOLUTION.   We set $t = (x, y, z)$, $x_1 = x/t$, $y_1 = y/t$, and $z_1 = z/t$. Then

$$t^2 x_1^2 + t^2 y_1^2 = t^2 z_1^2,$$

which upon division by $t^2 \neq 0$ gives

$$x_1^2 + y_1^2 = z_1^2, \tag{32}$$

where $(x_1, y_1, z_1) = 1$ (see, e.g., 5.1). The numbers x_1, y_1, z_1 are actually pairwise relatively prime: Indeed, if a prime p divides two of the numbers x_1, y_1, z_1, then by (32), it would also divide the third one, but because of $(x_1, y_1, z_1) = 1$ this is impossible. Thus at most one of the numbers x_1, y_1 is even. Let us assume that both are odd. Then it follows from the congruence

$$z_1^2 \equiv x_1^2 + y_1^2 \equiv 1 + 1 \pmod{8}$$

that z_1^2 is an even number not divisible by 4, which is impossible. Therefore, exactly one of x_1, y_1 is even. Since the equation (32) is symmetric in x_1

and y_1, we may as well assume that the even one is $x_1 = 2r$, $r \in \mathbb{N}$. Then it follows from (32) that

$$4r^2 = z_1^2 - y_1^2,$$

and thus

$$r^2 = \frac{z_1 + y_1}{2} \cdot \frac{z_1 - y_1}{2}.$$

If we set $u = (z_1 + y_1)/2$, $v = (z_1 - y_1)/2$, then $z_1 = u + v$, $y_1 = u - v$. Since y_1, z_1 are relatively prime, then so are u, v. Now it follows from the equation

$$r^2 = u \cdot v$$

and from 2.6.(iii) that there exist relatively prime integers a, b such that $u = a^2$, $v = b^2$, and furthermore from $u > v$ it follows that $a > b$. Altogether we therefore have

$$x = tx_1 = 2tr = 2tab,$$
$$y = ty_1 = t(u - v) = t(a^2 - b^2),$$
$$z = tz_1 = t(u + v) = t(a^2 + b^2),$$

which indeed satisfies the given equation for any $t \in \mathbb{N}$ and for arbitrary relatively prime $a, b \in \mathbb{N}$ such that $a > b$. The remaining solutions could be obtained by changing the order of x and y (in the course of solving this problem we assumed that only x_1 is even):

$$x = t(a^2 - b^2), \quad y = 2tab, \quad z = t(a^2 + b^2),$$

where again $t, a, b \in \mathbb{N}$ are any numbers such that $a > b$, $(a, b) = 1$.   □

We remark that the Diophantine equation $x^2 + y^2 = z^2$ that we just solved is called the *Pythagorean equation* and describes all right-angled triangles with integer sides.

5.20 Exercises

Solve the Diophantine equations (i)–(iv), where p is a fixed prime number:

(i) $xy = x + y$.
(iii) $x^2 - y^2 = 105$.

(ii) $x(x + 1) = 4y(y + 1)$.
(iv) $2x^2 - x - 36 = p^2$.

Solve the equations (v)–(viii) in natural numbers:

(v) $2x^2 + 5xy - 12y^2 = 28$.
*(vii) $2^{x-1} + 1 = y^{z+1}$.

(vi) $2^{x^2 + x - 2} - 2^{x^2 - 4} = 992$.
*(viii) $2^{x+1} - 1 = y^{z+1}$.

(ix) Show that no number of the form 2^n, $n \in \mathbb{N}_0$, can be written as a sum of two or more consecutive natural numbers.

*(x) Show that any natural number that is not of the form 2^n, $n \in \mathbb{N}_0$, can be written as a sum of at least two consecutive natural numbers.

Solve the systems of equations (xi), (xii) over the integers:

(xi) $x + y + z = 14$,
$x + yz = 18$.

*(xii) $x + y + z = 3$,
$x^3 + y^3 + z^3 = 3$.

*(xiii) Show that for natural numbers x, y, z such that $(x, y) = 1$ and $x^2 + y^2 = z^4$, the product xy is always divisible by 7. Show that this is generally not true without the condition $(x, y) = 1$.

6 Solvability of Diophantine Equations

In the previous section we saw that solving Diophantine equations is, in most cases, no easy matter. Although we have learned several methods, there are many specific examples of Diophantine equations where none of these methods of solution can be successfully used. Nevertheless, it may be possible in these cases to obtain some information about the solutions. For example, we may be able to find an infinite set of solutions, which would show that the set of all solutions is infinite, even though we may not be able to determine them all. Or on the other hand, we may be able to show that the set of solutions is empty (which would actually solve the given equation), or that it is finite.

6.1   The Nonexistence of Solutions

To prove that a Diophantine equation has no solutions at all, it is often possible to successfully use congruences. Indeed, if the Diophantine equation $L = R$ has solutions (where L, R are expressions that contain the variables, and take on integer values for integer values of the variables), then also the congruence $L \equiv R \pmod{m}$ must have solutions for any $m \in \mathbb{N}$, since, for instance, the solutions of the original equation are also solutions of all these congruences. This means that if we find a natural number m such that the congruence $L \equiv R \pmod{m}$ has no solution, then the original equation $L = R$ cannot have a solution either. However, it is important to note that the converse is generally not true: If the congruence $L \equiv R$ $\pmod{m}$ has solutions for each natural number m, this does not yet imply that the Diophantine equation $L = R$ has solutions (we will show this in 6.2.(v)).

6.2   Examples

(i) Solve the Diophantine equation

$$x_1^4 + x_2^4 + \cdots + x_{14}^4 = 15999.$$

SOLUTION.   We show that the congruence

$$x_1^4 + x_2^4 + \cdots + x_{14}^4 \equiv 15999 \pmod{16}$$

has no solution, which would mean that the given equation is also not solvable. Indeed, if an integer n is even, then $n = 2k$ for $k \in \mathbb{Z}$, and thus $n^4 = 16k^4 \equiv 0 \pmod{16}$. If n is odd, then

$$n^4 - 1 = (n - 1)(n + 1)(n^2 + 1) \equiv 0 \pmod{16},$$

since the numbers $n - 1$, $n + 1$ and $n^2 + 1$ are even and one of the integers $n - 1$, $n + 1$ must even be divisible by 4. This means that n^4 is congruent to 0 modulo 16 for even n, and congruent to 1 modulo 16 for odd n. Therefore, if exactly r of the numbers $x_1, x_2, \ldots, x_{14}$ are odd, then

$$x_1^4 + x_2^4 + \cdots + x_{14}^4 \equiv r \pmod{16}.$$

Now $15999 = 16000 - 1 \equiv 15 \pmod{16}$, and since $0 \leq r \leq 14$, the congruence

$$x_1^4 + x_2^4 + \cdots + x_{14}^4 \equiv 15 \pmod{16}$$

cannot have a solution, and thus neither can the given equation be solvable.   □

(ii) Find integer solutions of the system of equations

$$x^2 + 2y^2 = z^2,$$
$$2x^2 + y^2 = u^2.$$

SOLUTION.   It is easy to see that from $x = y = 0$ it follows that $z = u = 0$, and this is a solution of the given system. We will now show that there are no other solutions. We assume that x, y, z, u is a solution and that $x \neq 0$ or $y \neq 0$, and we denote by $d = (x, y) > 0$ the greatest common divisor of x and y. From the first equation it follows that $d \mid z$, and from the second one that $d \mid u$. If we set $x_1 = x/d$, $y_1 = y/d$, $z_1 = z/d$, and $u_1 = u/d$, then it follows from 1.14.(iii) that $(x_1, y_1) = 1$, and upon dividing the two original equations by d^2 we obtain

$$x_1^2 + 2y_1^2 = z_1^2,$$
$$2x_1^2 + y_1^2 = u_1^2.$$

By adding, we get $3x_1^2 + 3y_1^2 = z_1^2 + u_1^2$, and thus $3 \mid z_1^2 + u_1^2$. By 3.14.(vi) we have $3 \mid z_1$, $3 \mid u_1$ and thus $9 \mid z_1^2 + u_1^2$. But then $9 \mid 3(x_1^2 + y_1^2)$, and thus $3 \mid x_1^2 + y_1^2$. Again by 3.4.(vi) we have $3 \mid x_1$, $3 \mid y_1$, which is a contradiction to $(x_1, y_1) = 1$. The system has therefore the unique solution $x = y = z = u = 0$.   □

(iii) Find all positive integer solutions of the equation

$$1! + 2! + 3! + \cdots + x! = y^2.$$

SOLUTION.   By direct computation we convince ourselves that for $x < 5$ the equation is satisfied only by $x = y = 1$ and $x = y = 3$. We will now show that the equation has no solution for $x \geq 5$. Since $n!$ is divisible by 5 for any $n \geq 5$, we have

$$1! + 2! + 3! + \cdots + x! \equiv 1! + 2! + 3! + 4! = 33 \equiv 3 \quad (\text{mod } 5).$$

Now, the square of a natural number is always congruent to 0, 1, or 4 modulo 5. Hence the congruence $1! + 2! + \cdots + x! \equiv y^2$ (mod 5) has no solution for $x \geq 5$, which means that the given equation also has no solution for $x \geq 5$.   □

(iv) Find all positive integer solutions of the equation

$$x^2 - y^3 = 7.$$

SOLUTION.   We will show that the given equation has no solutions. Let us assume to the contrary that for appropriate $x, y \in \mathbb{Z}$ we have $x^2 - y^3 = 7$. If y were even, we would have $x^2 \equiv 7$ (mod 8), which is impossible. Hence y is odd, $y = 2k + 1$ for $k \in \mathbb{Z}$. Then

$$x^2 + 1 = y^3 + 2^3 = (y + 2)(y^2 - 2y + 4)$$
$$= (y + 2)((y - 1)^2 + 3) = (2k + 3)(4k^2 + 3).$$

The number $4k^2 + 3$ has to be divisible by some prime $p \equiv 3$ (mod 4). Otherwise, since $4k^2 + 3$ is odd, its prime decomposition would contain only primes congruent to 1 modulo 4; the product of these primes, namely $4k^2 + 3$, would then also be congruent to 1 modulo 4, which is clearly not the case. Therefore, $4k^2 + 3$ is divisible by a prime $p \equiv 3$ (mod 4), and thus

$$x^2 + 1 \equiv 0 \quad (\text{mod } p).$$

By 3.14.(vi) this implies $x \equiv 1 \equiv 0$ (mod p), which is a contradiction.   □

(v) Show that the congruence

$$6x^2 + 5x + 1 \equiv 0 \quad (\text{mod } m)$$

has solutions for every natural number m, while the Diophantine equation

$$6x^2 + 5x + 1 = 0$$

has no solution.

SOLUTION.   Since $6x^2+5x+1 = (3x+1)(2x+1)$, the equation $6x^2+5x+1 = 0$ has no integer solution. Now let m be any natural number, $m = 2^n \cdot k$, where $n \in \mathbf{N}_0$ and k is an odd number. Since $(3, 2^n) = (2, k) = 1$, both congruences of the system

$$3x \equiv -1 \pmod{2^n},$$
$$2x \equiv -1 \pmod{k}$$

have solutions by Theorem 4.2, and since $(2^n, k) = 1$, by Theorem 4.6 the whole system has solutions. Hence for any x satisfying this system, $3x + 1$ is divisible by 2^n and $2x + 1$ is divisible by k, and therefore the product $(3x + 1)(2x + 1)$ is divisible by $2^n \cdot k = m$. Thus x solves the congruence

$$6x^2 + 5x + 1 \equiv 0 \pmod{m}. \qquad \square$$

6.3 Exercises

Solve the Diophantine equations (i)–(v):

(i) $3x^2 - 4y^2 = 13$. **(ii)** $2x^2 - 5y^2 = 7$.

(iii) $3x^2 + 8 = y^2$. ***(iv)** $3x^2 + 3x + 7 = y^3$.

(v) $(x + 1)^3 + (x + 2)^3 + (x + 3)^3 + (x + 4)^3 = (x + 5)^3$.

Solve the equations (vi)–(viii) in natural numbers:

(vi) $2^x + 7^y = 19^z$. ***(vii)** $2^x + 5^y = 19^z$.

*(viii) $1! + 2! + 3! + \cdots + x! = y^{z+1}$.

(ix) Show that for any given odd numbers r, s, the Diophantine equation $x^{10} + rx^7 + s = 0$ has no solutions.

(x) Given a prime $p \equiv 3 \pmod 4$, show that $x^2 + y^2 = pz^2$ has no solution in natural numbers.

Find all integer solutions of the systems of equations (xi), (xii), and of the equations (xiii), (xiv):

(xi) $x^2 + 5y^2 = z^2$,
$5x^2 + y^2 = u^2$.

(xii) $x^2 + 6y^2 = z^2$,
$6x^2 + y^2 = u^2$.

*(xiii) $x^2 - y^3 + 1 = (4z + 2)^3$. ***(xiv)** $x^2 + 5 = y^3$.

*(xv) Show that the equation $4xy - x - y = z^2$ has no solution in natural numbers.

6.4 Reductio ad Absurdum

The method of *reductio ad absurdum* is a method for proving the nonexistence of solutions of a Diophantine equation. To carry out this method of proof, we characterize a supposed solution of the given Diophantine equation by some natural number (for instance, the greatest common divisor of the values of some variables, or the square of the value of some variable, or something like that) and show that if there exists a solution characterized by the natural number d, then there must exist another solution characterized by the natural number $d' < d$. But then no solution can exist, of which we can easily convince ourselves by contradiction: If there were a solution, we could choose one that was characterized by the smallest natural number d; then, of course, there would exist another solution characterized by a natural number $d' < d$, which, however, would be a contradiction to the choice of d. (We note that we could solve, for instance, Example 6.2.(ii) in this way if d denoted the greatest common divisor of the values of the variables x, y, z, u, or the exponent of 3 in the prime decomposition of this greatest common divisor.)

6.5   Examples

(i) Solve the Diophantine equation $x^3 + 2y^3 + 4z^3 - 6xyz = 0$.

SOLUTION.   The equation is certainly satisfied by $x = y = z = 0$. We will show that there is no other solution. We set $d = x^2 + y^2 + z^2$ and assume that for some solution x, y, z of the given equation we have $d > 0$. It follows from the original equation that x^3 is an even number, and thus $x = 2x_1$ for an appropriate $x_1 \in \mathbb{Z}$. Substituting this into the equation, we get

$$8x_1^3 + 2y^3 + 4z^3 - 12x_1yz = 0,$$

and upon dividing by 2,

$$4x_1^3 + y^3 + 2z^3 - 6x_1yz = 0,$$

so y^3 is also an even number, and we write $y = 2y_1$, with an appropriate $y_1 \in \mathbb{Z}$. Substituting and dividing by 2, we get

$$2x_1^3 + 4y_1^3 + z^3 - 6x_1y_1z = 0,$$

which implies that z^3 is also even, and thus $z = 2z_1$ for some $z_1 \in \mathbb{Z}$. Substituting and dividing by 2 once more, we obtain

$$x_1^3 + 2y_1^3 + 4z_1^3 - 6x_1y_1z_1 = 0,$$

so x_1, y_1, z_1 is a solution of the original Diophantine equation, and furthermore

$$x_1^2 + y_1^2 + z_1^2 = \frac{x^2}{4} + \frac{y^2}{4} + \frac{z^2}{4} = \frac{d}{4} < d.$$

According to the method described in 6.4, the given equation does not have any solution with the property $d > 0$, and thus $x = y = z = 0$ is its unique solution.   □

(ii) Solve the equation $x^2 + y^2 = 4^z$ in natural numbers.

SOLUTION. We use the method of 6.4 with $d = z$. Let us first assume that x, y, z is a solution of the given equation. Then certainly $z \neq 1$, since for $x = y = 1$ we have $x^2 + y^2 = 2 < 4$, and if at least one of the numbers x, y is greater than 1, then $x^2 + y^2 > 4$. Thus $z > 1$, and we have $x^2 + y^2 = 4^z \equiv 0$ (mod 8). Since the square of an odd number is congruent to 1 modulo 8, and the square of an even number is congruent to 0 or 4 modulo 8, it follows from this congruence that both x and y are even, i.e., $x = 2x_1$, $y = 2y_1$ for appropriate $x_1, y_1 \in \mathbb{N}$. But then

$$x_1^2 + y_1^2 = \frac{x^2}{4} + \frac{y^2}{4} = 4^{z-1},$$

and thus, if we set $z_1 = z - 1 \in \mathbb{N}$, the numbers x_1, y_1, z_1 satisfy the given equation, with $z_1 < z$. Therefore, the equation has no solution. □

(iii) Solve the Diophantine equation $x^4 + y^4 + z^4 = 9u^4$.

SOLUTION. If $u = 0$, then necessarily $x = y = z = 0$, which is a solution of the given equation. We will show that there are no other solutions. Let us assume that the integers x, y, z, u satisfy the given equation and that $u \neq 0$; we set $d = u^4$. If the number u were not divisible by 5, then Fermat's theorem gives $u^4 \equiv 1$ (mod 5), and we would have

$$x^4 + y^4 + z^4 \equiv 4 \pmod 5.$$

This, however, is impossible since by Fermat's theorem the numbers x^4, y^4, z^4 are congruent to 0 or 1 modulo 5. Thus, u is divisible by 5, i.e., $u = 5u_1$ for an appropriate $u_1 \in \mathbb{Z}$, and we get

$$x^4 + y^4 + z^4 \equiv 0 \pmod 5,$$

which implies that x, y, z are divisible by 5, i.e., $x = 5x_1$, $y = 5y_1$, $z = 5z_1$ for appropriate $x_1, y_1, z_1 \in \mathbb{Z}$. Substituting this into the original equation and dividing by 5^4, we obtain

$$x_1^4 + y_1^4 + z_1^4 = 9u_1^4,$$

and thus x_1, y_1, z_1, u_1 satisfy the given equation, and

$$u_1^4 = \frac{u^4}{5^4} < u^4 = d. \qquad\qquad □$$

(iv) Solve the equation $x^2 - y^2 = 2xyz$ in natural numbers.

SOLUTION. We assume that some natural numbers x, y, z satisfy the given equation, and we set $d = xy$. If we let $d = 1$, then $x = y = 1$ and the equation would give $z = 0$, which is impossible. Hence $d > 1$. Let p be some prime dividing d. Since

$$(x + y)(x - y) = x^2 - y^2 = 2xyz \equiv 0 \pmod{p},$$

we have $x \equiv y \pmod{p}$ or $x \equiv -y \pmod{p}$. In view of the fact that the prime p divides the product xy, either x or y is congruent to 0 modulo p, and together $x \equiv y \equiv 0 \pmod{p}$. Hence $x_1 = x/p$ and $y_1 = y/p$ are natural numbers, and

$$(px_1)^2 - (py_1)^2 = 2(px_1)(py_1)z,$$

from which, upon dividing by p^2, we see that x_1, y_1, z satisfy the given equation, and that

$$x_1 y_1 = \frac{x}{p} \cdot \frac{y}{p} = \frac{d}{p^2} < d.$$

By the principle of 6.4 the given equation has no solution.   □

(v) Solve the Diophantine equation $x^2 + y^2 + z^2 = 2xyz$.

SOLUTION.   The equation is certainly satisfied by $x = y = z = 0$. We will show that it has no further solutions. In fact, we will prove the following stronger assertion: Given any $u \in \mathbb{N}$, the equation

$$x^2 + y^2 + z^2 = 2^u xyz, \tag{33}$$

where $x, y, z \in \mathbb{Z}$, has no solution other than $x = y = z = 0$. Let us assume that $x, y, z \in \mathbb{Z}$ and $u \in \mathbb{N}$ satisfy (33), and that $d = x^2 + y^2 + z^2 > 0$. Since $u \geq 1$, $2^u xyz$ is an even number, and thus $x^2 + y^2 + z^2$ is even. But this means that either exactly one of the numbers x, y, z or all three of them are even. However, in the first case we have

$$x^2 + y^2 + z^2 \equiv 1 + 1 + 0 = 2 \pmod{4},$$

while

$$2^u xyz \equiv 0 \pmod{4},$$

since $u \geq 1$ and one of the numbers x, y, z is even. Hence we consider the second case, where the numbers $x_1 = x/2$, $y_1 = y/2$, $z_1 = z/2$ are integers. We set $u_1 = u + 1$ and substitute everything into (33):

$$4x_1^2 + 4y_1^2 + 4z_1^2 = 2^{u_1 - 1} \cdot 2x_1 \cdot 2y_1 \cdot 2z_1,$$

and upon dividing by 4,

$$x_1^2 + y_1^2 + z_1^2 = 2^{u_1} \cdot x_1 y_1 z_1,$$

so that x_1, y_1, z_1, u_1 satisfy (33). In addition, $0 < x_1^2 + y_1^2 + z_1^2 = d/4 < d$, since $d > 0$. By 6.4, the equation (33) can only have a solution with the property $d = 0$, which, as we have already seen, leads to the solutions $x = y = z = 0$, with arbitrary $u \in \mathbb{N}$. In particular, the given equation has the unique solution $x = y = z = 0$.   □

6.6   Exercises

Solve the Diophantine equations (i)–(viii):

(i) $x^3 - 2y^3 - 4z^3 = 0$.   (ii) $x^4 + 4y^4 = 2(z^4 + 4u^4)$.

(iii) $x^2 + y^2 + z^2 = 7u^2$.   (iv) $x^3 + 3y^3 + 9z^3 = 9xyz$.

*(v) $x^2 + y^2 + z^2 = x^2y^2u^2$.   *(vi) $x^2 + y^2 = 11^{2z}$.

*(vii) $x^2 + y^2 + z^2 + u^2 = 2xyzu$.   (viii) $2x^2 + y^2 = 7z^2$.

We have now studied some of the most common approaches used for proving that a given Diophantine equation has no solution. We do not claim completeness of this list of methods. Indeed, even though there is no universal method for solving Diophantine equations, one can find in the mathematical literature a large number of methods; most of them, however, are useful only in very special cases. We conclude this overview of nonexistence proofs of solutions of Diophantine equations by presenting one of these special instances.

6.7   Example

Given an arbitrary prime p, solve the Diophantine equation

$$x^4 + 4^x = p.$$

SOLUTION.   For any integer $x < 0$, 4^x is not an integer, and neither is $x^4 + 4^x$. Thus for any prime p, the given equation has no negative solution. For $x = 0$ we have $0^4 + 4^0 = 1$, which is not a prime, and for $x = 1$ we have $1^4 + 4^1 = 5$, a prime. We will now show that for any integer $x \geq 2$, the number $x^4 + 4^x$ is composite. If $x = 2k$ is even, where $k \in \mathbb{N}$, then

$$x^4 + 4^x = 2^4k^4 + 4^{2k} = 16(k^4 + 4^{2(k-1)}),$$

which is a composite number. If $x = 2k + 1$ ($k \in \mathbb{N}$) is odd, then

$$\begin{aligned}
x^4 + 4^x &= x^4 + 4 \cdot 4^{2k} = [x^4 + 4x^2(2^k)^2 + 4(2^k)^4] - 4x^2(2^k)^2 \\
&= [x^2 + 2(2^k)^2]^2 - (2x \cdot 2^k)^2 \\
&= [x^2 + 2x \cdot 2^k + 2(2^k)^2][x^2 - 2x \cdot 2^k + 2(2^k)^2] \\
&= [(x + 2^k)^2 + 2^{2k}][(x - 2^k)^2 + 2^{2k}],
\end{aligned}$$

which is again composite, since $(x \pm 2^k)^2 + 2^{2k} \geq 2^{2k} \geq 2^2 > 1$. To summarize, for $p = 5$ the given equation has the unique solution $x = 1$, while there are no solutions for any prime $p \neq 5$.   □

6.8   The Cardinality of the Set of Solutions

In many cases when we are unable to find all solutions of a Diophantine equation, we may at least succeed in determining whether there are finitely or infinitely many solutions. Finiteness can, for instance, be established by showing that the values of the variables that yield solutions are in absolute value less than a certain number. If we can find this number and it is "reasonably small," we can then find all solutions with the method described in 5.11.

On the other hand, the infinitude of solutions of a given Diophantine equation can, for instance, be established as follows. For each variable we find an expression involving a parameter, in such a way that upon substituting, the equation becomes an equality, and that for infinitely many values of the parameter the variables have different values. Or we can find one solution of the equation and give a description of how to obtain a new solution from any known one. If we can guarantee that upon repeated application of this rule we always obtain different solutions (for instance, if the solutions thus obtained get larger and larger), then we have again proved that the set of solutions is infinite. It is clear that with both approaches there may exist more solutions than the ones detected.

6.9 Examples

(i) Show that the Diophantine equation $4xy - x - y = z^2$ has infinitely many solutions (it can be shown that this equation has no solution in positive integers; see 6.3.(xv)).

SOLUTION. If we choose $x = -1$, the equation becomes significantly simpler:

$$-5y = z^2 - 1,$$

and this is an equation that we know how to solve by the method described in 5.5. Hence we first solve the congruence

$$z^2 - 1 \equiv 0 \pmod{5},$$

which is satisfied by exactly those integers z that are congruent to 1 or -1 modulo 5. Thus $z = \pm 1 + 5t$, with an arbitrary $t \in \mathbb{Z}$, and we get

$$-5y = z^2 - 1 = 1 \pm 10t + 25t^2 - 1,$$

and therefore

$$y = \mp 2t - 5t^2.$$

By substituting we convince ourselves that the values $x = -1$, $y = 2t - 5t^2$, and $z = -1 + 5t$ indeed satisfy the given equation. It is clear that different values of t give different values of z; hence the above solutions are indeed infinite in number.   □

(ii) Show that the Diophantine equation

$$(x - 1)^2 + (x + 1)^2 = y^2 + 1$$

has infinitely many solutions.

SOLUTION. The equation can easily be rewritten as

$$y^2 - 2x^2 = 1.$$

We try to find some solution; it is easy see that the choice $y = 3$, $x = 2$ satisfies the given equation. Now we assume that we have an arbitrary solution $x, y \in \mathbb{Z}$, and we try to find another one. We have

$$(y + \sqrt{2}x)(y - \sqrt{2}x) = 1.$$

Substituting the values $y = 3$ and $x = 2$, we obtain the equality $(3 + 2\sqrt{2})$ $\cdot (3 - 2\sqrt{2}) = 1$, and by multiplying we get

$$[(y + \sqrt{2}x)(3 + 2\sqrt{2})] \cdot [(y - \sqrt{2}x)(3 - 2\sqrt{2})] = 1.$$

Manipulating both expressions in square brackets, we get

$$\begin{aligned}
(y + \sqrt{2}x)(3 + 2\sqrt{2}) &= 3y + 3\sqrt{2}x + 2\sqrt{2}y + 4x \\
&= (4x + 3y) + (3x + 2y)\sqrt{2}, \\
(y - \sqrt{2}x)(3 - 2\sqrt{2}) &= 3y - 3\sqrt{2}x - 2\sqrt{2}y + 4x \\
&= (4x + 3y) - (3x + 2y)\sqrt{2}.
\end{aligned}$$

If we set $u = 4x + 3y$, $v = 3x + 2y$, then

$$(u + \sqrt{2}v)(u - \sqrt{2}v) = 1,$$

which means that

$$u^2 - 2v^2 = 1,$$

and thus $u, v \in \mathbb{Z}$ is a new solution of the given equation. If we set $x_1 = 2$, $y_1 = 3$ and

$$x_{n+1} = 3x_n + 2y_n, \qquad y_{n+1} = 4x_n + 3y_n$$

for any $n \in \mathbb{N}$, then for all $n \in \mathbb{N}$ we have a solution x_n, y_n of the given equation. Furthermore, since $0 < x_1 < x_2 < \cdots$, $0 < y_1 < y_2 < \cdots$, different indices n give different solutions x_n, y_n. The given equation has therefore infinitely many solutions. $\qquad \square$

(iii) Show that for any integer k the equation

$$k + x^2 + y^2 = z^2$$

has infinitely many solutions in natural numbers.

SOLUTION. Factoring $z^2 - y^2$, we rewrite the given equation as

$$k + x^2 = (z - y)(z + y).$$

It is not necessary to find all solutions, and hence we may assume that

$$z - y = 1,$$
$$z + y = k + x^2.$$

Any solution of this system will also be a solution of the given equation (the converse, however, is not true; the reader is encouraged to find, for some fixed k, an example of natural numbers x, y, z that satisfy the given equation but do not satisfy the above system of equations). If we solve the system for z and y, we get

$$z = \frac{1}{2}(x^2 + k + 1),$$

$$y = \frac{1}{2}(x^2 + k - 1).$$

If we choose $x = |k| + 1 + 2t$, where $t \in \mathbb{N}$, then x is a natural number. Then

$$x^2 + k \equiv k + 1 + 2t + k \equiv 1 \pmod 2,$$

and thus $z = \frac{1}{2}((|k| + 1 + 2t)^2 + k + 1) > 0$, $y = \frac{1}{2}((|k| + 1 + 2t)^2 + k - 1) > 0$ are also natural numbers. Since for different t we obtain different x and thus different solutions, the equation has infinitely many solutions.   □

(iv) Show that for any natural number k the Diophantine equation

$$5x^2 - 8xy + 5y^2 - 4k^2 = 0$$

has only finitely many solutions.

SOLUTION.   We rewrite the given equation as $(2x - y)^2 + (2y - x)^2 = 4k^2$, which implies $(2x - y)^2 \le (2k)^2$ and $(2y - x)^2 \le (2k)^2$, and thus $-2k \le 2x - y \le 2k$ and $-2k \le 2y - x \le 2k$. If we add the first and twice the second equation, we get $-2k \le y \le 2k$, and in complete analogy $-2k \le x \le 2k$. Since x and y can take on only finitely many values for a fixed k, the given equation has only finitely many solutions.   □

6.10   Exercises

Show that the Diophantine equations (i)–(iv) have infinitely many solutions:

(i) $z^3 + 5xz^2 + 3y^2z + 2y^3 = 9$.   *(ii) $x^2 + (x + 1)^2 = y^2$.
*(iii) $(x + 1)^3 - x^3 = y^2$.   *(iv) $x^2 + y^2 + z^2 = 3xyz$.

(v) Show that for any integer n the Diophantine equation

$$x^2 + y^2 - z^2 = n$$

has infinitely many solutions.

(vi) Show that for any integer n the Diophantine equation

$$x^2 + y^2 + z^2 + n^2 = n^6 - n(x + y + z)$$

has only finitely many solutions.

6.11 Examples

We devote the remainder of this section to equations with an undetermined number of variables; as a rule, we will solve them by induction.

(i) Show that for any natural number s and any rational number w the equation

$$\frac{1}{x_1} + \frac{1}{x_2} + \cdots + \frac{1}{x_s} = w$$

has a finite number of solutions in natural numbers.

SOLUTION.   We use mathematical induction. For $s = 1$ the equation has at most one solution, and thus the assertion is true for $s = 1$. We now assume that the assertion holds for $s \in \mathbb{N}$, and we will prove it for $s+1$. We assume that the natural numbers $x_1, \ldots, x_s, x_{s+1}$, with $x_1 \leq x_2 \leq \cdots \leq x_s \leq x_{s+1}$, satisfy the equation

$$\frac{1}{x_1} + \frac{1}{x_2} + \cdots + \frac{1}{x_s} + \frac{1}{x_{s+1}} = u, \tag{34}$$

where u is a given rational number, which is obviously positive. But then

$$u = \frac{1}{x_1} + \frac{1}{x_2} + \cdots + \frac{1}{x_{s+1}} \leq \frac{1}{x_1} + \frac{1}{x_1} + \cdots + \frac{1}{x_1} = \frac{s+1}{x_1},$$

and thus $x_1 \leq (s+1)/u$. The number x_1 can therefore take on only finitely many values. If we choose one of these values, then the $x_2, \ldots, x_{s+1}$ satisfy

$$\frac{1}{x_2} + \cdots + \frac{1}{x_{s+1}} = u - \frac{1}{x_1},$$

where for a fixed x_1 we have a fixed rational number on the right-hand side, and therefore by the induction hypothesis this equation has only finitely many solutions $x_2, \ldots, x_{s+1}$. Since x_1 can take on only finitely many values, the equation (34) can have only finitely many solutions such that $x_1 \leq x_2 \leq \cdots \leq x_{s+1}$. Finally, since an arbitrary solution can be obtained by rearranging a solution with this property, equation (34) can have only finitely many solutions.   □

(ii) Show that for any natural number $s > 1$ the equation

$$\frac{1}{x_0^2} = \frac{1}{x_1^2} + \frac{1}{x_2^2} + \cdots + \frac{1}{x_s^2}$$

has infinitely many solutions in natural numbers such that $x_0 < x_1 < \cdots < x_s$.

SOLUTION.   We convince ourselves that if $x_0, x_1, \ldots, x_s$ is a solution of the given equation with the required condition, then $tx_0, tx_1, \ldots, tx_s$ is also a solution for any natural number t, and we have $tx_0 < tx_1 < \cdots < tx_s$. Therefore, it is sufficient if we can show that for any natural number $s > 1$ the given equation has *at least one* solution $x_0, x_1, \ldots, x_s$ such that $x_0 < x_1 < \cdots < x_s$. To do this, we use induction on s.

If $s = 2$, then $x_0 = 12$, $x_1 = 15$, $x_2 = 20$ is a solution, since it is easy to verify that

$$\frac{1}{12^2} = \frac{1}{15^2} + \frac{1}{20^2}.$$

We now assume that the assertion holds for some $s \geq 2$, i.e., there exist natural numbers $x_0 < x_1 < \cdots < x_s$ such that

$$\frac{1}{x_0^2} = \frac{1}{x_1^2} + \frac{1}{x_2^2} + \cdots + \frac{1}{x_s^2}.$$

We set $y_0 = 12x_0$, $y_1 = 15x_0$, and $y_i = 20x_{i-1}$ for $i = 2, 3, \ldots, s+1$. It is easy to see that $y_0 < y_1 < \cdots < y_{s+1}$. Furthermore, we have

$$\frac{1}{y_0^2} = \frac{1}{x_0^2} \cdot \frac{1}{12^2} = \frac{1}{x_0^2} \left(\frac{1}{15^2} + \frac{1}{20^2} \right) = \frac{1}{15^2} \cdot \frac{1}{x_0^2} + \frac{1}{20^2} \left(\frac{1}{x_1^2} + \cdots + \frac{1}{x_s^2} \right)$$
$$= \frac{1}{y_1^2} + \frac{1}{y_2^2} + \cdots + \frac{1}{y_s^2} + \frac{1}{y_{s+1}^2}.$$

This completes the proof by induction. $\square$

6.12   Exercises

(i) Show that for any natural number s the equation

$$x_0^2 = x_1^2 + \cdots + x_s^2$$

has infinitely many solutions in natural numbers.

(ii) Show that for any rational number w and any natural number s the equation

$$\frac{x_1}{1+x_1^2} + \frac{x_2}{1+x_2^2} + \cdots + \frac{x_s}{1+x_s^2} = w$$

has only finitely many solutions in natural numbers.

*(iii) Show that for any integer $s > 2$ the equation

$$\frac{1}{x_0^3} = \frac{1}{x_1^3} + \cdots + \frac{1}{x_s^3}$$

has infinitely many solutions in natural numbers.

7 Integer Part and Fractional Part

The integer part and the fractional part of a real number are encountered in many different situations in number theory. Therefore, we devote an entire section to these concepts; we will summarize some of their properties and uses, and solve several equations in which they occur.

7.1 Integer and Fractional Parts

Definition. The greatest integer not exceeding the real number x is called the *integer part* of x, and is denoted by $[x]$. This means that $[x]$ is the (unique) integer satisfying the inequalities

$$[x] \leq x < [x] + 1. \tag{35}$$

The difference $x - [x]$ is called the *fractional part* of x and is denoted by $\langle x \rangle$.

We note that both symbols could also be defined in a different way, for instance by the conditions

$$[x] + \langle x \rangle = x, \quad [x] \in \mathbb{Z}, \quad 0 \leq \langle x \rangle < 1;$$

these, as the reader can easily verify, are equivalent to Definition 7.1. It also follows from these conditions that for arbitrary real numbers x, y such that $x \leq y$, we have $[x] \leq [y]$, but not $\langle x \rangle \leq \langle y \rangle$. If, however, $[x] = [y]$, then $\langle x \rangle \leq \langle y \rangle$ if and only if $x \leq y$.

7.2 Examples

Show that (i)–(v) hold for any $x, y \in \mathbb{R}$, $m \in \mathbb{Z}$, and $n \in \mathbb{N}$:

(i) $[x + m] = [x] + m$; $\langle x + m \rangle = \langle x \rangle$.

SOLUTION.   From (35) it follows that

$$[x] + m \leq x + m < ([x] + m) + 1.$$

Since $[x] + m \in \mathbb{Z}$, this means, by (35), that

$$[x + m] = [x] + m,$$

which is the first of the desired equalities. Since

$$\langle x + m \rangle = x + m - [x + m] = x + m - [x] - m = x - [x] = \langle x \rangle,$$

the second one follows as a consequence.   □

(ii) $\frac{m+1}{n} \leq [\frac{m}{n}] + 1.$

SOLUTION. From the inequality

$$\frac{m}{n} < \left[\frac{m}{n}\right] + 1$$

(see (35), with $x = m/n$) we obtain upon multiplying by n,

$$m < n \cdot \left[\frac{m}{n}\right] + n.$$

This last inequality is between integers; hence

$$m + 1 \leq n \cdot \left[\frac{m}{n}\right] + n,$$

which, upon dividing by n, gives the desired inequality.   □

(iii) $[x] + [y] \leq [x + y] \leq [x] + [y] + 1$.

SOLUTION. Since $[x] + y \leq x + y$, by 7.1 we have $[[x] + y] \leq [x + y]$, and thus it follows from (i) that

$$[x] + [y] = [[x] + y] \leq [x + y] \leq x + y.$$

Since by (35) we have $x < [x] + 1$ and $y < [y] + 1$, then $x + y < [x] + [y] + 2$, and therefore

$$[x] + [y] \leq [x + y] < [x] + [y] + 2.$$

Now the desired inequality follows from the fact than on both sides of the last, strict, inequality we have integers.   □

(iv) If $n \in \mathbb{N}$ and $x \geq n$, then $\left[\frac{x}{n}\right]$ is the number of positive integers divisible by n and not exceeding x.

SOLUTION. By (35) we have

$$\left[\frac{x}{n}\right] \leq \frac{x}{n} < \left[\frac{x}{n}\right] + 1,$$

and multiplying by n we obtain

$$\left[\frac{x}{n}\right] \cdot n \leq x < \left(\left[\frac{x}{n}\right] + 1\right) n,$$

and thus the $\left[\frac{x}{n}\right]$th multiple of n does not yet exceed the number x, while the following $\left(\left[\frac{x}{n}\right] + 1\right)$th multiple is already larger than x. The multiples of n that do not exceed x are therefore exactly the numbers

$$n, 2n, \ldots, \left(\left[\frac{x}{n}\right] - 1\right) \cdot n, \left[\frac{x}{n}\right] \cdot n,$$

namely $\left[\frac{x}{n}\right]$ of them, which was to be shown.   □

(v)
$$[x] + \left[x + \frac{1}{n}\right] + \left[x + \frac{2}{n}\right] + \cdots + \left[x + \frac{n-1}{n}\right] = [nx],$$

$$\langle x \rangle + \left\langle x + \frac{1}{n}\right\rangle + \left\langle x + \frac{2}{n}\right\rangle + \cdots + \left\langle x + \frac{n-1}{n}\right\rangle = \langle nx \rangle + \frac{n-1}{2}.$$

SOLUTION.   We begin with the first identity. If we set $[x] = k$, $\langle x \rangle = r$, then we have $x = k + r$, and for any $i = 0, 1, \ldots, n-1$ by (i),

$$\left[x + \frac{i}{n}\right] = \left[k + r + \frac{i}{n}\right] = k + \left[r + \frac{i}{n}\right].$$

In view of the fact that $0 \le r < 1$ and $0 \le \frac{i}{n} < 1$, the second summand takes on only the values 0 or 1. It is equal to 0 in the case where $r + \frac{i}{n} < 1$; on the other hand, if $1 \le r + \frac{i}{n} < 2$, then $\left[r + \frac{i}{n}\right] = 1$. Hence

$$[x] + \left[x + \frac{1}{n}\right] + \cdots + \left[x + \frac{n-1}{n}\right]$$
$$= nk + \left([r] + \left[r + \frac{1}{n}\right] + \cdots + \left[r + \frac{n-1}{n}\right]\right).$$

It is clear that for $i = 0, 1, \ldots, n-1$ we have $r + \frac{i}{n} \ge 1$ if and only if $-i \le nr - n$, which occurs when $-i \in \{-(n-1), -(n-2), \ldots, [nr-n]\}$. This set has exactly $[nr - n] + (n-1) + 1 = [nr - n] + n$ elements, and thus the sum within parentheses satisfies

$$[r] + \left[r + \frac{1}{n}\right] + \cdots + \left[r + \frac{n-1}{n}\right] = [nr - n] + n = [nr - n + n] = [nr],$$

where we have used (i). Altogether, it follows that

$$[x] + \left[x + \frac{1}{n}\right] + \cdots + \left[x + \frac{n-1}{n}\right] = nk + [nr] = [nk + nr]$$
$$= [n(k + r)] = [nx]$$

(where (i) has been used again), and this proves the first identity. The second one follows as a consequence, since

$$\langle x \rangle + \left\langle x + \frac{1}{n}\right\rangle + \cdots + \left\langle x + \frac{n-1}{n}\right\rangle$$
$$= x - [x] + \left(x + \frac{1}{n}\right) - \left[x + \frac{1}{n}\right] + \cdots + \left(x + \frac{n-1}{n}\right) - \left[x + \frac{n-1}{n}\right]$$
$$= nx + \frac{1}{n} + \cdots + \frac{n-1}{n} - [nx]$$

by the first identity. Since $1 + 2 + \cdots + (n-1) = n(n-1)/2$, we have

$$\langle x \rangle + \left\langle x + \frac{1}{n}\right\rangle + \cdots + \left\langle x + \frac{n-1}{n}\right\rangle = nx - [nx] + \frac{n-1}{2} = \langle nx \rangle + \frac{n-1}{2},$$

which completes the proof of the second identity. $\square$

7.3 Exercises

(i) How many integers m with $10^6 < m < 10^7$ are divisible by 786?

(ii) How many natural numbers less than 1000 are divisible neither by 5 nor by 7?

(iii) How many natural numbers less than 100 are relatively prime to 36?

(iv) For how many natural numbers $m \leq 1000$ is $v_5(m) = 1$ (see 2.9)?

Prove (v)–(xxi) for any $x, y \in \mathbb{R}$ and $m, n \in \mathbb{N}$:

(v) $\left[\dfrac{[x]}{n}\right] = \left[\dfrac{x}{n}\right].$

(vi) $\left[\dfrac{n}{2}\right] - \left[-\dfrac{n}{2}\right] = n.$

(vii) the smallest integer not less than x is $-[-x]$.

*(viii) $[x + y] + [x] + [y] \leq [2x] + [2y]$.

*(ix) $[x] \cdot [y] \leq [xy] \leq [x] \cdot [y] + [x] + [y]$, if $x \geq 0$ and $y \geq 0$.

(x) If $x > y$, then $[x] - [y]$ is the number of integers r such that $y < r \leq x$.

(xi) $[x] + [-x]$ is 0 for $x \in \mathbb{Z}$ and -1 for $x \in \mathbb{R} \backslash \mathbb{Z}$.

(xii) $\langle x \rangle + \langle -x \rangle$ is 0 for $x \in \mathbb{Z}$ and 1 for $x \in \mathbb{R} \backslash \mathbb{Z}$.

(xiii) $0 \leq [2x] - 2[x] \leq 1$.

(xiv) $-1 \leq \langle 2x \rangle - 2\langle x \rangle \leq 0$.

(xv) $[\sqrt[n]{x}] = \left[\sqrt[n]{[x]}\right]$, if $x \geq 0$.

*(xvi) $[\sqrt{n} + \sqrt{n+1}] = [\sqrt{4n+2}]$.

(xvii) $\left[\dfrac{x}{n}\right] + \left[\dfrac{x+1}{n}\right] + \cdots + \left[\dfrac{x+n-1}{n}\right] = [x]$.

(xviii) $\left\langle \dfrac{x}{n} \right\rangle + \left\langle \dfrac{x+1}{n} \right\rangle + \cdots + \left\langle \dfrac{x+n-1}{n} \right\rangle = \langle x \rangle + \dfrac{n-1}{2}$.

*(xix) $[mx] + \left[mx + \dfrac{m}{n}\right] + \cdots + \left[mx + \dfrac{(n-1)m}{n}\right]$
$$= [nx] + \left[nx + \dfrac{n}{m}\right] + \cdots + \left[nx + \dfrac{(m-1)n}{m}\right].$$

*(xx) $\langle mx \rangle + \left\langle mx + \dfrac{m}{n} \right\rangle + \cdots + \left\langle mx + \dfrac{(n-1)m}{n} \right\rangle + \dfrac{m-n}{2}$
$$= \langle nx \rangle + \left\langle nx + \dfrac{n}{m} \right\rangle + \cdots + \left\langle nx + \dfrac{(n-1)n}{m} \right\rangle.$$

*(xxi) $[\tau^2 n] = [\tau[\tau n] + 1]$, where $\tau = (1 + \sqrt{5})/2$.

The following theorem illustrates one of the uses of the concept of integer part in number theory.

7.4   Prime Powers in a Factorial

Theorem. *For any prime p and natural number n the exponent in the power of p in the prime decomposition of $n!$ (see Section 2.9) is determined by*

$$v_p(n!) = \left[\frac{n}{p}\right] + \left[\frac{n}{p^2}\right] + \cdots + \left[\frac{n}{p^k}\right], \tag{36}$$

where k is any natural number such that $n < p^{k+1}$.

PROOF. By 7.2.(iv), among the integers $1, 2, \ldots, n$ there are exactly $[n/p]$ numbers divisible by p, among those exactly $[n/p^2]$ are divisible by p^2, of which again $[n/p^3]$ are divisible by p^3, etc. The largest power of p dividing $n!$ is therefore given by the formula (36), since from $n < p^{k+1}$ it follows that none of the numbers $1, 2, \ldots, n$ is divisible by p^{k+1}.   □

7.5   Exercises

   (i) Find out in how many zeros the number $\binom{125}{62}$ ends.
 *(ii) Find all $n \in \mathbb{N}$ for which $n!$ is divisible by 2^{n-1}.
*(iii) Show that for any natural numbers m, n the product $\binom{2m}{m}\binom{2n}{n}$ is divisible by $\binom{m+n}{n}$.
  (iv) Show that for infinitely many $n \in \mathbb{N}$ the number $x_n = \left[2^n \sqrt{2}\right]$ is even.
   (v) Let $m > 1$ be an integer. The sequence $(x_n)_{n=0}^\infty$ is determined by $x_0 = m$ and $x_n = \left[\frac{3}{2}x_{n-1}\right]$ for $n \in \mathbb{N}$. Show that this sequence contains infinitely many odd and infinitely many even numbers.
 *(vi) Define $f(n) = n + \left[\sqrt{n}\right]$ for natural numbers n. Show that for any $m \in \mathbb{N}$ the sequence $m,\, f(m),\, f(f(m)),\, f(f(f(m))),\, \ldots$ contains at least one square of a natural number.
*(vii) Suppose that several numbers are given, each of which is less than $2\,000$, and suppose that the least common multiple of any two of them is larger than $2\,000$. Show that the sum of the reciprocals of these numbers is less than 2.

7.6   Examples

We devote the remainder of this section to several equations that contain the integer part or the fractional part of a real number.

(i) Solve the equation $4[x] = 3x$ in real numbers.

SOLUTION. Since $[x] \in \mathbb{Z}$ for any real number x, we have

$$\frac{3}{4}x = [x] = m \in \mathbb{Z},$$

and thus $x = 4m/3$. But then

$$m = [x] = \left[\frac{4m}{3}\right] = \left[m + \frac{m}{3}\right] = m + \left[\frac{m}{3}\right],$$

which implies that $[m/3] = 0$. This condition, however, is satisfied by only three integers, 0, 1, 2, and so x can take on only the values $0, \frac{4}{3}, \frac{8}{3}$. Finally, we verify that these three values are indeed solutions of the given equation.   □

(ii) Solve the equation $[x^3] + [x^2] + [x] = \langle x \rangle - 1$ in real numbers.

SOLUTION.   We rewrite the equation as

$$\langle x \rangle = [x^3] + [x^2] + [x] + 1,$$

and see that $\langle x \rangle \in \mathbb{Z}$. Since $0 \le \langle x \rangle < 1$, we have $\langle x \rangle = 0$, i.e., $x = [x] \in \mathbb{Z}$. Then $[x^3] = x^3$, $[x^2] = x^2$, and it remains to solve the Diophantine equation

$$x^3 + x^2 + x + 1 = 0.$$

Factoring,

$$(x + 1)(x^2 + 1) = 0,$$

and since for any integer x we have $x^2 + 1 \ne 0$, we get $x + 1 = 0$, and thus $x = -1$. Finally, by substituting we verify that $x = -1$ is indeed a solution.   □

(iii) Find all positive integer solutions of the equation

$$\left[\frac{1}{10}\right] + \left[\frac{2}{10}\right] + \cdots + \left[\frac{n-1}{10}\right] + \left[\frac{n}{10}\right] = 217.$$

SOLUTION.   We first add the expression on the left. For $i = 1, 2, \ldots, 9$ we have $[i/10] = 0$. If we set $k = [n/10]$, then by the solution of 7.2.(iv) the number $10k$ is the greatest multiple of 10 not exceeding n, and therefore

$$\left[\frac{1}{10}\right] + \left[\frac{2}{10}\right] + \cdots + \left[\frac{n-1}{10}\right] + \left[\frac{n}{10}\right]$$
$$= \left(\left[\frac{10}{10}\right] + \cdots + \left[\frac{19}{10}\right]\right) + \cdots + \left(\left[\frac{10(k-1)}{10}\right] + \cdots + \left[\frac{10k-1}{10}\right]\right)$$
$$+ \left(\left[\frac{10k}{10}\right] + \cdots + \left[\frac{n}{10}\right]\right)$$
$$= 10 \cdot 1 + 10 \cdot 2 + \cdots + 10(k-1) + k(n - 10k + 1)$$
$$= 5k(k-1) + k(n - 10k + 1) = k(n - 5k - 4).$$

In view of the fact that $k \in \mathbb{N}$, we will solve the equation

$$k(n - 5k - 4) = 217$$

in natural numbers. Since $217 = 31 \cdot 7$, there are four possibilities: $k = 1$, $k = 7$, $k = 31$, $k = 217$, and always $n - 5k - 4 = 217/k$, i.e., respectively $n = 226$, $n = 70$, $n = 166$, $n = 1090$. The condition $k = [n/10]$, however, is satisfied only by the pair $k = 7$, $n = 70$. The problem has therefore the unique solution $n = 70$. $\square$

7.7 Exercises

(i) Show that for any natural numbers m, n we have

$$\left[\frac{1}{m}\right] + \left[\frac{2}{m}\right] + \cdots + \left[\frac{n-1}{m}\right] + \left[\frac{n}{m}\right] = \left[\frac{n}{m}\right]\left(n + 1 - \frac{m}{2}\left(1 + \left[\frac{n}{m}\right]\right)\right).$$

Solve the equations (ii)–(v) in real numbers:

(ii) $[x^2] = x$.

(iii) $[x^2] = 2$.

(iv) $[3x^2 - x] = x + 1$.

(v) $2\langle x \rangle = 3[x] - [x^2]$.

*(vi) Find all solutions in real numbers of the system of equations

$$\left[\sqrt{y - 1}\right]^2 = x - 1,$$

$$y - 1 = 2\left[\sqrt{y + 2\sqrt{x}}\right].$$

*(vii) Find all real solutions of the equation

$$\left[\frac{2x - 1}{3}\right] + \left[\frac{4x + 1}{6}\right] = \frac{5x - 4}{3}.$$

*(viii) Find all positive integer solutions of the equation

$$\left[\sqrt[3]{1}\right] + \left[\sqrt[3]{2}\right] + \cdots + \left[\sqrt[3]{x^3 - 1}\right] = 400.$$

*(ix) Show that the equation

$$[x] + [2x] + [4x] + [8x] + [16x] + [32x] = 12345$$

has no solutions in real numbers.

*(x) Given any $n \in \mathbb{N}$, find the number of solutions of the equation

$$x^2 - [x^2] = \langle x \rangle^2$$

satisfying $1 \le x \le n$.

*(xi) Determine the natural numbers that cannot be written as $\left[n + \sqrt{n} + \frac{1}{2}\right]$ for any $n \in \mathbb{N}$.

8 Base Representations

Number-theoretic problems in which base representations occur can be rather diverse, which means that their methods of solution are also very diverse. Nevertheless, we will attempt to describe some of the more unusual methods. A good number of problems can be solved with the help of Diophantine equations. This is mainly the case for problems of the type "find a number whose digits satisfy ...." In this case we can get a handle on the conditions of the problem by way of an equation in which the variables are the digits of the number we want to find. Thus we can use, for instance, the method of 5.11, i.e., use the fact that for such an unknown x we have $0 \leq x \leq 9$, and then check all ten possibilities in turn. Such an approach, however, is rather tedious, and it is therefore often better to use a different, faster, method of solving the Diophantine equation in question.

We note that in this section we will deal only with digit representations in base 10. As usual, we suppose that no representation starts with the digit 0.

8.1 Examples

(i) Find all three-digit numbers that when divided by 11 are equal to the sum of the squares of their digits.

SOLUTION. We have to find a three-digit number $x = 100a + 10b + c$ whose digits $a \neq 0, b, c$ satisfy

$$100a + 10b + c = 11(a^2 + b^2 + c^2).$$

By rewriting we get

$$a - b + c = 11(a^2 + b^2 + c^2 - 9a - b), \tag{37}$$

which implies that $a - b + c$ is divisible by 11. Since $1 \leq a \leq 9$, $0 \leq b \leq 9$, $0 \leq c \leq 9$, we have

$$-8 \leq a - b + c \leq 18,$$

and thus there are two possibilities: $a - b + c$ is either 0 or 11. Let us begin with the first case. We have $b = a + c$, and by substituting,

$$a^2 + (a + c)^2 + c^2 - 9a - (a + c) = 0,$$

from which we get

$$2a^2 + (2c - 10)a + 2c^2 - c = 0,$$

and this is a quadratic equation in a. For it to have a solution, it must have a nonnegative discriminant, thus

$$(2c - 10)^2 - 8(2c^2 - c) \geq 0,$$

and upon simplifying and dividing by 4 we get

$$-3c^2 - 8c + 25 \geq 0.$$

This means that $c \leq 1$, since for $c \geq 2$ we have

$$3c^2 + 8c \geq 3 \cdot 4 + 8 \cdot 2 = 28 > 25.$$

If $c = 0$, then $2a^2 - 10a = 0$, which implies $a = b = 5$, since $a = 0$ does not satisfy the given conditions. If $c = 1$, then $2a^2 - 8a + 1 = 0$; this equation, however, has no integer solution.

Let us now turn to the case $b = a + c - 11$. By substituting this into (37) we obtain

$$a^2 + (a + c - 11)^2 + c^2 - 9a - (a + c - 11) = 1,$$

which again gives a quadratic equation in a, namely

$$2a^2 + (2c - 32)a + 2c^2 - 23c + 131 = 0. \tag{38}$$

Its discriminant

$$4(c - 16)^2 - 8(2c^2 - 23c + 131) = 4(-3c^2 + 14c - 6)$$

is negative for $c \geq 5$, since

$$-3c^2 + 14c - 6 = c(-3c + 14) - 6 < -6,$$

and thus $c \leq 4$. Since $c = b - a + 11 \geq 11 - a \geq 2$, we must have $c = 2$, $c = 3$, or $c = 4$. For even c we have an odd number on the left-hand side of (38); hence $c = 3$. Substituting this into (38) and solving, we get $a = 8$, which implies $b = 0$, or $a = 5$ and thus $b = -3$, which is a contradiction. The given conditions are therefore satisfied only by the numbers 550 and 803; it is easy to verify that these are indeed solutions.   □

(ii) Find all four-digit numbers that are squares of natural numbers and whose first and second digits, as well as their third and fourth digits, are equal.

SOLUTION.   The desired number is of the form

$$x = 1000a + 100a + 10b + b = 11(100a + b),$$

where $0 < a \leq 9$, $0 \leq b \leq 9$. Since x is the square of a natural number, $100a + b$ has to be divisible by 11, which means that $a + b$ as well has to be divisible by 11. Since $0 < a + b \leq 18$, we have $a + b = 11$, and thus

$$x = 11(99a + 11) = 11^2(9a + 1).$$

Again, since x is a square, we must have $9a + 1 = n^2$ for an appropriate $n \in \mathbb{N}$, which means that $a = (n+1)(n-1)/9$. But the difference of the factors $n + 1$, $n - 1$ is 2, so both cannot at the same time be divisible by 3, and thus $9 \mid n + 1$ or $9 \mid n - 1$. Along with $0 < a \leq 9$ it follows that $1 < n^2 \leq 82$, and thus $n = 8$, so finally $x = 88^2 = 7744$. $\qquad \square$

(iii) Find all natural numbers s with leading digit 6 that upon removal of this digit become 25-times smaller.

SOLUTION.   We assume that the desired number x has $k + 1$ digits. Then

$$x = 6 \cdot 10^k + y,$$

where $y \in \mathbb{N}_0$, $y < 10^k$. The given condition implies that $x = 25y$, so

$$25y = 6 \cdot 10^k + y,$$

which means $y = \frac{1}{4}10^k$, and thus $k \geq 2$, $y = 25 \cdot 10^{k-2}$, $x = 625 \cdot 10^{k-2}$. The problem has therefore infinitely many solutions; the condition is satisfied exactly by the numbers $x = 625 \cdot 10^{k-2}$, where $k \in \mathbb{N}$, $k \geq 2$. $\qquad \square$

(iv) Find all three-digit numbers that are equal to the sum of the factorials of their digits.

SOLUTION.   We assume that the number $100a + 10b + c$, where $0 < a \leq 9$, $0 \leq b \leq 9$, $0 \leq c \leq 9$, satisfies the condition of the problem, namely

$$100a + 10b + c = a! + b! + c!.$$

Since $7! = 5040 > 1000$, the digits a, b, c are less than 7. Furthermore, none of the digits can be 6, since $6! = 720 > 100a + 10b + c$. At least one digit is equal to 5, since otherwise we would have

$$a! + b! + c! \leq 3 \cdot 4! = 72 < 100a + 10b + c.$$

Since $3 \cdot 5! = 360$, we have $a \leq 3$. But $a \neq 3$, since $3! + 5! + 5! = 246 < 300$, and neither is $a = 2$ possible, since in that case $100a + 10b + c$ would be at least 200, while $2! + b! + c! \geq 200$ only for $b = c = 5$, but 255 does not satisfy the given condition. Thus $a = 1$, and the desired number is of the form $150 + c$ (if $b = 5$) or $105 + 10b$ (if $c = 5$). In the first case we get

$$150 + c = 1! + 5! + c!,$$

and thus

$$29 = c((c-1)! - 1),$$

which, however, is not satisfied for any $c \leq 5$. In the second case,

$$105 + 10b = 1! + b! + 5!,$$

and hence

$$b(10 - (b-1)!) = 16,$$

which gives $b = 4$. It is now easy to verify that the number 145 does indeed satisfy the given condition. $\qquad \square$

8.2 Exercises

(i) Find all two-digit numbers that equal the sum of the first digit and the square of the second digit.

(ii) Find all four-digit numbers that are the square of a natural number and are such that the first digit is equal to the third digit, and the second digit exceeds the last digit by one.

(iii) The square of a natural number is equal to a four-digit number whose first digit is three and whose last digit is five. Find this number.

(iv) Find a three-digit number n^2 such that the product of its digits is equal to $n - 1$.

(v) Find a three-digit number that, multiplied by 2, is equal to 3 times the product of the factorials of its digits.

(vi) Find three-digit numbers x, y satisfying $y = 8x$ and such that both the difference $y - x$ and the six-digit number obtained by writing y after x are squares of natural numbers.

(vii) Suppose that the decimal representation of n^2, where $n \in \mathbb{N}$, ends in a 5. Show that the third-to-last digit (the hundreds digit) is even.

(viii) Find all natural numbers n such that the sum of the digits of n^2 is equal to n.

(ix) Find the smallest natural number s with the following property: Its decimal expansion ends in a six, and if this digit is moved to the front, we obtain 4 times the original number.

(x) Find all natural numbers n divisible by the number obtained when the last digit is removed from n.

In problems (xi)–(xiii), let $s(n)$ denote the sum of digits of the natural number n.

(xi) Is there an $n \in \mathbb{N}$ such that $s(n) = 1996$ and $s(n^2) = 1996^2$?

(xii) Show that for all $n \in \mathbb{N}$ we have $s(n) \leq 8s(8n)$. Does the sharp inequality hold in general?

(xiii) Show that for any $m \in \mathbb{N}$ there exists an $n \in \mathbb{N}$ such that either $n + s(n) = m$ or $n + s(n) = m + 1$.

There are many problems that concern the final digits in a decimal expansion. On the one hand, there are problems of determining the last few digits of "huge" numbers, and on the other hand those of determining whether a given pair of integers ends in the same group of digits. Such problems can be solved with the help of congruences, as we will see in the following examples.

8.3 Examples

(i) Find the final digit of the numbers 9^{9^9} and 2^{3^4}.

SOLUTION.   We have to find the remainders when the given numbers are divided by 10; hence we use congruences modulo 10. We have

$$9^{9^9} \equiv (-1)^{9^9} = -1 \equiv 9 \pmod{10},$$

where we have used the fact that 9^9 is an odd number. To determine the final digit of the second number, we use separately congruences modulo 2 and modulo 5. Certainly,

$$2^{3^4} \equiv 0 \pmod 2.$$

To find the number congruent to the given integer modulo 5, we use 3.14.(v). Clearly, the smallest natural number with the property $2^n \equiv 1$ (mod 5) is 4. Therefore, we determine the remainder of 3^4 divided by 4:

$$3^4 \equiv (-1)^4 \equiv 1 \pmod 4,$$

and thus

$$2^{3^4} \equiv 2^1 = 2 \pmod 5.$$

Since 2^{3^4} is congruent to 2 modulo 2 and 5, we have

$$2^{3^4} \equiv 2 \pmod{10}.$$

We therefore conclude that the first of the given numbers ends in 9, and the second one in 2.   ☐

(ii) Find the last four digits of the number 2^{999}.

SOLUTION.   We must find the integer x, $0 \le x < 10^4$, satisfying

$$x \equiv 2^{999} \pmod{10^4}.$$

Since $\varphi(5^4) = 4 \cdot 5^3 = 500$, by Euler's Theorem 3.13 we have

$$2^{500} \equiv 1 \pmod{5^4},$$

which means that

$$2x \equiv 2^{1000} = (2^{500})^2 \equiv 1^2 \equiv 1 - 5^4 \pmod{5^4}.$$

Since $1 - 5^4 = -624$, upon division by 2 we obtain

$$x \equiv -312 \pmod{5^4}. \tag{39}$$

Certainly,

$$x \equiv 2^{999} \equiv 0 \pmod{2^4}; \tag{40}$$

we now solve the system of congruences (39) and (40). From (40) we obtain $x = 16k$ for an appropriate integer k. Substituting this into (39), we get

$$16k \equiv -312 \pmod{5^4},$$

and dividing by 8,

$$2k \equiv -39 \equiv 5^4 - 39 \pmod{5^4}.$$

Now, $5^4 - 39 = 586$, and thus upon division by 2,

$$k \equiv 293 \pmod{5^4}.$$

Hence $k = 293 + 625s$ for an appropriate integer s, and so $x = 4688 + 10000s$. Since $0 \le x < 10^4$, we have $x = 4688$, and thus the number 2^{999} ends in the four digits 4688. $\square$

(iii) Given any natural number n, find the final two digits of the integer

$$n^4 + (n+1)^4 + (n+2)^4 + \cdots + (n+99)^4.$$

SOLUTION.   If we divide all of the 100 consecutive integers $n, n+1, \ldots,$ $n + 99$ by 100, we obtain one hundred different—and thus all possible—remainders. Hence

$$n^4 + (n+1)^4 + \cdots + (n+99)^4 \equiv 0^4 + 1^4 + \cdots + 99^4 \pmod{100}.$$

To determine the right-hand sum, we could use formula (15) from Chapter 1. However, if we did not remember this formula, we could also proceed as follows. By the binomial theorem, we have for all $a = 0, 1, \ldots, 9$,

$$(10a)^4 + (10a + 1)^4 + \cdots + (10a + 9)^4$$
$$\equiv 4 \cdot 10a \cdot (0^3 + 1^3 + 2^3 + \cdots + 9^3) + 0^4 + 1^4 + \cdots + 9^4 \pmod{100}.$$

If we add this congruence for $a = 0, 1, \ldots, 9$, we get

$$0^4 + 1^4 + \cdots + 99^4 \equiv 4 \cdot 10(1 + \cdots + 9)(1^3 + 2^3 + \cdots + 9^3)$$
$$+ 10(1^4 + 2^4 + \cdots + 9^4) \pmod{100}.$$

Since $1 + 2 + \cdots + 9 = 45$, we have $40(1 + \cdots + 9) \equiv 0 \pmod{100}$. It remains to determine the second summand. By Fermat's theorem, for $1 \le k \le 9$, $k \ne 5$, we have $k^4 \equiv 1 \pmod 5$, and thus

$$1^4 + 2^4 + \cdots + 9^4 \equiv 8 \equiv 3 \pmod 5.$$

An odd number of the integers $1, 2, \ldots, 9$ is odd, and therefore the sum of the fourth powers of these integers is congruent to 3 modulo 2. Together, we have

$$1^4 + 2^4 + \cdots + 9^4 \equiv 3 \pmod{10},$$

and finally

$$1^4 + 2^4 + \cdots + 99^4 \equiv 10(1^4 + 2^4 + \cdots + 9^4) \equiv 30 \pmod{100}.$$

The final two digits of the given sum are therefore 3 and 0 for any n.   □

(iv) Determine the natural numbers $m > n$ such that 1234^m and 1234^n end in the same three digits and such that the sum $m + n$ is minimal.

SOLUTION. The numbers 1234^m and 1234^n end in the same three digits if and only if $1234^m - 1234^n = 1234^n(1234^{m-n} - 1)$ is divisible by 1000, i.e., if it is divisible by both 8 and 125. This occurs if and only if $8 \mid 1234^n$ and $125 \mid 1234^{m-n} - 1$, because $(1234^{m-n} - 1, 2) = (1234^n, 5) = 1$. Since 1234 is divisible by 2 but not by 4, 1234^n is divisible by 8 if and only if $n \geq 3$. We now determine when $125 \mid 1234^{m-n} - 1$, i.e., when

$$1234^{m-n} \equiv 1 \pmod{125}.$$

By 3.14.(v) we have to find the smallest $k \in \mathbb{N}$ for which $1234^k \equiv 1 \pmod{125}$. We know that this k satisfies $k \mid \varphi(125) = 100$. Now, by Euler's Theorem 3.13,

$$1234^{50} \equiv (-16)^{50} = 4^{100} \equiv 1 \pmod{125}.$$

By 3.14.(v) this means that $k \mid 50$. If $k \neq 50$, we would have $k \mid 10$ or $k \mid 25$. But

$$1234^{10} \equiv (-16)^{10} = 256^5 \equiv 6^5 = 36 \cdot 216 \equiv 36 \cdot (-34)$$
$$= -1224 \equiv 26 \pmod{125},$$
$$1234^{25} \equiv (-16)^{25} = -16^{25} = -2^{100} \equiv -1 \pmod{125},$$

again by Euler's theorem. Hence neither $k \mid 10$, nor $k \mid 25$. Therefore, $k = 50$ and $1234^{n-m} \equiv 1 \pmod{125}$ if and only if $n - m \equiv 0 \pmod{50}$, i.e., $n \equiv m \pmod{50}$. The sum $m + n$ is therefore minimal when $n = 3$ and $m = 53$.   □

8.4 Exercises

(i) Show that for any integer $n \geq 2$ the number 2^{2^n} ends in a 6.

(ii) Find the last two digits of 7^{9^9}.

(iii) Find the last four digits of 3^{999}.

(iv) Find the last two digits of $14^{14^{14}}$.

*(v) Show that if the difference of the natural numbers n, k is divisible by 4, then the numbers

$$7^{7^{7^{7^{7^n}}}} \quad \text{and} \quad 7^{7^{7^{7^{7^k}}}}$$

end in the same six digits.

*(vi) Find the smallest natural number n for which the numbers 19^n and 313^n end in the same three digits. Find these three digits.

8.5 *Examples*

A popular kind of problem involves large numbers with special digit patterns, such as those consisting only of ones. Before considering several such problems, we note that the number written with k ones (also called a *repunit*) can be expressed as

$$\underbrace{11\ldots1}_{k} = \frac{1}{9}(10^k - 1),$$

which can be seen either by summing $10^{k-1} + 10^{k-2} + \cdots + 10 + 1$, or by noting that

$$10^k - 1 = \underbrace{99\ldots9}_{k},$$

and dividing this expression by 9.

(i) Suppose that the integer a consists of $2m$ ones, that b consists of $m+1$ ones, and c is written as m sixes, where m is an arbitrary natural number. Show that $a + b + c + 8$ is the square of some integer.

SOLUTION.   We have

$$a = \frac{1}{9}\left(10^{2m} - 1\right), \quad b = \frac{1}{9}\left(10^{m+1} - 1\right), \quad c = \frac{6}{9}\left(10^m - 1\right),$$

which implies

$$a + b + c + 8 = \frac{1}{9}\left(10^{2m} - 1 + 10^{m+1} - 1 + 6 \cdot 10^m - 6 + 72\right)$$
$$= \frac{1}{9}\left(10^{2m} + 16 \cdot 10^m + 64\right) = \left(\frac{1}{3}(10^m + 8)\right)^2.$$

Since $10^m + 8 \equiv 1 + 8 \equiv 0 \pmod 3$, the number $n = \frac{1}{3}(10^m + 8)$ is an integer, and we have

$$a + b + c + 8 = n^2,$$

which was to be shown.   □

(ii) Show that the number written as 3^n identical digits is divisible by 3^n, where n is an arbitrary natural number.

SOLUTION.   It suffices to show that the number written as 3^n ones is divisible by 3^n, i.e., that for any $n \in \mathbb{N}$ we have $3^n \mid \frac{1}{9}(10^{3^n} - 1)$, or $3^{n+2} \mid 10^{3^n} - 1$. But this can be written as the congruence

$$10^{3^n} \equiv 1 \pmod{3^{n+2}}.$$

We prove this assertion by induction on $n \in \mathbb{N}_0$. For $n = 0$ it is clear. We assume that for some $n \geq 0$ we have $10^{3^n} \equiv 1 \pmod{3^{n+2}}$. Then by 3.4.(vi) we also have

$$10^{3^{n+1}} = (10^{3^n})^3 \equiv 1 \pmod{3^{n+3}},$$

which is the desired statement for $n + 1$. The proof by induction is now complete.   □

(iii) Show that each odd integer m that is not a multiple of 5 divides some of the numbers $1, 11, 111, 1111, \ldots$.

SOLUTION.   An integer m that is relatively prime to 10 satisfies by Euler's theorem

$$10^{\varphi(9m)} \equiv 1 \pmod{9m},$$

which implies that there exists an integer k such that $9mk = 10^{\varphi(9m)} - 1$. Then mk is the number consisting of $\varphi(9m)$ ones. Thus m divides this number, which consists of $\varphi(9m)$ ones.   □

(iv) Find the digits a, b, c, given that for each natural number n we have

$$\underbrace{aa \ldots a}_{n} \underbrace{bb \ldots b}_{n} + 1 = (\underbrace{cc \ldots c}_{n} + 1)^2,$$

where $\overline{a_1 a_2 \ldots a_k}$ denotes $a_1 \cdot 10^{k-1} + a_2 \cdot 10^{k-2} + \cdots + a_k$, with $a_1 \neq 0$.

SOLUTION.   If we set $t_n = \frac{1}{9}(10^n - 1)$, we can rewrite the condition in the problem as

$$a t_n \cdot 10^n + b t_n + 1 = (c t_n + 1)^2. \tag{41}$$

By substituting $9t_n + 1$ for 10^n and rewriting, we obtain

$$9a t_n^2 + (a + b)t_n + 1 = c^2 t_n^2 + 2c t_n + 1,$$

from which upon subtracting 1 and dividing by t_n we get

$$t_n(9a - c^2) = 2c - a - b.$$

This equality has to hold for all n, in particular for

$$t_1(9a - c^2) = 2c - a - b,$$
$$t_2(9a - c^2) = 2c - a - b,$$

and by subtracting we obtain $(t_2 - t_1)(9a - c^2) = 0$. Since $t_2 - t_1 = 10 \neq 0$, this implies $c^2 = 9a$, $2c = a+b$. If, on the other hand, the digits a, b, c satisfy the conditions $c^2 = 9a$, $2c = a + b$, then the identity (41) is satisfied for all n. It remains to determine the a, b, c that satisfy these two conditions. From $c^2 = 9a$ it follows that c is divisible by 3. Since $c \neq 0$, we have $c \in \{3, 6, 9\}$. If $c = 3$, then $a = 1$, $b = 5$; if $c = 6$, then $a = 4$, $b = 8$, and finally, if $c = 9$ then $a = 9$, $b = 9$. Our problem has therefore three solutions altogether.   □

8.6 Exercises

(i) Show that for each natural number n, the difference between the number consisting of $2n$ ones and the number consisting of n twos is the square of an integer. Determine this integer.

(ii) Suppose that the number m_1 consists of n threes, and m_2 of n sixes, where $n \in \mathbb{N}$. What are the digits of the product $m_1 m_2$?

(iii) Show that for any $n, k \in \mathbb{N}$ the integer consisting of n ones is divisible by the integer consisting of k ones if and only if $k \mid n$.

(iv) Given a natural number k, find the smallest number consisting only of ones that is divisible by the number consisting of k threes.

(v) For which digits a is there a natural number $n > 3$ such that

$$1 + 2 + \cdots + n$$

has no other digits than a in its decimal expansion?

(vi) For $m > 1$ we denote by $M = 11 \ldots 1$ the number consisting of m ones. Does there exist a natural number n divisible by M such that its digit sum is less than the digit sum m of M?

To conclude this section, we describe a method that exemplifies the diversity of methods that can be used for solving problems on digit expansions. We will take advantage of the concept, introduced in the previous section, of the integer part of a real number.

8.7 Example

Show that for any digits $a_1, a_2, \ldots, a_s$, with $a_1 \neq 0$, there exists a natural number n such that the decimal representation of n^2 begins with $a_1, a_2, \ldots, a_s$.

SOLUTION. We set $m = a_1 \cdot 10^{s-1} + a_2 \cdot 10^{s-2} + \cdots + a_{s-1} \cdot 10 + a_s$. We choose a $k \in \mathbb{N}$ such that $2\sqrt{m} < 10^{k-1}$ and set $n = \left[10^k \cdot \sqrt{m} \right] + 1$. Then by (35) we have

$$10^k \cdot \sqrt{m} < n \le 10^k \cdot \sqrt{m} + 1,$$

which implies

$$10^{2k} m < n^2 \le 10^{2k} m + 1 + 2 \cdot 10^k \sqrt{m}$$
$$< 10^{2k} m + 1 + 10^{2k-1} < 10^{2k} \cdot m + 10^{2k} - 1,$$

and thus

$$10^{2k} m < n^2 < 10^{2k} m + (10^{2k} - 1).$$

Both integers $10^{2k} m$, $10^{2k} m + (10^{2k} - 1)$ have the same number of digits, and their decimal representations begin with $a_1, a_2, \ldots, a_s$. Hence the representation of n^2 begins with the digits $a_1, a_2, \ldots, a_s$ as well. $\square$

8.8 Exercise

Show that for any digits $a_1, a_2, \ldots, a_s$, where $a_1 \ne 0$, there exists a natural number n such that the decimal expansion of n^3 begins with $a_1, a_2, \ldots, a_s$.

9 Dirichlet's Principle

Dirichlet's principle (also known as the *pigeonhole principle*) is a simple combinatorial concept that we can use in a number of practical situations. For instance, if you have in your pocket 13 marbles of 3 different colors, then we can conclude that at least 5 of them are of the same color. Similarly easy arguments are also useful in more complicated mathematical situations.

9.1 Dirichlet's Principle

Theorem. *Let $n, k \in \mathbb{N}$. If at least $nk+1$ objects are divided into n groups, then at least one of these groups contains at least $k + 1$ objects.*

PROOF. We prove this by contradiction. For $i = 1, 2, \ldots, n$ we denote by m_i the number of objects in the ith group. Let us suppose that the conclusion does not hold, i.e., each group contains no more than k objects. Then $m_i \le k$ for all $i = 1, 2, \ldots, n$, which implies that

$$nk + 1 \le m_1 + m_2 + \cdots + m_n \le k + k + \cdots + k = nk.$$

But this is a contradiction, since $nk + 1 > nk$. $\square$

9.2   Examples

(i) Show that among 50 arbitrarily chosen distinct prime numbers one can always find 13 primes such that the difference between any two of them is divisible by 5.

SOLUTION. The difference between two integers is divisible by 5 if and only if they have the same remainder upon division by 5. Hence we divide the given set of primes into five groups according to their remainder upon division by 5. Since only the prime 5 is divisible by 5, the group belonging to remainder 0 contains at most one prime, while the other four groups together contain at least 49 primes. By Dirichlet's principle, at least one group contains at least 13 primes. These 13 primes satisfy the required condition. □

(ii) Show that any set of ten two-digit numbers has two nonempty disjoint subsets such that the sums of their elements are the same.

SOLUTION. A ten-element set has altogether $2^{10} - 1 = 1023$ nonempty subsets. The sum of at most ten two-digit numbers is less than $10 \cdot 100 = 1000 < 1023$. Hence there exist two different nonempty subsets of the given set of ten numbers such that the sums of their elements are equal. By removing, if necessary, any common elements, we obtain two disjoint nonempty subsets with the desired property. □

(iii) Show that there exists a number of the form

$$123456789123456789\ldots123456789$$

that is divisible by 987654321.

SOLUTION. We set $n = 123456789$, $m = 987654321$, and

$$a_i = \frac{(10^{9i} - 1)n}{10^9 - 1}$$

for $i = 1, 2, \ldots, m + 1$. Then clearly, the a_i are exactly the numbers of the given form. Any integer has one of the m remainders $0, 1, \ldots, m - 1$ when divided by m. By Dirichlet's principle at least two of the $m + 1$ numbers $a_1, a_2, \ldots, a_{m+1}$ must have the same remainder, and thus there exist i, j, $1 \le j < i \le m + 1$, such that $m \mid a_i - a_j$. Now,

$$a_i - a_j = ((10^{9i} - 1) - (10^{9j} - 1)) \cdot \frac{n}{10^9 - 1}$$

$$= (10^{9i} - 10^{9j}) \cdot \frac{n}{10^9 - 1} = 10^{9j} a_{i-j}.$$

Since $(m, 10^{9j}) = 1$, the number a_{i-j} is divisible by m, which was to be shown. □

(iv) Show that from among any fifteen natural numbers one can choose eight such that their sum is divisible by 8.

SOLUTION.   From Dirichlet's principle it follows that from any set of at least three integers one can choose two numbers of the same parity (i.e., either both even or both odd). Let us choose two such numbers from among the given fifteen, and denote their sum by a_1. From the remaining thirteen numbers we choose again two of the same parity, and denote their sum by a_2. Repeating this procedure, we obtain seven even numbers $a_1, \ldots, a_7$. Again from Dirichlet's principle it follows that from any set of at least three even numbers we can choose two that have the same remainder when divided by 4. We choose two such numbers from among $a_1, \ldots, a_7$ and denote their sum, which must be a multiple of 4, by b_1. From the remaining five elements we choose in the same way b_2, and finally b_3 from the remaining triple. Once again it follows from Dirichlet's principle that from the three numbers b_1, b_2, b_3, which are divisible by 4, we can choose two that have the same remainder when divided by 8. Their sum, which is the sum of eight of the original 15 numbers, must therefore be divisible by 8.   □

9.3   Exercises

(i) Show that for any natural number n there exists a number k, divisible by n, whose decimal expansion begins with ones only, and then has only zeros.

*(ii) Let $a > 1$ be a natural number. Find all natural numbers that divide at least one of the integers $a_n = a^n + a^{n-1} + \cdots + a + 1$, where $n \in \mathbb{N}$.

*(iii) Given a natural number n, show that from among $2^{n+1} - 1$ natural numbers one can choose 2^n numbers such that their sum is divisible by 2^n.

*(iv) Given a natural number n, show that from among $2 \cdot 3^n - 1$ natural numbers one can choose 3^n numbers such that their sum is divisible by 3^n.

9.4   Examples

(i) Suppose that the product of nine distinct natural numbers is divisible by exactly three primes. Show that from among these nine numbers we can choose two distinct ones whose product is the square of an integer.

SOLUTION.   Let us denote the given primes by p_1, p_2, p_3 and let $A = \{a_1, a_2, \ldots, a_9\}$ be the set of the nine given numbers. From the given condition it follows that none of the numbers $a_1, \ldots, a_9$ is divisible by any prime other than p_1, p_2, p_3. Each element of A can therefore be written in the form $p_1^{n_1} \cdot p_2^{n_2} \cdot p_3^{n_3}$, where $n_1, n_2, n_3 \in \mathbb{N}_0$. It is clear that the product $p_1^{n_1} p_2^{n_2} p_3^{n_3} \cdot p_1^{m_1} p_2^{m_2} p_3^{m_3}$ is the square of an integer if and only if the sums $n_1 + m_1$, $n_2 + m_2$, $n_3 + m_3$ are even, i.e., the exponents n_i and m_i have the same parity for each $i = 1, 2, 3$. We now consider all eight triples $(\varepsilon_1, \varepsilon_2, \varepsilon_3)$ of zeros and ones, and for each such triple we consider the set of numbers $p_1^{n_1} \cdot p_2^{n_2} \cdot p_3^{n_3}$ from A such that all numbers $n_i + \varepsilon_i$, for $i = 1, 2, 3$, are even. Since in this way we distribute nine numbers over eight sets, one of these sets must contain at least two numbers $p_1^{n_1} \cdot p_2^{n_2} \cdot p_3^{n_3}$ and $p_1^{m_1} \cdot p_2^{m_2} \cdot p_3^{m_3}$. But then the exponents n_i, m_i have the same parity for $i = 1, 2, 3$, and thus the product of these numbers is the square of an integer. □

(ii) Suppose that four distinct natural numbers are divisible by exactly three primes. Show that for some k, $1 \le k \le 4$, one can choose k numbers from among these four such that their product is the square of an integer.

SOLUTION.   Let us again denote the given primes by p_1, p_2, p_3, and the set of given numbers by $A = \{a_1, a_2, a_3, a_4\}$. The set A has $2^4 - 1 = 15$ nonempty subsets. For each of these subsets we consider the product of its elements; each of these 15 products is of the form $p_1^{n_1} p_2^{n_2} p_3^{n_3}$. Just as in (i) we group these products into eight groups according to the parity of their exponents n_1, n_2, n_3. By Dirichlet's principle there exist two different subsets of A whose products are in the same group. The product of these two products is the square of an integer. Therefore, also the product of all numbers of the set A that belong to exactly one of the two subsets (i.e., the product of the elements of the symmetric difference of the two subsets) is the square of an integer. □

(iii) Suppose that the product of 48 distinct natural numbers is divisible by exactly ten primes. Show that one can choose four of these 48 numbers such that their product is the square of an integer.

SOLUTION.   We denote the given primes by $p_1, \ldots, p_{10}$ and form all two-element subsets of the given set A of 48 numbers. The number of these subsets is

$$\binom{48}{2} = \frac{48 \cdot 47}{2} = 1128.$$

We form the products of the elements of these subsets; they can be written as $p_1^{n_1} p_2^{n_2} \cdots p_{10}^{n_{10}}$. We consider all 10-tuples $(\varepsilon_1, \ldots, \varepsilon_{10})$ of zeros and ones; there are $2^{10} = 1024$ of them. We distribute the 1128 subsets over the 1024 groups determined by the 10-tuples $(\varepsilon_1, \ldots, \varepsilon_{10})$ such that the subset with product of elements $p_1^{n_1} \cdots p_{10}^{n_{10}}$ belongs to the 10-tuple $(\varepsilon_1, \ldots, \varepsilon_{10})$ for which all the numbers $n_1 + \varepsilon_1, \ldots, n_{10} + \varepsilon_{10}$ are even. By Dirichlet's principle there exists at least one group containing at least two such subsets $A_1 = \{a, b\}$, $A_2 = \{c, d\}$. The product $abcd$ is then the square of an integer. If A_1, A_2 are disjoint, then a, b, c, d are the desired four numbers. Otherwise, A_1, A_2 have exactly one common element, since $A_1 \ne A_2$. Without loss of generality we assume that $a = c$, $b \neq d$. Then the product bd is the square of an integer.

Now we repeat the previous argument with the 46-element set $B = A \setminus \{b, d\}$. This time the number of two-element subsets is

$$\binom{46}{2} = \frac{46 \cdot 45}{2} = 1035,$$

which is again more than 1024. As a consequence there are two subsets $B_1 = \{e, f\}$, $B_2 = \{g, h\}$ such that the product $efgh$ is the square of an integer. If B_1, B_2 are disjoint, then e, f, g, h are the desired numbers. If B_1 and B_2 have an element in common, say $e = g$, $f \neq h$, then the product fh is a square, and the numbers b, d, f, h have the desired property.   □

(iv) Suppose that the product of 55 distinct natural numbers has exactly three prime divisors. Show that the product of some three of these 55 numbers is the third power of an integer.

SOLUTION.   We denote the given primes by p_1, p_2, p_3 and consider all triples $(\varepsilon_1, \varepsilon_2, \varepsilon_3)$, where $\varepsilon_1, \varepsilon_2, \varepsilon_3 \in \{0, 1, 2\}$; their number is $3^3 = 27$. We divide the given 55 numbers, which are of the form $p_1^{n_1} p_2^{n_2} p_3^{n_3}$, into 27 groups by associating to each triple $(\varepsilon_1, \varepsilon_2, \varepsilon_3)$ those numbers $p_1^{n_1} p_2^{n_2} p_3^{n_3}$ that satisfy $n_1 \equiv \varepsilon_1 \pmod 3$, $n_2 \equiv \varepsilon_2 \pmod 3$, $n_3 \equiv \varepsilon_3 \pmod 3$. By Dirichlet's principle, at least one of these groups contains at least three numbers, say $a = p_1^{m_1} p_2^{m_2} p_3^{m_3}$, $b = p_1^{n_1} p_2^{n_2} p_3^{n_3}$, $c = p_1^{k_1} p_2^{k_2} p_3^{k_3}$. Then for all $i = 1, 2, 3$ we have $m_i \equiv n_i \equiv k_i \pmod 3$, and thus $m_i + n_i + k_i \equiv 3k_i \equiv 0 \pmod 3$, i.e., $m_i + n_i + k_i = 3r_i$ for appropriate $r_i \in \mathbb{N}_0$. This implies that

$$abc = p_1^{n_1 + m_1 + k_1} p_2^{n_2 + m_2 + k_2} p_3^{n_3 + m_3 + k_3} = (p_1^{r_1} p_2^{r_2} p_3^{r_3})^3.$$

The product abc is therefore the third power of an integer.   □

(v) Given the natural numbers $a_1, a_2, \ldots, a_{25}$, whose product is divisible by exactly two primes, show that there exist i, j, $1 \leq i \leq j \leq 25$, such that $a_i \cdot a_{i+1} \cdots a_j$ is the fifth power of an integer.

SOLUTION.   We denote the given primes by p_1, p_2 and consider all pairs $(\varepsilon_1, \varepsilon_2)$, where $\varepsilon_1, \varepsilon_2 \in \mathbb{N}_0$ satisfy $\varepsilon_1 < 5$, $\varepsilon_2 < 5$. There are $5^2 = 25$ such pairs. We divide the numbers $j = 1, 2, \ldots, 25$ into 25 groups such that j belongs to the group corresponding to $(\varepsilon_1, \varepsilon_2)$ if and only if for $a_1 \cdot a_2 \cdots a_j = p_1^{n_1} p_2^{n_2}$ we have $n_1 \equiv \varepsilon_1 \pmod 5$, $n_2 \equiv \varepsilon_2 \pmod 5$. If some number j belongs to the group associated with $(0, 0)$, it suffices to set $i = 1$; then these numbers i, j satisfy the given conditions, since $a_1 \cdot a_2 \cdots a_j = p_1^{n_1} p_2^{n_2}$, where $n_1 \equiv n_2 \equiv 0 \pmod 5$, and thus for appropriate $r_1, r_2 \in \mathbb{N}_0$ we have $n_1 = 5r_1$, $n_2 = 5r_2$, which means that $a_1 \cdot a_2 \cdots a_j = (p_1^{r_1} p_2^{r_2})^5$.

Now we assume that none of the numbers j belongs to the group corresponding to the pair $(0, 0)$. Then the 25 numbers j are distributed over the remaining 24 groups, and by Dirichlet's principle, at least one of these groups will contain at least two numbers $j_1, j_2, j_1 < j_2$. If we set $i = j_1 + 1$, $j = j_2$, then clearly $1 \leq i \leq j \leq 25$, and we have

$$a_1 \cdot a_2 \cdots a_{i-1} = p_1^{n_1} p_2^{n_2}, \quad a_1 \cdot a_2 \cdots a_j = p_1^{m_1} p_2^{m_2},$$

where $n_1 \equiv m_1 \pmod 5$, $n_2 \equiv m_2 \pmod 5$. From the construction of the numbers n_1, n_2, m_1, m_2 it follows that $n_1 \leq m_1$, $n_2 \leq m_2$, and so there exist $r_1, r_2 \in \mathbb{N}_0$ such that $n_1 + 5r_1 = m_1$, $n_2 + 5r_2 = m_2$. This implies that

$$a_i \cdot a_{i+1} \cdots a_j = p_1^{m_1 - n_1} p_2^{m_2 - n_2} = (p_1^{r_1} \cdot p_2^{r_2})^5,$$

which means that the numbers i, j have the desired property. $\qquad\square$

9.5 Exercises

(i) Suppose that the product of five distinct natural numbers is divisible by exactly three primes. Show that from these five one can choose an even number of integers whose product is the square of an integer.

(ii) Suppose that the product of 45 distinct natural numbers is divisible by exactly ten primes. Show that for some k, $1 \leq k \leq 4$, one can choose k of those 45 numbers such that their product is the square of an integer.

*(iii) Prove the following strengthening of Example 9.4.(iv): Suppose that the product of 29 distinct natural numbers has exactly three prime divisors. Show that the product of some three of these 29 numbers is the third power of an integer.

*(iv) Suppose that the product of 27 distinct natural numbers is divisible by exactly three primes. Show that for some k, $1 \leq k \leq 3$, one can choose k of these 27 numbers such that their product is the third power of an integer.

(v) Given the natural numbers s, n, with $s > 1$, show that for any natural numbers $a_1, a_2, \ldots, a_{s^n}$ whose product is divisible by exactly n primes there exist i, j, $1 \leq i \leq j \leq s^n$, such that the product $a_i \cdot a_{i+1} \cdots a_j$ is the sth power of an integer.

To conclude this section, we consider another problem that can be solved by way of an easy combinatorial concept. This time we will not resort to Dirichlet's principle, but instead use the following consideration: If A is a subset of the set B and if A has fewer elements than B, then there must exist an element of B not belonging to A.

9.6 Example

Show that for any integer $n > 1$ there exist natural numbers m, k such that the number $x^n + y^n + k$ is not divisible by m for any integers x, y.

SOLUTION.   We choose $m = n^2$. An integer x is congruent to one of the numbers $0, 1, 2, \ldots, n - 1$ modulo n. By 3.4.(vi), x^n is then congruent to one of the numbers $0, 1^n, 2^n, \ldots, (n-1)^n$ modulo n^2. The same is also true for y^n.

The sum $x^n + y^n$ is therefore congruent modulo n^2 to one of the values $r^n + s^n$, where $0 \le r \le s < n$, of which there are at most $\frac{1}{2}n(n + 1)$. Therefore, the set of remainders of $x^n + y^n$ modulo m has at most $\frac{1}{2}n(n+1)$ elements, and since from $n > 1$ it follows that $\frac{1}{2}n(n + 1) < n^2 = m$, there has to exist $0 \le k < m$ such that $x^n + y^n$ is not congruent to $-k$ modulo m for any $x, y \in \mathbb{Z}$. The number $x^n + y^n + k$ is therefore not divisible by m for any x, y.   □

9.7 Exercise

Prove the following supplement to Theorem 4.14: For any prime $p > 2$ there exists an integer t, $0 < t < p$, such that the congruence $x^2 \equiv t \pmod{p}$ has no solution.

10 Polynomials

We return now to the topic of the third section of Chapter 1, namely polynomials, but this time from a number-theoretic point of view. We will be mainly interested in polynomials with integer coefficients, or polynomials that take on integer values at integers. First we recall Theorem 4.10, which will later enable us to prove several negative statements about polynomials with integer coefficients.

10.1 A Substitution Result

Theorem. Let $F(x)$ be a polynomial with integer coefficients, and let m be a natural number. If a and b are integers such that $a \equiv b \pmod{m}$, then $F(a) \equiv F(b) \pmod{m}$.

10.2 Examples

(i) Show that there is no polynomial $F(x)$ with integer coefficients such that $F(10) = 12$, $F(11) = 10$, $F(12) = 11$.

SOLUTION.   We show that that there does not even exist a polynomial $F(x)$ with integer coefficients such that $F(10) = 12$, $F(12) = 11$. Indeed, by 10.1 it follows from $10 \equiv 12 \pmod{2}$ that such a polynomial would satisfy $12 = F(10) \equiv F(12) = 11 \pmod{2}$, which is a contradiction. □

(ii) Show that a polynomial $F(x)$ with integer coefficients and such that $F(1989)$ and $F(1990)$ are odd numbers cannot have an integer zero.

SOLUTION. We assume to the contrary that there exists an integer n such that $F(n) = 0$. If n is odd, then $n \equiv 1989 \pmod{2}$, which means that $0 = F(n) \equiv F(1989) \pmod{2}$; but this is a contradiction. If n is even, then $n \equiv 1990 \pmod{2}$, and thus $0 = F(n) \equiv F(1990) \pmod{2}$, which is again a contradiction. Hence there is no integer n for which $F(n) = 0$. □

(iii) Show that a polynomial $F(x)$ with integer coefficients that takes on the values 1 or -1 at three different integers has no integer zero.

SOLUTION. Let us suppose that for three distinct integers a, b, c we have

$$|F(a)| = |F(b)| = |F(c)| = 1,$$

and that at the same time there exists an integer n such that $F(n) = 0$. Certainly, $n \neq a$. We set $|n - a| = m \in \mathbb{N}$. Since $n \equiv a \pmod{m}$, by Theorem 10.1 we have $F(n) \equiv F(a) \pmod{m}$, i.e., $1 \equiv 0 \pmod{m}$, which means that $m = 1$, and thus $a = n + 1$ or $a = n - 1$. In exactly the same way we show that $b = n + 1$ or $b = n - 1$, and that $c = n + 1$ or $c = n - 1$. But this means that at least two of the numbers a, b, c must be identical (this is really Dirichlet's principle), which is a contradiction. □

(iv) Show that there does not exist a polynomial $F(x)$ with integer coefficients such that

$$F(10) + F(10^2) + \cdots + F(10^9) = 10^{10}.$$

SOLUTION. For any $k = 1, 2, \ldots, 9$ we have $10^k \equiv 1 \pmod{9}$, and thus by 10.1 also $F(10^k) \equiv F(1) \pmod{9}$. We have therefore

$$F(10) + F(10^2) + \cdots + F(10^9) \equiv 9 \cdot F(1) \equiv 0 \pmod{9},$$

while $10^{10} \equiv 1 \pmod{9}$. □

10.3 Exercises

(i) Show that there does not exist a polynomial $F(x)$ with integer coefficients such that $F(19) = 15$ and $F(62) = 70$.

(ii) Show that if the polynomial $F(x)$ with integer coefficients satisfies $F(r) = F(s) + 1$ for some integers r, s, then $r = s + 1$ or $r = s - 1$.

   (iii) Suppose that for the polynomial $F(x)$ with integer coefficients there exist distinct integers a_1, a_2, a_3, a_4, a_5 such that $|F(a_i)| = 73$ for all $i = 1, \ldots, 5$. Show that $F(x)$ has no integer zero.

   (iv) Given an arbitrary integer r, prove the following: If the polynomial $F(x)$ with integer coefficients takes on the values r or $r + 1$ for each of the five distinct integers $a_1, a_2, \ldots, a_5$, then $F(a_1) = F(a_2) = F(a_3) = F(a_4) = F(a_5)$.

   *(v) Find all polynomials $F(x)$ with integer coefficients such that for each integer n there exists an integer k with $F(k) = n$.

10.4 Integer-Valued Polynomials

In contrast to the above we will now study polynomials $F(x)$ with real coefficients and with the property that $F(x) \in \mathbb{Z}$ for all $x \in \mathbb{Z}$. Apart from polynomials with integer coefficients, these include others, for instance $F(x) = \frac{1}{2}x(x-1)$, since $x(x-1)$ is even for all $x \in \mathbb{Z}$. Note that for integers $k \geq 2$ the value $F(k)$ is equal to the binomial coefficient $\binom{k}{2}$. This example was not arbitrarily chosen; indeed, in order to be able to easily write down polynomials with the desired property, it will be convenient to extend the concept of binomial coefficient. For any real number t and any natural number k we define the *generalized binomial coefficient* $\binom{t}{k}$ by

$$\binom{t}{k} = \frac{t(t-1)\cdots(t-k+1)}{k!}.$$

We remark that if t is a natural number and $t \geq k$, then this new definition coincides with the original one.

Theorem 1. *For any integer t and any natural number k the generalized binomial coefficient $\binom{t}{k}$ is an integer.*

PROOF.   Given $t \in \mathbb{Z}$ we find a rational number r such that $r \cdot k! > k - t$, and we set $n = t + r \cdot k!$. By 10.1, the polynomial

$$F(x) = x(x-1)\cdots(x-k+1)$$

satisfies $F(t) \equiv F(n) \pmod{k!}$, since $t \equiv n \pmod{k!}$. But then

$$F(n) = \binom{n}{k} \cdot k!,$$

where the binomial coefficient $\binom{n}{k}$ is a natural number (since $n \in \mathbb{N}$ and $n > k$, and so $\binom{n}{k}$ is a coefficient in the binomial expansion of $(A + B)^n$). Together,

$$F(t) \equiv F(n) = \binom{n}{k}k! \equiv 0 \pmod{k!},$$

and thus the integer $F(t)$ is divisible by $k!$. Therefore, $\binom{t}{k} = \frac{F(t)}{k!}$ is an integer. $\qquad\Box$

Theorem 2. *Given any polynomial $F(x) = b_n x^n + b_{n-1} x^{n-1} + \cdots + b_1 x + b_0$, where $b_0, b_1, \ldots, b_n$ are real numbers, there exist real numbers $a_0, a_1, \ldots, a_n$ such that*

$$F(x) = a_0 + a_1 \binom{x}{1} + \cdots + a_{n-1} \binom{x}{n-1} + a_n \binom{x}{n}.$$

PROOF. We use induction on the degree of the polynomial $F(x)$. If $F(x)$ is the zero polynomial or $\deg F(x) = 0$, then the assertion is clear. We will now assume that given any natural number n, the assertion holds for all polynomials of degree less than n. Our aim is to prove it for a polynomial $F(x) = b_n x^n + \cdots + b_1 x + b_0$ of degree n. We set $a_n = b_n \cdot n!$ and consider the polynomial $G(x) = F(x) - a_n \binom{x}{n}$. Since the polynomials $F(x)$ and $a_n \binom{x}{n}$ have the same degree n and the same coefficient b_n of x^n, $G(x)$ is either the zero polynomial or has smaller degree than $F(x)$. It follows from the induction hypothesis that there exist real numbers $a_0, \ldots, a_{n-1}$ such that $G(x) = a_0 + a_1 \binom{x}{1} + \cdots + a_{n-1} \binom{x}{n-1}$. But then

$$F(x) = G(x) + a_n \binom{x}{n} = a_0 + a_1 \binom{x}{1} + \cdots + a_{n-1} \binom{x}{n-1} + a_n \binom{x}{n},$$

which was to be shown. $\qquad\Box$

Theorem 3. *For any polynomial $F(x) = a_0 + a_1 \binom{x}{1} + \cdots + a_n \binom{x}{n}$, where $a_0, \ldots, a_n$ are real numbers, the following holds: $a_0, a_1, \ldots, a_n$ are integers if and only if $F(t)$ is an integer for any integer t.*

PROOF. If $a_0, a_1, \ldots, a_n$ are integers, then by Theorem 1, $F(t)$ must be an integer for any $t \in \mathbb{Z}$.

Let us now assume that $F(t)$ is an integer for any integer t; we will prove by induction that $a_0, a_1, \ldots, a_n$ are integers. Certainly, $a_0 = F(0)$ is an integer. Next we assume that for some natural number $k \leq n$, the coefficients $a_0, a_1, \ldots, a_{k-1}$ are integers, and we aim to show that a_k is an integer as well. We have

$$F(k) = a_0 + a_1 \binom{k}{1} + \cdots + a_{k-1} \binom{k}{k-1} + a_k,$$

since $\binom{k}{k} = 1$ and $\binom{k}{m} = 0$ for any integer $m \geq k + 1$. Hence

$$a_k = F(k) - a_0 - a_1 \binom{k}{1} - \cdots - a_{k-1} \binom{k}{k-1},$$

and thus a_k is an integer, which completes the proof. $\qquad\Box$

We note that in the second part of the proof we used only the fact that $F(0), F(1), \ldots, F(n)$ are integers. Then it followed that $a_0, a_1, \ldots, a_n$ are integers, and therefore $F(t)$ has to be an integer for any integer t. This fact is worth being formulated as a separate result.

Consequence. *If a polynomial $F(x)$ with real coefficients and degree n is such that $F(0), F(1), \ldots, F(n)$ are integers, then $F(t)$ is an integer for any integer t.*

PROOF. This follows from the above remark and the fact that by Theorem 2 we can express the polynomial $F(x)$ in the form $a_0 + a_1\binom{x}{1} + \cdots + a_n\binom{x}{n}$. $\quad\square$

10.5 Examples

(i) Find all real numbers a, b, c with the property that for each integer t the number $at^2 + bt + c$ is an integer.

SOLUTION. Using Theorem 2 in 10.4, we express the trinomial $ax^2 + bx + c$ as

$$ax^2 + bx + c = 2a\binom{x}{2} + ax + bx + c = 2a\binom{x}{2} + (a+b)\binom{x}{1} + c.$$

Hence by Theorem 3 in 10.4, $at^2 + bt + c$ is an integer for all integers t if and only if $2a$, $a + b$, c are integers. $\quad\square$

(ii) Decide whether the polynomial $F(x) = \frac{1}{25}(x^9 - 2x^5 + x)$ takes on integer values $F(t)$ for all integers t.

SOLUTION. By the consequence in 10.4 it suffices to establish whether $F(0), F(1), \ldots, F(9)$ are integers. It is easy to check that $F(0)$, $F(1)$, $\ldots$, $F(4)$ are indeed integers. However, $F(5)$ is not an integer, since $5^9 - 2 \cdot 5^5 + 5 \equiv 5 \not\equiv 0 \pmod{25}$. $\quad\square$

(iii) Decide whether there exists a polynomial $F(x)$ with real coefficients and degree $n \in \mathbb{N}$ such that $F(0), F(1), \ldots, F(n-1)$, and $F(n+1)$ are integers, while $F(n)$ is not an integer.

SOLUTION. By Theorem 2 in 10.4 we can express $F(x)$ as

$$F(x) = a_0 + a_1\binom{x}{1} + \cdots + a_n\binom{x}{n}.$$

The numbers $F(0), F(1), \ldots, F(n-1)$ must be integers, and thus $a_0, a_1, \ldots$, $a_{n-1} \in \mathbb{Z}$, while $F(n)$ cannot be an integer, which means that $a_n \notin \mathbb{Z}$ (see the proof of Theorem 3 in 10.4). Let us now evaluate $F(n+1)$:

$$F(n+1) = a_0 + a_1\binom{n+1}{1} + \cdots + a_{n-1}\binom{n+1}{n-1} + a_n\binom{n+1}{n};$$

thus $F(n + 1)$ will be an integer if and only if $a_n \binom{n+1}{n} = (n + 1)a_n$ is an integer. It is easy to see that if we choose, for instance, $a_0 = a_1 = \cdots = a_{n-1} = 0$, $a_n = 1/(n + 1)$, then all conditions are satisfied. The conditions of the problem are therefore satisfied by the polynomial

$$F(x) = \frac{1}{n + 1} \binom{x}{n} = \frac{x(x - 1) \cdots (x - n + 1)}{(n + 1)!}. \qquad \square$$

10.6 Exercises

(i) Find all real numbers a, b, c, d with the property that $at^3 + bt^2 + ct + d$ is an integer for all $t \in \mathbb{Z}$.

(ii) Decide whether the polynomial $F(x) = \frac{1}{15}(x^7 - x^5 - x^3 + x)$ takes on an integer value $F(t)$ for any integer t.

*(iii) For each $n \in \mathbb{N}$ decide whether there exists a polynomial $F(x)$ with real coefficients and degree n such that $F(0), F(1), \ldots, F(n - 1)$, $F(n + 1)$, $F(n + 2)$ are integers, while $F(n)$ is not an integer.

*(iv) Determine the $n \in \mathbb{N}$ for which there exists a polynomial $F(x)$ with real coefficients and degree n such that $F(0), F(1), \ldots, F(n - 1)$, $F(n + 1)$, $F(n + 2)$, $F(n + 3)$ are integers, while $F(n)$ is not an integer.

(v) Let p be a prime and suppose that the polynomial $F(x)$ of degree $k < p$ and with integer coefficients has the following property: For each integer x the value $F(x)$ is an integer multiple of p. Show that then all coefficients of $F(x)$ are integer multiples of p.

We will now deal with the decomposition of a polynomial with integer coefficient into a product of two polynomials with integer or rational coefficients.

10.7 Example

Show that there do not exist polynomials $F(x), G(x)$ with integer coefficients and degrees at least one that satisfy the identity

$$F(x) \cdot G(x) = x^{10} + 2x^9 + 2x^8 + 2x^7 + 2x^6 + 2x^5 + 2x^4 + 2x^3 + 2x^2 + 2x + 2.$$

SOLUTION.   We assume to the contrary that such polynomials $F(x), G(x)$ exist, and we write $F(x) = a_s x^s + \cdots + a_1 x + a_0$, $G(x) = b_t x^t + \cdots + b_1 x + b_0$, where $a_0, a_1, \ldots, a_s, b_0, b_1, \ldots, b_t$ are integers. Then $a_0 b_0 = 2$, so exactly one of the numbers a_0, b_0 is even. We assume that it is a_0 (if it were b_0, we could simply interchange the polynomials $F(x), G(x)$). Hence a_0 is even and b_0 is odd. Equating the coefficients of x, we obtain $2 = a_0 b_1 + a_1 b_0$; thus $a_1 b_0 = 2 - a_0 b_1$, which is an even number. Since b_0 is odd, a_1 must be even. Comparing coefficients of x^2 and simplifying, we find that $a_2 b_0 = 2 - a_1 b_1 - a_0 b_2$ is even, and thus a_2 is even. If we continue in this way, we find after s steps that a_s is also an even number. But this is a contradiction to the fact that $a_s \cdot b_t = 1$; hence the polynomials $F(x)$, $G(x)$ with the given properties do not exist.   □

With a similar argument one can obtain a proof of the following useful criterion.

10.8 Eisenstein's Irreducibility Criterion

Theorem. *Let p be a prime and*

$$F(x) = a_n x^n + a_{n-1} x^{n-1} + \cdots + a_1 x + a_0$$

a polynomial with integer coefficients such that $a_0, a_1, \ldots, a_{n-1}$ are divisible by p, the coefficient a_n is not divisible by p, and a_0 is not divisible by p^2. Then there do not exist polynomials $G(x)$, $H(x)$ with rational coefficients and with degrees at least one such that $F(x) = G(x) \cdot H(x)$.

A proof can be found, for instance, in [4].

10.9 Examples

(i) Show that the polynomial $4x^5 + 7x^4 - 14x^3 + 49x^2 - 28$ cannot be written as the product of two polynomials with rational coefficients and degrees at least one.

SOLUTION.   The assertion follows from 10.8 with $p = 7$.   □

(ii) Find a polynomial with rational coefficients and smallest degree that has the irrational number $\alpha = \sqrt[1987]{2}$ as a zero.

SOLUTION.   The number α is certainly a zero of the polynomial $x^{1987} - 2$. Let $F(x)$ denote the desired polynomial, which has the smallest degree among all polynomials with rational coefficients and that have α as a zero. We will show that $F(x) = c \cdot (x^{1987} - 2)$, where $c \in \mathbb{Q} \setminus \{0\}$. To do this, we divide the polynomial $x^{1987} - 2$ by $F(x)$ with remainder:

$$x^{1987} - 2 = H(x) \cdot F(x) + R(x)$$

(see Theorem 3.9 in Chapter 1), where $R(x)$ is either the zero polynomial or $\deg R(x) < \deg F(x)$. If we think about the division algorithm, we easily see that both $H(x)$ and $R(x)$ have rational coefficients. Also,

$$R(\alpha) = (\alpha^{1987} - 2) - H(\alpha) \cdot F(\alpha) = 0$$

(since $F(\alpha) = 0$), and thus the polynomial $R(x)$ with rational coefficients has α as a zero. From the definition of $F(x)$ it follows that the degree of $R(x)$ (if it is defined) cannot be less than that of $F(x)$. Hence $R(x)$ is the zero polynomial, and therefore

$$x^{1987} - 2 = H(x) \cdot F(x).$$

By 10.8, however, the polynomial $x^{1987} - 2$ cannot be written as a product of two polynomials with rational coefficients and of degrees at least one. Since $F(x)$ has α as a zero, we have $\deg F(x) \geq 1$, and thus $\deg H(x) = 0$, i.e., $H(x) = 1/c$, where $c \neq 0$. Hence we have indeed $F(x) = c \cdot (x^{1987} - 2)$.   □

10.10   Exercises

Show that the polynomials (i)–(iv) cannot be decomposed into a product of two polynomials (with rational coefficients and degrees at least one).

(i) $x^5 - 5x^4 + 25x - 5$.   **(ii)** $x^6 - 6x^3 + 12$.

(iii) $3x^7 - 21x^4 - 63x^2 - 21$.   **(iv)** $5x^5 + 25x^4 - 100x^2 + 125x - 25$.

(v) Find a polynomial with rational coefficients and smallest degree that has $\sqrt[125]{75}$ as a zero.

In the concluding example of this chapter we will present a result that simplifies the search for rational roots of a polynomial with integer coefficients. We have already illustrated its use in 3.27.(ii) of Chapter 1.

10.11   Example

Suppose that the polynomial

$$F(x) = a_n x^n + a_{n-1} x^{n-1} + \cdots + a_1 x + a_0$$

with integer coefficients has a zero $\alpha = r/s$, where $r \in \mathbb{Z}$, $s \in \mathbb{N}$, and r, s are relatively prime. Show that then for each integer k the number $F(k)$ is divisible by $r - ks$.

SOLUTION.   We set $G(x) = s^n F(x/s)$. Then $G(x)$ is a polynomial with integer coefficients:

$$G(x) = a_n x^n + a_{n-1} s x^{n-1} + \cdots + a_1 s^{n-1} x + a_0 s^n.$$

We have $G(r) = s^n F(\alpha) = 0$ and $G(ks) = s^n F(k)$. Let us set $m = |r - ks|$. If $r - ks = 0$, then $k = \alpha$, and so $F(k) = 0$ is indeed divisible by $r - ks = 0$. We now assume that $r - ks \neq 0$, i.e., $m \in \mathbb{N}$. Since $ks \equiv r \pmod{m}$, we have by 10.1,

$$G(ks) \equiv G(r) = 0 \pmod{m},$$

which implies $m \mid G(ks)$, i.e., $m \mid s^n F(k)$. Now, m and s are relatively prime integers (if there were a prime p dividing both m and s, it would also have to divide r, which contradicts $(r, s) = 1$), and thus by 1.14.(ii) we have $m \mid F(k)$, or $r - ks \mid F(k)$, which was to be shown. $\square$

4

Hints and Answers

1   Hints and Answers to Chapter 1

1.4 (i) Add identities (3) and (4).
(ii) Subtract (4) from (3).
(iii) After multiplying by 2, use (1).
(iv) The summand $\binom{50}{22}(\sqrt{2})^{22}(\sqrt{3})^{28}$.
(v) $m = 3$, $n = 6$.

1.7 Instead of B in (5) write $-B$ and set $n = 2m$ in (i) and $n = 2m - 1$ in (ii).

2.3 (i) n^2.   (ii) For $n = 2k$: $S = -k$; for $n = 2k + 1$: $S = k + 1$.
(iii) $5n(n + 1)/2$.   (iv) 17.   (v) $2^{n+1} - 2$.
(vi) $\frac{2^n + (-1)^{n+1}}{3 \cdot 2^{n-1}}$.   (vii) $3(2^n - 1) - n$.   (viii) $2n + \frac{(x^{2n} - 1) \cdot (x^{2n+2} + 1)}{x^{2n} \cdot (x^2 - 1)}$.

2.8 (i) $(n + 2) \cdot 2^{n-1}$.
(ii) 1. [Compare with the sum in 2.6 and use (4).]
(iii) $n(n - 1) \cdot 2^{n-2}$.
(iv) $n(n + 1)2^{n-2}$. [Note that $k^2 = k(k - 1) + k$ and use (iii).]
(v) $\frac{n}{n+1}$. [Consider $(n + 1)S$.]
(vi) $\frac{n \cdot 2^{n+1} + 1}{(n+1)(n+2)}$. [Consider $(n + 1)(n + 2)S$.]
(vii) $\frac{3^{n+1} - 2n - 3}{n+1}$. [Consider $(n + 1)S$; after manipulations as in 2.5 (ii) use 1.2.]
(viii) $2^{2n-1}/(2n)!$. [Consider $(2n)!S$.]

2.10 (i) Consider the coefficient of x^p in the polynomial $(1 + x)^{n+m} = (1 + x)^n \cdot (1 + x)^m$.

(ii) Follows from (i) by choosing $m = n$, $p = n - r$.

(iii) $\frac{(2n-1)!}{[(n-1)!]^2}$. [Use the method of 2.5.]

2.15 Determine the sums $S_k(n)$ successively from the formulas $B(3, n) = S_3(n) + 3 \cdot S_2(n) + 2 \cdot S_1(n)$, $B(4, n) = S_4(n) + 6 \cdot S_3(n) + 11 \cdot S_2(n) + 6 \cdot S_1(n)$, $B(5, n) = S_5(n) + 10 \cdot S_4(n) + 35 \cdot S_3(n) + 50 \cdot S_2(n) + 24 \cdot S_1(n)$.

2.22 (i) $(n^3 + 2n)/3$.   (ii) n^4.   (iii) $n^2(n^2 + 2n + 2)$.

2.25 (i) $n(2n - 1)(2n + 1)/3$.   (ii) $k(k + 1)(k + 2)(3k + 1)/12$.

(iii) $n^2(2n^2 - 1)$.

(iv) For $n = 2k$: $S = -k^2(4k + 3)$; for $n = 2k + 1$: $S = (k + 1)^2 \cdot (4k + 1)$.

(v) $n(4n^2 + 15n + 17)/36$. [First evaluate each term in the sum S.]

2.30 (ii) $[n(n - 1)q^{n+1} - 2(n^2 - 1)q^n + n(n + 1)q^{n-1} - 2]/[(q - 1)^3]$.

(iii) For $t = 1$: $S(1) = n(n - 1)/2$; for $t \neq 1$: $S(t) = \frac{(n-1)t^n - n \cdot t^{n-1} + 1}{(t-1)^2}$.

(iv) For $v = 1$: $S(1) = n(n + 1)(2n + 1)/6$; for $v \neq 1$: $S(v) = R_2(v, n)/v$.

(v) For $x = 1$: $S(1) = n^2$; for $x \neq 1$: $S(x) = \frac{(x^n - 1)^2}{x^{n-1}(x-1)^2}$.

(vi) For $q = 1$: $R_3(1, n) = S_3(n) = [n(n + 1)/2]^2$; for $q \neq 1$:

$$R_3(q, n) = \frac{q}{(q-1)^4}[n^3 q^{n+3} - (3n^3 + 3n^2 - 3n + 1)q^{n+2}$$
$$+ (3n^3 + 6n^2 - 4)q^{n+1} - (n + 1)^3 q^n + q^2 + 4q + 1].$$

(vii) $R_k(q, n) = \frac{1}{q-1}\left[n^k \cdot q^{n+1} + \sum_{i=1}^{k}(-1)^i \binom{k}{i}R_{k-i}(q, n)\right]$.

2.33 (i) $(n - 1) \cdot 2^{n+1} + 2$.   (ii) $(n^2 - n + 2) \cdot 2^{n+1} - 4$.

(iii) $\frac{3(2^n - 1) - 2n}{2^{n-1}}$.   (iv) $\frac{2^n + (-1)^{n+1}(6n+1)}{9 \cdot 2^{n-1}}$.

2.38 Try to find the sums in 2.33 in the form

(i) $d + (a + bn) \cdot 2^n$.   (ii) $d + (a + bn + cn^2) \cdot 2^n$.

(iii) $d + (a + bn) \cdot \left(\frac{1}{2}\right)^n$.   (iv) $d + (a + bn) \cdot \left(-\frac{1}{2}\right)^n$.

2.40 (i) $\frac{1}{k} - \frac{1}{n+1}$.   (ii) $\frac{n}{2n+1}$.   (iii) $\frac{n(3n+5)}{4(n+1)(n+2)}$.

(iv) $\frac{n}{3n+1}$.   (v) $\frac{n(n+3)}{4(n+1)(n+2)}$.   (vi) $\frac{n(n+2)}{3(2n+1)(2n+3)}$.

2.42 (i) $\frac{n(n+2)}{n+1}$.   (ii) $\frac{n(n^2+2n+3)}{2(n+1)}$.   (iii) $\frac{n(2n+3)}{n+1}$.

2.44 (i) $\frac{n(n+1)}{2(2n+1)(2n+3)}$. [Consider $2S$ and use the method of 2.43.]

(ii) $\frac{n(n+1)(n^2+n+1)}{6(2n+1)}$. [Use the equalities $\frac{k^4}{(2k-1) \cdot (2k+1)} = \frac{k^2}{4} + \frac{1}{16} + \frac{1}{32(2k-1)} - \frac{1}{32(2k+1)}$ for $k = 1, 2, \ldots, n$.]

3.3 (i) $F(x) = \sum_{i=0}^{k}\left(2\binom{2k}{2i} - \binom{k}{i}\right)x^{2k}$. [Use the binomial theorem.]

(ii) $G(x) = 1 + x + x^2 + \cdots + x^{2^k - 1}$. [Manipulate $(1 - x)F(x)$.]

(iii) $F(x)$. [The polynomial $H(x) = F(-x) = (1 + x^2 + x^3)^{1000}$ has the same coefficients at even powers of x as $F(x)$. The coefficient of x^{20} of the polynomial $(1 + yx^2 + x^3)^{1000}$ is a polynomial $P(y)$ in the variable y of the form $c_1 y + c_2 y^4 + c_3 y^7 + c_4 y^{10}$ with positive coefficients c_i. Hence $P(1) - P(-1) = 2c_1 + 2c_3 > 0$.]

(iv) Verify by induction that $(x^2 + x + 1)F_k(x) = x^{2^{k+1}} - x^{2^k} + 1$.

3.6 (i) $S = F(1) = 1$.

(ii) After multiplication by $a^2(x - p)(x - q)$ and simplification, find a quadratic polynomial with discriminant $D = (p - q)^2 + 4a^4 > 0$ and with nonzero values for $x = p$ and for $x = q$.

(iii) Does not exist: For $a = 0$ it has at most one zero, or infinitely many zeros; for $a > 0$, $\lambda > -c + b^2/(4a)$ it has no real zeros; for $a < 0$, $\lambda > -c$ it has one negative zero.

(iv) $a = \pm \sqrt[20]{2^{20} - 1}$, $b = -\frac{a}{2}$, $p = -1$, $q = \frac{1}{4}$. [Comparing coefficients of x^{20}, you obtain $2^{20} - a^{20} = 1$. Substituting $x = \frac{1}{2}$ leads to the equation $\left(\frac{a}{2} + b\right)^{20} + \left(\frac{1}{4} + \frac{p}{2} + q\right)^{10} = 0$, which implies $b = -\frac{a}{2}$. Then the original equation of polynomials can be rewritten as $\left(x - \frac{1}{2}\right)^{20} = \left(x^2 + px + q\right)^{10}$, which implies $x^2 + px + q = \left(x - \frac{1}{2}\right)^2$.]

3.8 (i) $F(x) = x^2 + x$.

(ii) No such polynomial exists: $F(x^2 + 1)$ has even degree.

(iii) $F(x) = x^2 - 3x$. [Since $\deg F(x) = 2$, set $F(x) = ax^2 + bx + c$.]

(iv) Set $y = x - 1$ and $G(y) = F(y - 1)$, and compare with 3.7; apart from the zero polynomial, the solutions are the polynomials $F(x) = (x + 1)^n$, where $n \in \mathbb{N}_0$.

(v) $F(x) = x^n$. [From a comparison of the degrees of $F(F(x))$ and $(F(x))^n$ it follows that $\deg F(x) = n$; therefore, $F(x) = a_n x^n + a_{n-1} x^{n-1} + \cdots + a_0$. Comparing the coefficients of the powers x^{n^2}, you then obtain $a_n^{n+1} = a_n^n$, that is, $a_n = 1$. Assume that $a_k \neq 0$ for some $k < n$ and take the largest such k. The condition $F(F(x)) = (F(x))^n$ can be rewritten as

$$a_k(x^n + a_k x^k + \cdots + a_0)^k + a_{k-1}(x^n + a_k x^k + \cdots + a_0)^{k-1} + \cdots + a_0 = 0,$$

which is a contradiction, since the polynomial on the left has degree nk.]

3.13 (i) (a) 6; (b) $6x$. [Use (25), and replace x by 1 and -1.]

(ii) 0 is a zero only for even n, and it has multiplicity 2. [Use 1.2.]

(iii) For even n. [Use (5) and determine when $x + 1$ divides $x^{n+1} + 1$.]

(iv) $F(x) = 4x^4 - 27x^3 + 66x^2 - 65x + 24$. [From the equality of the polynomials

$$F(x) = (x - 1)^2 P(x) + 2x = (x - 2)^3 Q(x) + 3x$$

it follows that you must find a polynomial of smallest degree such that $(x - 1)^2$ divides $(3x - 4)Q(x) + x$. From this you can see that $\deg Q(x) \geq 1$.

For $\deg Q(x) = 1$, that is, $Q(x) = ax + b$, you obtain the condition

$$(3x - 4)(ax + b) + x = R(x)(x - 1)^2,$$

which implies $R(x) = 3a$, and comparing coefficients, $a = 4$, $b = -3$.]

(v) The desired remainder is $F(1) = 2\,000$. By Bézout's theorem there exists a polynomial H such that $F(x) = F(1) + (x - 1)H(x)$, which implies $F(x^5) = F(1) + (x^5 - 1)H(x^5) = F(1) + (1 + x + x^2 + x^3 + x^4)(x - 1)H(x^5)$.

3.17 (i),(ii) Use the method of 3.16 (ii); verify the equality of the polynomials by substituting 0, b, $-b$, c, $-c$ for x (in (i), four of them suffice). Special care must be taken in the case where some of these quantities coincide.

(iii) Set $x = a$, $x = b$, $x = c$.

(iv) Set $x = a$, $x = b$, $x = c$.

(v) Start with $2x^3 + 3x^2 + x = x(x + 1)(2x + 1)$, show that $0, -1, -\frac{1}{2}$ are zeros of the given polynomial, and use 3.15 (ii).

(vi) Show that $F(x)$ has the zeros $0, 1, \ldots, 25$; therefore, $F(x) = x(x - 1) \cdots (x - 25) \cdot G(x)$ for an appropriate polynomial $G(x)$. By substituting you see that $G(x) = G(x - 1)$, and therefore $G(x)$ and the constant polynomial $G(0)$ have the same values at all integers; thus $G(x) = G(0)$ by 3.15 (i). The given equation is therefore satisfied exactly by the polynomials $a \cdot x \cdot (x - 1) \cdots (x - 25)$, where a is an arbitrary number.

3.20 (i) $p = q = 0$; $p = 1$ and $q = -2$. [Use (28).]

(ii) Proceed as in 3.19.

(iii) Add the equations $x_i^k(ax_i^2 + bx_i + c) = 0$ for both roots x_i.

3.22 (i) If $x_1 + x_2 = 1$, then by (29) you have $x_3 = -\frac{1}{2}$, which implies $\lambda = -3$ and $x_{1,2} = (1 \pm \sqrt{13})/2$.

(ii) From (29) and from the hypothesis it follows that $-q = x_1 x_2 x_3 = x_1 + x_2 = -x_3$; hence $q = x_3$ is a root, and therefore $q^3 + pq + q = 0$. On the other hand, if the numbers p, q are such that at least one of them is nonzero and $q^3 + pq + q = 0$, then one zero of the given polynomial is the sum of the reciprocals of the other two zeros.

(iii) From (29) and from the hypothesis it follows that $-p/2$ is a zero of the polynomial, and therefore $8r = 4pq - p^3$. On the other hand, if the numbers p, q, r satisfy $8r = 4pq - p^3$, then one zero of the given polynomial is the sum of the other two.

(iv) $G(x) = x^3 - qx^2 + prx - r^2$; $H(x) = x^3 + 2px^2 + (p^2 + q)x + (pq - r)$.

(v) $r = (9pq - 2p^3)/27$. [Show that one zero is $-p/3$.]

(vi) Use (29) and substitute.

(vii) $x + y + z = p + q + r - b - c$. Consider the monic cubic polynomial

$$F(t) = \left(1 - \frac{x}{t} - \frac{y}{t - b} - \frac{z}{t - c}\right)t(t - b)(t - c)$$

with three different zeros p, q, and r. Vieta's relation gives that $-p - q - r$ equals the coefficient of t^2 in $F(t)$, which is $-b - c - x - y - z$, and the result follows. Let us add that the numbers x, y, z always exist and are unique:

$$x = \frac{pqr}{bc}, \quad y = \frac{(p-b)(q-b)(r-b)}{b(b-c)}, \quad z = \frac{(p-c)(q-c)(r-c)}{c(c-b)}.$$

3.25 Use the method of 3.24 to show that 4 is the unique real zero of the polynomial $x^3 - 6x - 40$ in (i), resp. $\sqrt{5}$ is the unique real zero of $x^3 - 3x - 2 \cdot \sqrt{5}$ in (ii).

3.28 (i) $F(x)$ has a double zero at 1 and the two simple complex zeros $-2 + i$, $-2 - i$.

(ii) $F(x)$ has the zeros $-3, \frac{1}{2}, (-1 + \sqrt{29})/2, (-1 - \sqrt{29})/2$.

(iii) $F(x)$ has the double zero -1 and simple zeros $1, -\frac{2}{3}, -2 + \sqrt{3}, -2 - \sqrt{3}$.

(iv) Let α denote the common irrational zero of $F(x)$ and $G(x)$. Substituting $\alpha^2 = -a\alpha - b$ into the equation $F(\alpha) = 0$, you obtain $(a^2 - b + p)\alpha + (ab + q) = 0$, which implies $p = b - a^2$, $q = -ab$, and this leads to $F(x) = (x - a)G(x)$.

(v) The root in question is $x = 1$.

4.7 (i) $x^2 + (p^3 - 3pq)x + q^3$.

(ii) $x^2 + (-p^2 + 2q + p)x + (-p^3 + 3pq + q^2 + q)$.

(iii) $b \in \{0, \sqrt{\sqrt{5} - 1}, -\sqrt{\sqrt{5} - 1}\}$.

4.9 (i) $x^2 - 6x + 8$ or $x^2 + (13/3)x - 52/9$.

(ii) $x^2 + 8x + 15$ or $x^2 - 8x + 15$.

(iii) $x^2 - 5x + 6$ or $x^2 + (3/2)x - 19/12$.

(iv) For the numbers $p = a + b$ and $q = ab$, find two "independent" equations; from those you will realize by excluding the number p that q is a root of the given equation of degree 6. Since $a^3(a+1) = b^3(b+1) = 1$, you have $1 = a^3(a+1)b^3(b+1) = q^3(p+q+1)$. From the equation $a^4 + a^3 = b^4 + b^3$, upon dividing by $a - b$, obtain the second equation for the numbers p and q:

$$0 = a^3 + a^2 b + ab^2 + b^3 + a^2 + ab + b^2$$
$$= (a^3 + a^2) + (b^3 + b^2) + q(p + 1)$$
$$= \frac{1}{a} + \frac{1}{b} + q(p + 1) = \frac{p}{q} + q(p + 1).$$

4.14 (i) $x^3 - 2x^2 - 47x - 144$.

(ii) Use the fact that a, b, c are three distinct zeros of a cubic polynomial with vanishing quadratic term.

(iii) Use the formula for $s_4 = x_1^4 + x_2^4 + x_3^4$ from 4.12.

(iv) $x^3 + 2px^2 + (p^2 + q)x + (pq - r)$.

(v) Transform the supposed equation into $(a + b)(a + c)(b + c) = 0$ and then discuss (for example) the case $a + b = 0$, in which $a^n + b^n = 0$ for any odd n.

4.16 (i) $(x_1 + x_2 + x_3)(x_1 x_2 + x_1 x_3 + x_2 x_3)$.
   (ii) $(x_1 + x_2 + x_3)(2x_1^2 + 2x_2^2 + 2x_3^2 - x_1 x_2 - x_1 x_3 - x_2 x_3)$.
   (iii) $(x_1 x_2 + x_1 x_3 + x_2 x_3)^2$.
   (iv) $24 x_1 x_2 x_3$.

5.3 (i) $\{(x_1, x_2, \ldots, x_n)\}$, where $x_k = k - n/2$ for $k = 1, 2, \ldots, n$.
   (ii) $\{(x_1, x_2, \ldots, x_n)\}$, where $x_k = (n - 2k + 2)/2$.
   (iii) $\{(x_1, x_2, \ldots, x_n)\}$, where $x_1 = (n+1)(2-n)/2$ and $x_k = 1$ $(k > 1)$.

5.6 (i) $\{((a_1 + a_2 + a_3 + a_4)/2, (a_1 + a_2 - a_3 - a_4)/2, (a_1 - a_2 + a_3 - a_4)/2, (a_1 - a_2 - a_3 + a_4)/2)\}$.
   (ii) For $p = -3$: no solutions; for $p = 0$: infinitely many solutions of the form $(r, s, 1 - r - s)$, where $r, s \in \mathbb{R}$; for $p \in \mathbb{R} \setminus \{0, -3\}$: the unique solution $x_1 = x_2 = x_3 = 1/(p+3)$.
   (iii) Necessary and sufficient condition for solvability: $a_1 a_2 + a_3 a_4 = a_1 a_3 + a_2 a_4 = a_1 a_4 + a_2 a_3$, which holds if and only if at least three of the numbers a_1, a_2, a_3, a_4 are equal. If $\{1, 2, 3, 4\} = \{p, q, r, s\}$ and $a_p = a_q = a_r$, there exists the unique solution $x_p = x_q = x_r = \frac{1}{2} a_p^2$, $x_s = a_p(a_s - \frac{1}{2} a_p)$.
   (iv) $\{(\frac{a_3 - a_2}{a}, \frac{a_1 - a_3}{a}, \frac{a_2 - a_1}{a})\}$, where $a = a_1 + a_2 + a_3$.
   (v) If $a_1 = a_2 = a_3$, then there are infinitely many solutions of the form $(r, s, 3 - r - s)$, where $r, s \in \mathbb{R}$; otherwise, there is the unique solution $x_1 = x_2 = x_3 = 1$.
   (vi) $\{(x_1, x_2, \ldots, x_n)\}$, where $x_k = a \cdot \frac{2^{n+1} - (n \cdot 2^k + 1)}{2^{n+1} - 1}$ for $k = 1, 2, \ldots, n$.
   (vii) $\{(x_1, x_2, x_3, x_4)\}$, where $x_1 = (4a_1 + 3a_2 + 2a_3 + a_4)/5$, $x_2 = (3a_1 + 6a_2 + 4a_3 + 2a_4)/5$, $x_3 = (2a_1 + 4a_2 + 6a_3 + 3a_4)/5$, $x_4 = (a_1 + 2a_2 + 3a_3 + 4a_4)/5$.
   (viii) Show that $x_1 + x_2 + \cdots + x_{100} = 0$; comparing this result with the sum of the 1st, 4th, $\ldots$, 97th equation, you obtain $x_{100} = 0$, and from the sum of the 2nd, 5th, $\ldots$, 98th equation you get $x_1 = 0$. With this, the last equation immediately gives $x_2 = 0$, the first one gives $x_3 = 0$, etc.

5.9 (i) $(x_1, x_2) \in \{(-6, -2), (-4, -4)\}$.
   (ii) $(x_1, x_2) \in \{(3, 2), (2, 3)\}$.
   (iii) $(x_1, x_2) \in \{(1, 2), (\frac{7}{3}, \frac{4}{3})\}$. [By subtracting you obtain a linear equation.]
   (iv) $(x_1, x_2) \in \{(-3, 4), (4, -3)\}$.
   (v) $(x_1, x_2) \in \{(4, 3), (4, -3)\}$. [Solve the first equation for x_1 and substitute into the second.]
   (vi) $(x_1, x_2) \in \{(-7, -3), (7, 3)\}$. [Add the two equations.]
   (vii) By adding five times the first and seven times the second equation, you obtain the homogeneous equation $5x_1^2 - 19 x_1 x_2 + 12 x_2^2 = 0$. Upon dividing by x_2^2 $(x_2 \neq 0)$ you get a quadratic equation in x_1/x_2, which leads to the solutions $(x_1, x_2) \in \{(-4, -5), (-3\sqrt{3}, -\sqrt{3}), (3\sqrt{3}, \sqrt{3}), (4, 5)\}$.

(viii) $(x_1, x_2) \in \{(-1, -2), (-\sqrt{2}, -\sqrt{2}), (1, 2), (\sqrt{2}, \sqrt{2})\}$. [Proceed as in (vii).]

(ix) $(x_1, x_2) \in \{(1, -1), (3, -3), (\sqrt{157} - 13, (\sqrt{157} - 13)/2), (-\sqrt{157} - 13, (-\sqrt{157} - 13)/2)\}$. [The second equation is homogeneous; hence use it to determine x_1/x_2.]

(x) $(x_1, x_2) \in \{(2, -?)\} \cup \{(c, 1) \mid c \in \mathbb{R}\}$. [Subtract twice the first equation from the second one.]

(xi) $(x_1, x_2, x_3, x_4) \in \{(1 + \sqrt{2}, 1 - \sqrt{2}, 2 - 3\sqrt{2}, 2 + 3\sqrt{2}), (1 - \sqrt{2}, 1 + \sqrt{2}, 2 + 3\sqrt{2}, 2 - 3\sqrt{2})\}$. We describe a procedure that applies to a more general system

$$x_1 + x_2 = a_1,$$
$$x_1 x_3 + x_2 x_4 = a_2,$$
$$x_1 x_3^2 + x_2 x_4^2 = a_3,$$
$$x_1 x_3^3 + x_2 x_4^3 = a_4,$$

with arbitrary parameters a_i. Let $x^2 + px + q = 0$ be the quadratic equation with zeros x_3 and x_4, i.e., $x^2 + px + q$ is the polynomial $(x - x_3)(x - x_4)$. Since

$$x_1(x_3^2 + px_3 + q) + x_2(x_4^2 + px_4 + q) = a_3 + pa_2 + qa_1,$$
$$x_1 x_3(x_3^2 + px_3 + q) + x_2 x_4(x_4^2 + px_4 + q) = a_4 + pa_3 + qa_2,$$

the coefficients p and q satisfy the two equations

$$a_3 + pa_2 + qa_1 = 0, \quad a_4 + pa_3 + qa_2 = 0,$$

which in our case imply that $p = -4$ and $q = -14$ and hence $x_{3,4} = 2 \pm 3\sqrt{2}$. The computation of the unknowns $x_{1,2}$ is now trivial.

5.12 (i) $(x_1, x_2) \in \{(2, 3), (3, 2)\}$.

(ii) $(x_1, x_2) \in \{(3, 4), (4, 3)\}$.

(iii) $(x_1, x_2) \in \{((-16 + 8\sqrt{10})/7, (-16 - 8\sqrt{10})/7), ((-16 - 8\sqrt{10})/7, (-16 + 8\sqrt{10})/7), (3, 4), (4, 3)\}$.

(iv) $(x_1, x_2) \in \{(3, -2), (-2, 3), (1 + \sqrt{6}, 1 - \sqrt{6}), (1 - \sqrt{6}, 1 + \sqrt{6})\}$.

(v) $(x_1, x_2) \in \{(6, 6), ((-3 + 3\sqrt{5})/2, (-3 - 3\sqrt{5})/2), ((-3 - 3\sqrt{5})/2, (-3 + 3\sqrt{5})/2)\}$.

(vi) Eliminating σ_2 gives $\sigma_1(7\sigma_1^4 - 43\sigma_1^2 + 36) = 0$. Since $\sigma_1 \neq 0$, the expression in parentheses is zero; solve it as a quadratic equation in σ_1^2. The real solutions are $(x_1, x_2) \in \{(2, -1), (-1, 2), (-2, 1), (1, -2)\}$. In addition, there are the four complex solutions $(x_1, x_2) \in \{((3 + i\sqrt{6})/\sqrt{7}, (3 - i\sqrt{6})/\sqrt{7}), ((3 - i\sqrt{6})/\sqrt{7}, (3 + i\sqrt{6})/\sqrt{7}), ((-3 + i\sqrt{6})/\sqrt{7}, (-3 - i\sqrt{6})/\sqrt{7}), ((-3 - i\sqrt{6})/\sqrt{7}, (-3 + i\sqrt{6})/\sqrt{7})\}$.

(vii) First, deal with the cases $x_1 = x_2$, $x_1 = -x_2$. Then, in the case $x_1^2 \neq x_2^2$, divide the first equation by $x_1 - x_2$, and the second one by $x_1 + x_2$. The system has the following nine solutions: $(x_1, x_2) \in$

$\{(0,0), (\sqrt{7}, \sqrt{7}), (-\sqrt{7}, -\sqrt{7}), (\sqrt{19}, -\sqrt{19}), (-\sqrt{19}, \sqrt{19}), (2,3), (3,2),$
$(-2,-3), (-3,-2)\}$.

(viii) $(x_1, x_2) \in \{(1,4), (-1,4), (2,1), (-2,1)\}$. [Solve as a system of symmetric equations in x_1^2 and x_2.]

(ix) $(x_1, x_2) \in \{(a,0), (0,a)\}$, and also $(x_1, x_2) \in \{((1 + i\sqrt{7})a/2, (1 - i\sqrt{7})a/2), ((1 - i\sqrt{7})a/2, (1 + i\sqrt{7})a/2)\}$.

(x) $(x_1, x_2) \in \{(1,2), (2,1)\}$, and also $(x_1, x_2) \in \{((3 + i\sqrt{19})/2, (3 - i\sqrt{19})/2), ((3 - i\sqrt{19})/2, (3 + i\sqrt{19})/2)\}$.

(xi) Using $\sigma_1 = x_1 + x_2$, $\sigma_2 = x_1 x_2$, set up a new system in the unknowns σ_1, σ_2, x_3. Use the first equation to eliminate x_3 from the remaining two, and then eliminate σ_2 from the second equation of the new system. This system has the unique solution $\sigma_1 = 19$, $\sigma_2 = 90$, $x_3 = 12$, and thus the original system has the two solutions $(x_1, x_2, x_3) \in \{(9,10,12), (10,9,12)\}$.

5.14 (i) $(x_1, x_2, x_3) = (3,3,3)$.

(ii) $(x_1, x_2, x_3) \in \{(\frac{1}{3},1,3), (\frac{1}{3},3,1), (3,\frac{1}{3},1), (3,1,\frac{1}{3}), (1,\frac{1}{3},3), (1,3,\frac{1}{3})\}$.

(iii) $(x_1, x_2, x_3) \in \{(0,-1,3), (0,3,-1), (3,0,-1), (3,-1,0), (-1,3,0), (-1,0,3)\}$.

(iv) $(x_1, x_2, x_3) \in \{(2,2,-1), (2,-1,2), (-1,2,2)\}$.

(v) $(x_1, x_2, x_3) \in \{(-1,-1,2), (-1,2,-1), (2,-1,-1)\}$.

(vi) $(x_1, x_2, x_3) \in \{(0,0,a), (0,a,0), (a,0,0)\}$.

(vii) $(x_1, x_2, x_3) \in \{(-1,2,-2), (-1,-2,2), (2,-1,-2), (2,-2,-1), (-2,-1,2), (-2,2,-1), (1,2,-2), (1,-2,2), (2,1,-2), (2,-2,1), (-2,1,2), (-2,2,1)\}$.

(viii) Substitute $y_2 = 2x_2$, $y_3 = -x_3$ and solve for x_1, y_2, y_3. In real numbers, the system has the unique solution $x_1 = 1$, $x_2 = \frac{1}{2}$, $x_3 = -1$; in addition, it has the following six complex solutions: $(x_1, x_2, x_3) \in \{(1, w/2, -w^2), (1, w^2/2, -w), (w, 1/2, -w^2), (w, w^2/2, -1), (w^2, 1/2, -w), (w^2, w/2, -1)\}$, where $w = (-1 + i\sqrt{3})/2$.

(ix) Solve the last equation for x_4 and substitute into the remaining equations. Solve the new system using the method of symmetric polynomials, where the auxiliary polynomial has the solutions $\sigma_1 = 2$, $\sigma_2 = -3$, $\sigma_3 = 0$ and $\sigma_1 = -2$, $\sigma_2 = -3$, $\sigma_3 = 12$. The first one of these gives six solutions of the original system: $x_4 = 2$ and x_1, x_2, x_3 are all permutations of the numbers $-1, 0, 3$. The second one gives six more solutions: $x_4 = -2$ and x_1, x_2, x_3 are all permutations of the zeros of the polynomial $x^3 + 2x^2 - 3x - 12$; these, however, are not rational.

(x) Although the polynomial on the left of the third equation is not symmetric, the method of symmetric polynomials can be used. Rewrite the third equation as $x_1^2 + x_2^2 + x_3^2 = 5 + 2x_3^2$ and express the polynomials on the left sides by way of elementary symmetric polynomials. Solution: $(x_1, x_2, x_3) \in \{(1,2,0), (2,1,0)\}$.

6.3 (i) $\frac{1}{3}$.   (ii) $-\frac{1}{2}$.   (iii) $\frac{1}{5}(15 + 2\sqrt{5})$.   (iv) 1.

6.5 (i) $D = \emptyset$.

(ii) $L > 0$, $R = 0$.

(iii) For $x \geq 0$ you have $\sqrt{x+9} \geq 3$, $\sqrt{x} \geq 0$, so $L \geq 3 > 2 = R$.
(iv) $L \geq 0 > R$.

6.8 (i) 5.   (ii) $2\sqrt{2}$, $-2\sqrt{2}$.   (iii) 2.   (iv) 4, -4.   (v) 3, $\frac{5}{4}$.
(vi) 4.   (vii) $\frac{1}{2}$.

6.10 (i) 1, $-\frac{8}{3}$.   (ii) 1.   (iii) 2, $-\frac{9}{2}$.

6.12 (i) $\frac{5}{2}$.   (ii) $\frac{5}{3}$.   (iii) 2.   (iv) -2, $\frac{7}{2}$.   (v) 2.
(vi) 0, -5.   (vii) 8.   (viii) $[1, 10]$.   (ix) $[2, \infty)$.   (x) 0, -1.
(xi) 1, -2.   (xii) 5.   (xiii) 1.

6.14 (i) -2. [Rewrite $\sqrt{\frac{2(x+1)}{x}} = 2 \cdot \sqrt{\frac{x+1}{2x}}$.]   (ii) 15.   (iii) 1.
(iv) $-(5 + \sqrt{73})/14$, $\frac{3}{5}$, $\frac{4}{5}$. [Square the given equation and use the substitution $u = 1/x$, $v = 1/\sqrt{1 - x^2}$.]

6.16 (i) -15, 1.   (ii) 1, 3.   (iii) $\frac{1}{2}$, $-\frac{1}{2}$.   (iv) 0.   (v) 2, 6.

6.18 (i) 13, -15.   (ii) 1, 20.   (iii) 1.   (iv) -2.   (v) 1, 3.
(vi) 8, $8 \pm 12\sqrt{21}/7$.

6.21 (i) For $a < 0$ there is no root; for $a = 0$, every $x \in [0, \infty)$ is a solution; for $a > 0$ there is the unique root $x = 0$.
(ii) For $a \leq -1$ there is the unique root $x = ((a+1)/2)^2$; no solution for $a > -1$.
(iii) For $a \in [0, 1)$: the unique root $x = a/(1-a^2)$; otherwise, no solutions.
(iv) For $a \leq -1$, one root $x = -1$; for $a = 0$, every $x \in \mathbb{R}_0^+$ is a solution; no solutions for $a \in (-1; 0) \cup (0; \infty)$.
(v) Denote $x_1 = (1-\sqrt{4a+1})/2$; $x_2 = (1+\sqrt{4a+1})/2$. Then for $a < -\frac{1}{4}$ there are no solutions, for $a = -\frac{1}{4}$ there is one root $x = \frac{1}{2}$, for $a \in (-\frac{1}{4}, 0]$ there are two roots x_1, x_2, and for $a > 0$ there is the unique root x_2. [Use the method of Example 6.20.]

6.23 (i) $(x, y) \in \{(27, 1), (1, 27)\}$.
(ii) $(x, y) \in \{(2, 8), (8, 2)\}$.
(iii) $x = 2$, $y = 1$.
(iv) $(x, y) \in \{(10 + 6\sqrt{3}, 10 - 6\sqrt{3}), (10 - 6\sqrt{3}, 10 + 6\sqrt{3})\}$.
(v) $(x, y) \in \{(8, 2), (-2, -8)\}$.
(vi) $(x, y) \in \{(16, 4), (5, 15)\}$.
(vii) $(x, y) \in \{(5, 4), (-5, -4), (15, -12), (-15, 12)\}$. [In the first equation, clear the denominator; in the second one use the substitution $t = \sqrt{x^2 + xy + 4}$.]
(viii) $x = \frac{a+b\sqrt{a^2-b^2}}{\sqrt{1-a^2+b^2}}$, $y = \frac{b+a\sqrt{a^2-b^2}}{\sqrt{1-a^2+b^2}}$. Condition for solvability: $a^2 \geq b^2 > a^2 - 1$. [Add and subtract the given equations, then multiply the new equations together, thus obtaining $x^2 - y^2 = a^2 - b^2$. Substituting this into the original equations, you obtain two linear equations in x, y. Since the final formulas are obtained from the original system by interchanging the pairs (a, b) and $(x, -y)$, it is not necessary to check the result.]

7.3 Set $w = (-1 + i\sqrt{3})/2$.

(i) $[2^n + 2\cos(n\pi/3)]/3$. [Rewrite $(1 + 1)^n + (1 + w)^n + (1 + w^2)^n$ with the binomial theorem and de Moivre's theorem.]

(ii) $[2^n + 2\cos((n-2)\pi/3)]/3$. [Rewrite $(1+1)^n + w^2(1+w)^n + w(1+w^2)^n$.]

(iii) $[2^n + 2\cos((n-4)\pi/3)]/3$. [Rewrite $(1+1)^n + w(1+w)^n + w^2(1+w^2)^n$.]

(iv) $2^n \cos\frac{n\pi}{3}$. [Rewrite w^n with the binomial theorem and de Moivre's theorem, and equate real parts.]

(v) $\frac{2^n}{\sqrt{3}} \sin\frac{n\pi}{3}$. [Rewrite w^n and equate imaginary parts.]

7.5 (i) $(6\tan\alpha - 20\tan^3\alpha + 6\tan^5\alpha)/(1 - 15\tan^2\alpha + 15\tan^4\alpha - \tan^6\alpha)$. [Use 7.4.(i) for $n = 6$.]

(ii) Express the real part of the sum of the complex roots of the equation $x^{2n+1} = 1$. The result also follows from the formula for the sum S_2 in 7.4.(ii).

(iii) For $\varepsilon = \cos\alpha + i\sin\alpha$ rewrite $(\varepsilon + \varepsilon^{-1})^n$.

(iv) $S_1 = 2^n \cos^n(\alpha/2)\cos(\beta + (n\alpha/2))$; $S_2 = 2^n \cos^n(\alpha/2)\sin(\beta + (n\alpha/2))$. [Rewrite $S_1 + iS_2$ with the help of the binomial theorem and de Moivre's theorem, and use the identities $\cos\alpha = 2\cos^2(\alpha/2) - 1$; $\sin\alpha = 2\cos(\alpha/2)\sin(\alpha/2)$.]

(v) Proceed as in 7.4.(ii); in order to make the denominator real, multiply numerator and denominator of $((r \cdot \varepsilon)^{n+1} - 1)/(r \cdot \varepsilon - 1)$ by $r\varepsilon^{-1} - 1$, where $\varepsilon = \cos\alpha + i\sin\alpha$.

(vi) Begin with the formula

$$\cos 2n\alpha = \sum_{j=0}^{n}(-1)^j\binom{2n}{2j}\cos^{2(n-j)}\alpha\sin^{2j}\alpha$$

(see 7.4.(i)), factor out $\cos^{2n}\alpha$ on the right, set $\alpha = \alpha_k = \frac{(2k-1)\pi}{4n}$, where $k = 1, 2, \ldots, n$, and thus determine that the desired numbers $x_k = \tan^2\alpha_k$ are roots of the equation $\sum_{j=0}^{n}(-1)^j\binom{2n}{2j}x^j = 0$. Then by Vieta's formula you have $\sum_{k=1}^{n} x_k = \binom{2n}{2n-2} = n(2n-1)$.

(vii) First verify that $|1+\varepsilon^k| = 2|\cos\frac{k\pi}{n}|$ for all $k = 1, 2, \ldots, n$. Substitute this expression, for $n = 2m+1$, into $1 = |1+\varepsilon| \cdot |1+\varepsilon^2| \cdots |1+\varepsilon^{2m}|$, which is a consequence of (65) for $x = -1$. Then you easily obtain (66), if you use the relation $\left|\cos\frac{(m+k)\pi}{2m+1}\right| = \cos\frac{(m+1-k)\pi}{2m+1}$ for $k = 1, 2, \ldots, m$.

(viii) The polynomials $A_m(x)$ and $B_m(y)$ are (up to a constant factor) determined as follows:

$$A_m(x) = \sum_{j=0}^{m}(-1)^j\binom{2m+1}{2j+1}(1-x)^{m-j}x^j,$$

$$B_m(y) = \sum_{j=0}^{m}(-1)^j\binom{2m+2}{2j+1}(1-y)^{m-j}y^j.$$

The constant coefficient of $A_m(x)$ is obviously $2m + 1$, and the coefficient of x^m is equal to

$$\sum_{j=0}^{m}(-1)^j \binom{2m+1}{2j+1}(-1)^{m-j} = (-1)^m \cdot \sum_{j=0}^{m}\binom{2m+1}{2j+1} = (-1)^m \cdot 2^{2m}$$

by 1.4.(ii). Therefore, $x_1 x_2 \cdots x_m = (2m+1)/2^{2m}$, which implies

$$\sin\frac{\pi}{2m+1}\sin\frac{2\pi}{2m+1}\cdots\sin\frac{m\pi}{2m+1} = \frac{\sqrt{2m+1}}{2^m},$$

and this proves (63) for odd n. Similarly, from the coefficient $(-1)^m \cdot 2^{2m+1}$ of y^m and the coefficient $2m + 2$ of y^0 in the polynomial $B_m(y)$ it follows that

$$y_1 y_2 \cdots y_m = \frac{2m+2}{2^{2m+1}};$$

hence

$$\sin\frac{\pi}{2m+2}\sin\frac{2\pi}{2m+2}\cdots\sin\frac{m\pi}{2m+2} = \frac{\sqrt{m+1}}{2^m},$$

so (63) holds also for even n. Finally, from the equation

$$(-1)^m \cdot 2^{2m}(1 - x_1)(1 - x_2)\cdots(1 - x_m) = A_m(1) = (-1)^m,$$

in view of the fact that $1 - x_k = \cos^2\frac{k\pi}{2m+1}$, you obtain (66).

(ix) The equation $\frac{x^9-1}{x^3-1} = x^6 + x^3 + 1 = 0$ has the six roots $\varepsilon = \cos\frac{2\pi}{9} + i\sin\frac{2\pi}{9}$, ε^2, ε^4, ε^5, ε^7, ε^8. Substituting $y = x + \frac{1}{x}$, you obtain the cubic equation $y^3 - 3y + 1 = 0$ with roots

$$y_1 = 2\cos\frac{2\pi}{9}, \quad y_2 = 2\cos\frac{4\pi}{9}, \quad y_3 = 2\cos\frac{8\pi}{9}.$$

To determine the sum $\sqrt[3]{y_1} + \sqrt[3]{y_2} + \sqrt[3]{y_4}$, use the same approach as in 7.4.(v). This time you get $C = -1$, the pair of equations $A^3 = 3AB + 3$ and $B^3 = -3AB - 6$, and $(w + 3)^3 = 9$, which implies $A^3 = 3\sqrt[3]{9} - 6$ and $B^3 = 3 - 3\sqrt[3]{9}$.

(x) Set $\varepsilon = \cos\frac{2\pi}{5} + i\sin\frac{2\pi}{5}$. Then $\varepsilon^5 = 1$ and $\varepsilon \neq 1$, so ε, ε^2, ε^3, ε^4 are roots of $\frac{x^5-1}{x-1} = x^4 + x^3 + x^2 + x + 1$. Substituting $y = x + \frac{1}{x}$ you find that $y_1 = \varepsilon + \frac{1}{\varepsilon} = 2\cos\frac{2\pi}{5}$ is the positive root of $y^2 + y - 1$. Hence $2\cos\frac{2\pi}{5} = \frac{-1+\sqrt{5}}{2}$. The other root is $y_2 = \varepsilon^2 + \frac{1}{\varepsilon^2} = 2\cos\frac{4\pi}{5} = -2\cos\frac{\pi}{5}$, so $-2\cos\frac{\pi}{5} = \frac{-1-\sqrt{5}}{2}$.

7.7 (i) Proceed as in 7.6.(i), or use the fact that for an arbitrary root α of the polynomial $x^2 + x + 1$ you have $\alpha^2 + \alpha + 1 = 0$, and thus $(\alpha + 1)^{2n+1} = (\alpha + 1)(\alpha^2 + 2\alpha + 1)^n = -\alpha^2 \cdot \alpha^n$.

(ii) Substitute i and $-i$, and use de Moivre's theorem.

(iii) Substitute i, -1 and $-i$.

(iv) $a \in \{0, \pm 1, \pm\sqrt{3}, \pm\frac{1+\sqrt{5}}{2}, \pm\frac{1-\sqrt{5}}{2}\}$. Each $a \in (-2, 2)$ is of the form $a = 2\cos\beta$, where $\beta \in (0, \pi)$. The polynomial $g(x) = x^2 - ax + 1$ has roots $\cos\beta \pm i\sin\beta$, and hence the polynomial $f(x) = x^{14} - ax^7 + 1$ has 14 different zeros,

$$\cos\frac{\pm\beta + 2k\pi}{7} + i\sin\frac{\pm\beta + 2k\pi}{7}, \qquad k \in \{0, 1, \ldots, 5, 6\}.$$

So the polynomial $f(x)$ divides the polynomial $h(x) = x^{154} - ax^{77} + 1$ if and only if each root of $f(x)$ is also a root of $h(x)$, i.e., it is of the form

$$\cos\frac{\pm\beta + 2l\pi}{77} + i\sin\frac{\pm\beta + 2l\pi}{77}$$

with an integer l. Especially,

$$\frac{\beta}{7} = \frac{\pm\beta + 2l\pi}{77}$$

for a suitable $l \in \mathbb{Z}$, which gives $\beta = \frac{l\pi}{6}$ or $\beta = \frac{l\pi}{5}$. Hence

$$\beta \in \left\{\frac{\pi}{6}, \frac{\pi}{3}, \frac{\pi}{2}, \frac{2\pi}{3}, \frac{5\pi}{6}, \frac{\pi}{5}, \frac{2\pi}{5}, \frac{3\pi}{5}, \frac{4\pi}{5}\right\}.$$

For the values of $\cos\frac{l\pi}{5}$, see Exercise 7.5.(x).

7.9 (i) From 3.14 derive $(x_1^2 + x_1 + 1)\cdots(x_n^2 + x_n + 1) = F(w)\cdot F(w^2)$, where $w = (-1 + i\sqrt{3})/2$; show that $F(w) = Cw^2 + Bw + A$, $F(w^2) = Cw + Bw^2 + A$. Multiply and simplify.

(ii) From 3.14 derive $(x_1^4 + 1)\cdots(x_4^4 + 1) = G(\tau)G(\tau^3)G(\tau^5)G(\tau^7)$, where $\tau = (1+i)\sqrt{2}/2$; show that $G(\tau)\cdot G(\tau^5) = -q^2 - i(pi + r)^2$, $G(\tau^3)\cdot G(\tau^7) = -q^2 + i(-pi + r)^2$, and multiply these two expressions together.

2   Hints and Answers to Chapter 2

2.2 (i) $\sqrt[n]{n!} < \sqrt[n+1]{(n+1)!}$ for all $n \in \mathbb{N}$.

(ii) $2^{700} > 5^{300}$, since $2^7 > 5^3$.

(iii) The first number is larger. For all $x \in \mathbb{R}$,

$$\frac{10^x + 1}{10^{x+1} + 1} > \frac{10^{x+1} + 1}{10^{x+2} + 1} \iff (10^x + 1)(10^{x+2} + 1) > (10^{x+1} + 1)^2.$$

Expanding, you get $10^{x+2} - 2\cdot 10^{x+1} + 10^x > 0$, that is, $81\cdot 10^x > 0$.

(iv) $L - R = (\sqrt{b} - a)\cdot\frac{a + 1 - \sqrt{b}}{1 + 2a} > 0$.

(v) Better than squaring: Multiply the inequality by $\sqrt{a+1} + \sqrt{a}$.

(vi) By squaring you obtain

$$\frac{(c+a)^2}{c^2+a^2} > \frac{(c+b)^2}{c^2+b^2} \iff 1 + \frac{2ac}{c^2+a^2} > 1 + \frac{2bc}{c^2+b^2}$$

$$\iff \frac{a}{a^2+c^2} > \frac{b}{b^2+c^2}.$$

The last inequality can be rewritten as $(a-b)c^2 > ab(a-b)$.

(vii) This follows from 2.1.(v) for $r = \frac{1}{3}$ and $s = \frac{1}{4}$.

(viii) Upon dividing by $(n!)^2$ you get $1 < \frac{2n-k}{n} \cdot \frac{2n-k-1}{n-1} \cdot \frac{2n-k-2}{n-2} \cdots \frac{n+1}{k+1}$. Each term on the right-hand side is greater than 1.

2.4 (i) Each term on the left-hand side is greater than $1/(a+b+c)$.

(ii) Rewrite this as $2(x_1 + x_2 + \cdots + x_6) \leq 3(x_7 + x_8 + x_9 + x_{10})$. The left-hand side is at most $12x_6$, and the right-hand side at least $12x_7$.

(iii) $\frac{1}{L} = \frac{1}{1+a} + \frac{1}{1+b} > \frac{1}{1+a+b} + \frac{1}{1+a+b} = \frac{1}{R} \implies L < R$.

(iv) Adding $a + b > c$ and $2\sqrt{ab} > 0$, you obtain $(\sqrt{a} + \sqrt{b})^2 > (\sqrt{c})^2$, and thus $\sqrt{a} + \sqrt{b} > \sqrt{c}$. The inequality $(a+b)^{-1} + (b+c)^{-1} > (c+a)^{-1}$ follows from $a + b < 2(a+c)$ and $b + c < 2(a+c)$. Changing the order of a, b, c gives the remaining triangle inequalities.

(v) Distinguish between two cases: For $x \geq 1$ you have $L = (x^8 - x^5) + (x^2 - x) + 1 \geq 1$; for $x < 1$ it is $L = x^8 + (x^2 - x^5) + (1 - x) > 0$.

(vi) Upon rewriting, $(n-1)(a^n + 1) \geq 2(a + a^2 + \cdots + a^{n-1})$. Therefore, it suffices to prove the inequalities $a^n + 1 \geq a^k + a^{n-k}$ and add them for $k = 1, 2, \ldots, n-1$.

(vii) Prove the inequalities

$$\left(m + 2j - \frac{1}{2}\right)^2 > (m + 2j - 1)(m + 2j),$$

multiply them for $j = 1, 2, \ldots, k$, and take the square root of the result.

(viii) Without loss of generality you may take $a \geq b \geq c$. Show that then $a^3 + b^3 + 2abc \geq ab(a+b) + (a^2 + b^2)c$ and $c^3 + abc \geq (a+b)c^2$. Adding these two, you obtain the desired inequality.

(ix) The left-most term on the right-hand side is smallest. Hence

$$1 + (n+1)a > \left[\frac{1 + (n+1)a}{1+na}\right]^{n+1}.$$

(x) $31^{11} < 17^{14}$, since $31^{11} < 32^{11} = 2^{55} < 2^{56} = 16^{14} < 17^{14}$.

(xi) The first number is greater. Both can be rewritten as $2^{9 \cdot 3^{98}}$ and $3^{4 \cdot 2^{148}}$, and in view of the fact that $2^9 > 3^4$, it remains to show that $3^{98} > 2^{148}$. This last inequality is a product of $3^9 > 2^{12}$ (which follows from $3^3 > 2^4$) and $3^{89} > 2^{136}$ (which follows from $3^2 > 2^3$ and $89/2 > 136/3$).

(xii) Rewrite the equation as $1 = F(x)$, where

$$F(x) = \frac{a_1}{x} + \frac{a_2}{x^2} + \cdots + \frac{a_n}{x^n}.$$

It is easy to see that $F(x) > F(y)$ if $0 < x < y$ and $a_i \neq 0$ for some i.

(xiii) $L - R = (c - a - b)[c(c - a - b) - \frac{a^3 + b^3}{a + b}] > 0$.

(xiv) Multiplying the first, resp. the last two inequalities together, and simplifying, you obtain $a^2 + b^2 < cd - ab$, resp. $(a^2 + b^2)cd < ab(ab - cd)$. However, not both numbers $cd - ab$, $ab - cd$ can be positive.

(xv) By the inequality given in the hint you have

$$\frac{A + B + a + b}{A + B + s} > \frac{A + a}{A + s - b} \quad \text{and} \quad \frac{B + C + b + c}{B + C + s} > \frac{C + c}{C + s - b}.$$

If you replace the denominators on the right-hand sides by the larger number $A + C + s$ and subsequently add both inequalities, you obtain one of the desired inequalities; in view of symmetry, the remaining triangle inequalities are also true.

(xvi) By expanding both sides $A(r)A(u)$, $A(s)A(t)$ you will find that it suffices to show that $a_j^r a_k^u + a_k^r a_j^u > a_j^s a_k^t + a_k^s a_j^t$ whenever $a_j \neq a_k$. Upon dividing by $(a_j a_k)^r$ you obtain the inequality from 2.1.(v), with the numbers r, s replaced by $s - r, t - r$.

2.6 (i) Add $\sqrt{2} \leq \sqrt{k} \leq \sqrt{n}$ for $k = 2, 3, \ldots, n$ and exclude the cases of equality.

(ii) Multiply $(k + 1)(n - k) > n$ for $k = 1, 2, \ldots, n - 2$.

(iii) Add the inequalities

$$\sum_{j=1}^{n} \frac{1}{kn + j} > \frac{n}{(k + 1)n} = \frac{1}{k + 1}, \quad \text{resp.} \quad \sum_{j=1}^{n} \frac{1}{kn + j - 1} < \frac{n}{kn} = \frac{1}{k}.$$

(iv) $S(n) > 2\sqrt{n + 1} - 2\sqrt{2} + 1$. From this and from 2.5.(iv) it follows that $1998 < S(10^6) < 1999$.

(v) Use the estimate $\frac{1}{2k^2 + k} < \frac{1}{2k^2}$ for $k = 2, 3, \ldots, n$.

(vi) Add $\frac{1}{k^3} < \frac{1}{k^3 - k}$ for $k = 2, 3, \ldots, n$.

(vii) Use the inequality

$$\frac{k}{(2k - 1)(2k + 1)(2k + 3)} > \frac{k}{(2k + 1)(4k^2 + 4k)} = \frac{1}{4(k + 1)(2k + 1)}.$$

(viii) Add $4k - 1 \leq \frac{(2k - 1)(2k + 1)}{k} < 4k$ for $k = 1, 2, \ldots, n$.

(ix) Use the method of 2.5.(vii) with the inequalities $\frac{4k - 1}{4k + 1} < \frac{4k + 1}{4k + 3}$, $k \geq 1$.

(x) Estimate each term on the right of

$$Q(n) = Q(p) \cdot \frac{2p + 1}{2p + 2} \cdot \frac{2p + 3}{2p + 4} \cdots \frac{2n - 1}{2n}$$

as in Example 2.5.(vii).

2.8 (i) In view of the symmetry you may assume that $0 < a \leq b \leq c$. Then $c(c - a) \geq b(b - a)$, so that $L = a(b - a)(c - a) + (c - b)[c(c - a) - b(b - a)] \geq 0$.

(ii) Let $0 < a \leq b \leq c$; then $b = a + u$ and $c = a + u + v$, where $u, v \in \mathbb{R}_0^+$. Substituting, you get $L = u^2(a - v) + (u + v)^2(a + v) + v^2(a + 2u + v)$. The only term with negative coefficient, namely $-u^2 v$, clearly disappears upon simplification, and so $L \geq 0$.

(iii) The inequality

$$\frac{t^2 + 1}{2} - t \geq \frac{(t^2 - 1)^2(t^2 + 3)(1 + 3t^2)}{8(t^2 + 1)(t^4 + 6t^2 + 1)}$$

can be transformed into

$$\frac{(t - 1)^8}{8(t^2 + 1)(t^4 + 6t^2 + 1)} \geq 0.$$

3.2 (i) $(n^2 + n)(L - R) = (1 - a)(1 + a + \cdots + a^{n-1} - na^n) > 0$, since $(1 - a) > 0$ and $a^k > a^n$ for $k = 0, 1, \ldots, n - 1$.

(ii) You obtain the inequality $x - y < a - b$, with $x = \sqrt[n]{a^n + c}$ and $y = \sqrt[n]{b^n + c}$, from the equation $x^n - y^n = a^n - b^n$. To see this, expand both sides according to formula (14) and use the fact that the inequalities $x > a$ and $y > b$ imply the estimate

$$x^{n-1} + x^{n-2}y + \cdots + xy^{n-2} + y^{n-1} > a^{n-1} + a^{n-2}b + \cdots + ab^{n-2} + b^{n-1}.$$

(iii) Each inequality on the right-hand side of the implication is of the form $\frac{k}{a^k - a^{-k}} > \frac{k+1}{a^{k+1} - a^{-k-1}}$, where $k \in \mathbb{N}$. Under the assumption $a > 1$ you can use formula (14), with $n = 2k$ and $n = 2k + 2$, to change this inequality into the equivalent form $k(a^{2k+1} + 1) > a + a^2 + \cdots + a^{2k}$. But this is inequality (6).

(iv) It suffices to show that $2(a^2 - ab + b^2) \geq a^2 + b^2$, since

$$L = \frac{a^5 + b^5}{a + b} = \frac{a^2 \cdot a^3 + b^2 \cdot b^3}{a + b} > \frac{2a^3 + 2b^3}{a + b} = 2(a^2 - ab + b^2).$$

(v) The equation $a^3 + b^3 = a - b$ gives the expression $a^2 + ab + b^2 = \frac{a^3 - b^3}{a^3 + b^3}$; from this obtain the stronger inequality $a^2 + ab + b^2 < 1$.

(vi) Since $\frac{a^3 - b^3}{a^2 + ab + b^2} + \frac{b^3 - c^3}{b^2 + bc + c^2} + \frac{c^3 - a^3}{c^2 + ca + a^2} = (a - b) + (b - c) + (c - a) = 0$, you have

$$L = \frac{1}{2}\left(\frac{a^3 + b^3}{a^2 + ab + b^2} + \frac{b^3 + c^3}{b^2 + bc + c^2} + \frac{c^3 + a^3}{c^2 + ca + a^2}\right).$$

Therefore, it suffices to show that $\frac{x^3 + y^3}{x^2 + xy + y^2} \geq \frac{x+y}{3}$ for all $x, y \in \mathbb{R}^+$. Upon dividing by $(x + y)$ and multiplying by $3(x^2 + xy + y^2)$, this inequality changes to $2(x - y)^2 \geq 0$.

3.4 (i) $(a + 1)^n - 2^n > n(a - 1)2^{n-1}$, by the left part of (16).

(ii) $2^n - (1+a)^n > n(1-a)(1+a)^{n-1} = n(1-a^2)(1+a)^{n-2} > n(1-a^2)1^{n-2}$.

(iii) Rewrite as $(1-a)^n - a^n \geq na^{n-1}(1-2a)$. If $0 < a < \frac{1}{2}$, use (16) with $A = 1-a$ and $B = a$; if $\frac{1}{2} < a < 1$, substitute in the opposite order, namely $A = a$ and $B = 1-a$; if $a = \frac{1}{2}$, you obtain equality.

(iv) Since $A_n - A_{n-1} = (a_n - A_{n-1})/n$, you have $A_n > A_{n-1}$, unless $a_1 = a_2 = \cdots = a_n$. Then by (16), $A_n^n - A_{n-1}^n > n(A_n - A_{n-1})A_{n-1}^{n-1} = (a_n - A_{n-1})A_{n-1}^{n-1}$.

(v) Substitute $n = k$ and $x = 1/k$ in (17).

(vi) Rewrite the left inequality as $\left[1 - \frac{1}{(k+1)^2}\right]^k > 1 - \frac{1}{k+2}$; this follows from (17) for $n = k$ and $x = -\frac{1}{(k+1)^2}$. The right inequality can be written in the form (17) for $n = k+2$ and $x = \frac{1}{k^2+2k}$.

(vii) With $x = ab^r - 1$ in (17), obtain $a^n + (n-1)b^{-nr} \geq nab^{r(1-n)}$. Therefore, choose $r = -1/n$. The inequality (18) is strict if $x \neq 0$, that is, $a \neq \sqrt[n]{b}$.

(viii) Use (18) with $a = 1$ and $n = k+1$, $b = 1 + \frac{1}{k}$, resp. $n = k+2$, $b = 1 - \frac{1}{k+1}$.

(ix) The inequality $d_i^n \leq D^{n-1}d_i$ implies $L \leq D^{n-1} \cdot (d_1 + \cdots + d_k)/k$. Hence it suffices to show that $aD^{n-1} \leq a^n + \alpha_n D^n$; but this is (18) with $b = \alpha_n D^n/(n-1)$.

3.6 (i) Use $(n+2k)! > (n+2)!(n+3)^{2k-2}$ for $k \geq 2$, and (19) with $q = \frac{1}{(n+3)^2}$.

(ii) The numbers $a_k = \frac{1 \cdot 3 \cdots (2k-1)}{3^k \cdot k!}$ satisfy $\frac{a_{k+1}}{a_k} < \frac{2}{3}$.

(iii) Let $2^s a_s^k$ be the largest of the numbers $2a_1^k, 2^2 a_2^k, \ldots, 2^n a_n^k$. For $j = 1, 2, \ldots, n$ sum the inequalities $a_j \leq 2^{\frac{s-j}{k}} a_s$ and then use (19) with $q = 1/\sqrt[k]{2}$.

3.8 (i) Since $a < c$ and $b < c$, it suffices to show that $a + b > c$. Let us assume the opposite: $a + b \leq c$; then by (22), $a^n + b^n < (a+b)^n \leq c^n$, which is a contradiction.

(ii) In the expansion of $(1+a)^n$, use the inequality $a^k > a^{n-1}$ for $k = 1, 2, \ldots, n-2$.

(iii) $\frac{(a+1)^n - (a-1)^n}{a^n} = 2\left[\binom{n}{1}\frac{1}{a} + \binom{n}{3}\frac{1}{a^3} + \cdots\right]$. Estimate the term in square brackets from below by its first term, and from above by the sum $\frac{n}{a} + (\frac{n}{a})^3 + (\frac{n}{a})^5 + \cdots$, and use (19) with $q = (\frac{n}{a})^2$.

(iv) The expression (23) follows from (21) for $A = 1$, $B = 1/n$, and exponent $n = k$. Since $1 - \frac{j}{k} < 1 - \frac{j}{k+1}$ for $j = 1, 2, \ldots, k-1$, you get the inequality $c_{k+1} > c_k$ by comparing the two right-hand sides (23). The left inequality in (24) follows from (23) and from the estimates $1 - \frac{j}{k} < 1$ for $j > 0$; the right inequality of (24) is obtained by summing $\frac{1}{j!} < \frac{1}{2^{j-1}}$ for $j = 3, 4, \ldots, k$.

4.2 (i) $R - L = (x - yz)^2$.

(ii) $L - R = (x - y)^2[(x + y)^2 - 4xy] = (x - y)^4$.

(iii) $L - R = (x - y)^2(x^2 + xy + y^2)$, by 4.1.(i) you have $x^2 + xy + y^2 \geq 0$.

(iv) Multiply by $\sqrt{ab}$, to obtain $(\sqrt{a} + \sqrt{b})(\sqrt{a} - \sqrt{b})^2 \geq 0$.

(v) Rewrite as $(x^2 - 1)^2 \geq 0$.

(vi) Divide by $(1 + x^4)(1 + y^4)$; then this is the sum of two inequalities (v).

(vii) $L - R = (x - 1)^2 + (2y - 3)^2 + 3(z - 1)^2 + 1$.

(viii) $L = (x + y)^2 + (y + \frac{1}{2})^2 + \frac{3}{4}$.

(ix) $L - R = (xy - 1)^2 + (x - 1)^2$.

(x) $L - R = (x^2 - y^2)^2 + (x - z)^2 + (x - 1)^2$.

(xi) Since $R = (\sqrt{a} - \sqrt{b})^2 + \sqrt{2ab} > 0$, you can square the inequality. Upon simplification you obtain $L^2 - R^2 = 2(2 - \sqrt{2})\sqrt{ab}(\sqrt{a} - \sqrt{b})^2$.

(xii) $L - R = (x - 2y)^2 + (x - z)^2$.

(xiii) $L = 2(x^2 - y^2)^2 + (2xy - 3)^2 + 1$.

(xiv) $R - L = (1 - \sqrt[4]{a})^2 + (1 - \sqrt[4]{b})^2 + (1 - \sqrt[4]{c})^2$.

(xv) $L - R = (\sqrt{a} - \sqrt{bc})^2 + (\sqrt{b} - \sqrt{ac})^2 + (\sqrt{c} - \sqrt{ab})^2$.

(xvi) $L - R = (a - b)^2 + (a - \sqrt{b})^2 + (b - \sqrt{a})^2$.

(xvii) $L - R = (a^2 - b^2)^2 + 2ab(a - b)^2$.

(xviii) $L = (z^2 - x^2 - y^2)^2$.

(xix) $L = (x^2 + p)^2 + a(x + q)^2 + b$ for $p = -1$, $q = -\frac{3}{2}$, $a = 1$ and $b = \frac{3}{4}$.

(xx) $L = (a^2 + p)^2 + a(a + q)^2 + b = (a^2 - 2)^2 + a(a - 2)^2$. The numbers $(a^2 - 2)$ and $(a - 2)$ cannot be both zero at the same time.

(xxi) $L - R = (\frac{1}{2}x - y + z)^2$.

(xxii) $L - R = (\sqrt{a} + \sqrt{b} - \sqrt{c})^2$.

(xxiii) $L - R = (\frac{1}{2}a - b - c)^2 + (a^3 - 36abc)/(12a)$.

(xxiv) Rewrite as $1 + x^2 + y^2 \geq x + y + xy$, which follows from (27) with $Z = 1$.

(xxv) Squaring and simplifying, you get $a^2b^2 + b^2c^2 + c^2a^2 \geq a^2bc + ab^2c + abc^2$, which follows from (27); see the proof of (28) in 4.1.(viii).

(xxvi) Applying (27) three times, you obtain $a^8 + b^8 + c^8 \geq a^4b^4 + b^4c^4 + c^4a^4 \geq a^2b^4c^2 + b^2c^4a^2 + c^2a^4b^2 \geq a^2b^3c^3 + b^2c^3a^3 + c^2a^3b^3$. Divide by the expression $a^3b^3c^3$.

(xxvii) Show that $xy + yz + zx \geq -(x^2 + y^2 + z^2)/2$ for $x, y, z \in \mathbb{R}$.

(xxviii) $L = [2(x^2 + xy + xz) + yz]^2$.

(xxix) Assume that you have at the same time $pz + rx = 2qy$, $pr > q^2$ and $xz > y^2$. Multiplying these inequalities together, obtain $4prxz > 4q^2y^2 = (2qy)^2 = (pz + rx)^2$. The inequality $4prxz > (pz + rx)^2$ leads to a contradiction: $0 > (pz - rx)^2$.

(xxx) For $x \neq 0$,

$$F(x) > x^4 + ax^3 + \frac{c^2 + a^2d}{4d}x^2 + cx + d = \left(x^2 + \frac{ax}{2}\right)^2 + \left(\frac{cx}{2\sqrt{d}} + \sqrt{d}\right)^2 \geq 0.$$

(xxxi) First estimate the right-hand side from above:

$$R \leq a^p \cdot \frac{b^2 + c^2}{2} + b^p \cdot \frac{a^2 + c^2}{2} + c^p \cdot \frac{a^2 + b^2}{2}.$$

Hence it suffices to prove the inequality

$$2(a^{p+2} + b^{p+2} + c^{p+2}) \geq a^p \cdot (b^2 + c^2) + b^p \cdot (a^2 + c^2) + c^p \cdot (a^2 + b^2).$$

Write this as the sum of three inequalities of the type $a^{p+2} + b^{p+2} \geq a^p b^2 + b^p a^2$, which are easy to prove; see Example 2.1.(v).

(xxxii) $A \geq B$, with equality only when $a = b = c = 1$. Use the identity

$$abc(abc+1)(A-B) = ab(b+1)(ca-1)^2 + bc(c+1)(ab-1)^2 + ca(a+1)(bc-1)^2.$$

(xxxiii) The desired largest p is $\frac{2}{3}$. For $x = y = z \neq 0$ you obtain the necessary condition $6 \geq 9p$; for $p = \frac{2}{3}$ the difference $L - R$ is equal to one-twelfth of the sum

$$5(x^2-y^2)^2 + 5(y^2-z^2)^2 + 5(z^2-x^2)^2 + 2(x^2-yz)^2 + 2(y^2-xz)^2 + 2(z^2-xy)^2.$$

(xxxiv) Answer: For $p = 1$ all n; for $p = \frac{4}{3}$ only $n \in \{2,3\}$; for $p = \frac{6}{5}$ only $n \in \{2,3,4\}$. In the case $p = 1$ the given inequality is equivalent to

$$x_1^2 + (x_1 - x_2)^2 + (x_2 - x_3)^2 + \cdots + (x_{n-1} - x_n)^2 + x_n^2 \geq 0,$$

in the case $p = \frac{4}{3}$ and $n = 3$ to $(3x_1 - 2x_2)^2 + x_2^2 + (2x_2 - 3x_3)^2 \geq 0$, and finally in the case $p = \frac{6}{5}$ and $n = 4$ it is equivalent to $(5x_1 - 3x_2)^2 + x_2^2 + 15(x_2 - x_3)^2 + x_3^2 + (3x_3 - 5x_4)^2 \geq 0$. (If n is less than mentioned in the last sentence, fill up the set of numbers $x_1, x_2, \ldots, x_n$ with an appropriate number of zeros.) If $p = \frac{4}{3}$ and $n \geq 4$, consider the example $x_1 = x_4 = 2$, $x_2 = x_3 = 3$, and $x_i = 0$ for $i > 4$, and if $p = \frac{6}{5}$ and $n \geq 5$, consider $x_1 = x_5 = 9$, $x_2 = x_4 = 15$, $x_3 = 16$, and $x_i = 0$ for $i > 5$.

4.4 (i) Follows from (29) with $A = a^2b$ and $B = 1/b$.

(ii) By (29), $L \geq 2\sqrt{6(c+2)(2c+3)}ab$. Prove the inequality $\sqrt{6(c+2)(2c+3)} > 3(c+2)$ by equivalent transformations.

(iii) Multiply the inequalities $a + b \geq 2\sqrt{ab}$ and $a^3 + b \geq 2\sqrt{a^3b}$. One could also use the method of squares: $L = (a^2 - b)^2 + ab(a-1)^2$.

(iv) Use (29) to estimate from below each of the four expressions in parentheses on the left.

(v) Use (29) to estimate from below each of the six expressions in parentheses under the square roots.

(vi) By (29), $\frac{1}{ab} + \frac{1}{cd} \geq 2 \cdot \frac{1}{\sqrt{ab}} \cdot \frac{1}{\sqrt{cd}} \geq 2 \cdot \frac{2}{a+b} \cdot \frac{2}{c+d}$.

(vii) Use (29) with $A = a + b$ and $B = 2\sqrt{ab}$.

(viii) Since, by (29), $L \geq 4\sqrt{ab}(a+b) + 2\sqrt{ab}$ and $R = 4\sqrt{ab}(\sqrt{a} + \sqrt{b})$, it suffices to verify the inequality

$$4(a+b) + 2 \geq 4(\sqrt{a} + \sqrt{b}), \quad \text{i.e.,} \quad (2\sqrt{a} - 1)^2 + (2\sqrt{b} - 1)^2 \geq 0.$$

(ix) The proof follows from adding the inequalities (see also (vi))

$$\frac{1}{a_i a_j} \geq \frac{4}{(a_i + a_j)^2}, \quad \frac{1}{a_i a_k} + \frac{1}{a_j a_k} \geq \frac{8}{(a_i + a_k)(a_j + a_k)},$$

$$\frac{1}{a_i a_k} + \frac{1}{a_j a_m} \geq \frac{8}{(a_i + a_k)(a_j + a_m)}.$$

4.6 (i) Substitute $A = 1 + 2a$ and $B = 1 + 2b$ in (31), or show that $L - R = (a - b)^2$.

(ii) $\sqrt{b} - \left(a + \frac{b-a^2}{2a+1}\right) = \frac{(\sqrt{b}-a)(1+a-\sqrt{b})}{2a+1}$. Use (31) with $A = \sqrt{b} - a$ and $B = 1 - A$. Alternative solution: Rewrite $R = A + B$, where $A = \frac{2a+1}{4}$ and $B = \frac{b}{2a+1}$, and use (29). (The condition $a < \sqrt{b} < a + 1$ is not necessary.)

(iii) By (31), with $A = a + b$ and $B = a^2 - ab + b^2$,

$$a^3 + b^3 = (a + b)(a^2 - ab + b^2) \leq \left(\frac{a + b + a^2 - ab + b^2}{2}\right)^2.$$

From this you obtain the result by taking the square root and setting $ab = 1$.

(iv) Use (31) to estimate from below the products ab, bc, and ca on the left-hand side.

(v) Using (31) for the products ab, cd, and $(a + b)(c + d)$, you get

$$R^3 = \frac{ab(c + d) + cd(a + b)}{4} \leq \frac{(\frac{a+b}{2})^2(c + d) + (\frac{c+d}{2})^2(a + b)}{4}$$

$$= (a + b)(c + d) \cdot \frac{a + b + c + d}{16}$$

$$\leq \left[\frac{(a + b) + (c + d)}{2}\right]^2 \cdot \frac{a + b + c + d}{16} = L^3.$$

(vi) Choosing $x_1 = x_2 = 1$ and $x_k = 0$ for the remaining indices k, obtain the necessary condition $p_n \leq 4$. On the other hand, show that for $p_n = 4$ the given inequality holds everywhere: In view of its homogeneity it suffices to consider the case where the sum of all x_k is 1. (If $x_k = 0$ for all k, then the inequality is clearly satisfied.) Then the inequality (with $p_n = 4$) can be rewritten as $\sum_{k=1}^{n} x_k x_{k+1}(1 - 4x_k x_{k+1}) \geq 0$. But this holds, since by (31) for $i \neq j$,

$$x_i x_j \leq \left(\frac{x_i + x_j}{2}\right)^2 \leq \frac{(x_1 + x_2 + \cdots + x_n)^2}{4} = \frac{1}{4},$$

so that $1 - 4x_i x_j \geq 0$.

(vii) If n is even, then $L \leq (a_1 + a_3 + \cdots + a_{n-1})(a_2 + a_4 + \cdots + a_n)$; apply (31) to this product. For odd n, use the previous case for the even number of elements $a_1, a_2, \ldots, a_{k-1}, a_k + a_{k+1}, a_{k+2}, \ldots, a_n$, where the index k is chosen such that the inequality

$$a_{k-1}a_k + a_k a_{k+1} + a_{k+1}a_{k+2} \leq a_{k-1}(a_k + a_{k+1}) + (a_k + a_{k+1})a_{k+2}$$

holds (one can usually take $a_0 = a_n$, $a_{n+1} = a_1$ and $a_{n+2} = a_2$). For this it suffices, for example, that $a_{k+1} \leq a_{k+2}$.

4.8 (i) Follows from (32) with $A = \sqrt{a^2 + 2} \neq 1$.

(ii) Follows from (32) with $A = \sqrt{4a^2 + 1} \neq 1$.

(iii) $L = \left(\frac{b}{c} + \frac{c}{b}\right) + \left(\frac{a}{d} + \frac{d}{a}\right) \geq 2 + 2 = 4$.

(iv) Upon dividing by $2ab$ rewrite as $2\left(b + \frac{1}{b}\right) + \left(2a + \frac{1}{2a}\right) \geq 6$.

(v) Divide by $a\sqrt{b}$ and rewrite as $\left(\frac{a}{\sqrt{b}} + \frac{\sqrt{b}}{a}\right) + \left(\frac{1}{\sqrt{ab}} + \sqrt{ab}\right) \geq 4$.

5.3 (i) From $p^2 - 4q \leq 0$ and $r^2 - 4s \leq 0$ it follows that $p^2 \leq 4q$ and $r^2 \leq 4s$, which implies $p^2 r^2 \leq 16qs$. This means that also the third polynomial has a nonpositive discriminant.

(ii) If you set $q = 1 - p$, you obtain the equivalent inequality

$$c^2 p^2 + p(a^2 - b^2 - c^2) + b^2 > 0.$$

This holds for all $p \in \mathbb{R}$ if and only if $D = (a^2 - b^2 - c^2)^2 - 4c^2 b^2 < 0$. Use the factorization $D = -(a + b + c)(a + b - c)(b + c - a)(c + a - b)$.

(iii) Let $a \geq b \geq c$. Set $F(x) = x^2 - 2x(b + c) + (b - c)^2$ and show that $F(a) < 0$. Since $b \leq a < b + c$, by 5.1 it suffices to show that $F(b) < 0$ and $F(b + c) < 0$. But this is easy.

(iv) $L - R = x^2 + 2(y - 3z + u)x + (y + z + u)^2 - 8yu$. This polynomial has discriminant $D = 4(y - 3z + u)^2 - 4(y + z + u)^2 + 32yu = 32(y - z)(u - z)$. Clearly, $D < 0$. This implies the assertion also under the following weaker assumption: The number z lies between y and u, and x is arbitrary.

(v) Since $x^2 - 2x + 2 > 0$ for all $x \in \mathbb{R}$, the inequalities are equivalent to the pair

$$3x^2 + 2(p - 2)x + 2 > 0 \quad \text{and} \quad x^2 - 2(p + 2)x + 6 > 0.$$

This system is satisfied for all $x \in \mathbb{R}$ if and only if the discriminants of both polynomials are negative, that is, $|p - 2| < \sqrt{6}$ and $|p + 2| < \sqrt{6}$. The common solution of the last two inequalities is $2 - \sqrt{6} < p < -2 + \sqrt{6}$.

5.6 (i) Follows from (39) with $n = 2$, $u_1 = \sqrt{a}$, $u_2 = \sqrt{b}$, $v_1 = \sqrt{c}$, and $v_2 = \sqrt{d}$.

(ii) By (39), $b^2 = (a_1 x_1 + a_2 x_2 + \cdots + a_n x_n)^2 \leq (a_1^2 + a_2^2 + \cdots + a_n^2)(x_1^2 + x_2^2 + \cdots + x_n^2)$, which implies $x_1^2 + \cdots + x_n^2 \geq b^2/(a_1^2 + \cdots + a_n^2)$. Equality occurs for $x_k = (a_k b)/(a_1^2 + \cdots + a_n^2)$, $1 \leq k \leq n$.

(iii) Use (39) with $u_k = \sqrt{a_k}$ and $v_k = x_k \sqrt{a_k}$ for $k = 1, 2, \ldots, n$.

(iv) Follows from (39) for $u_k = \sqrt{a_k}$ and $v_k = x_k/\sqrt{a_k}$, $1 \leq k \leq n$.

(v) By (39) with $n = 2$: $(1 + p^4)(1 + q^4) \geq (1 + p^2 q^2)^2$, $(1 + r^4)(1 + s^4) \geq (1 + r^2 s^2)^2$, $(1 + p^2 q^2)(1 + r^2 s^2) \geq (1 + pqrs)^2$.

(vi) Use (v) with $p = a^{3/4}$, $q = b^{3/4}$, $r = c^{3/4}$, and $s = (abc)^{1/4}$.

(vii) Apply (39) three times: for the pairs (u_k, v_k) equal to $(p_k q_k, r_k s_k)$, (p_k^2, q_k^2), and (r_k^2, s_k^2).

(viii) Use (46) with $n = 4$, $a_1 = a+b+c$, $a_2 = a+b+d$, $a_3 = a+c+d$, and $a_4 = b+c+d$.

(ix) In (vii), substitute $p_k = x_k$ and $q_k = r_k = s_k = 1$, or use (48) twice, with $u_k = x_k$, resp. $u_k = x_k^2$.

(x) Use inequality (48) with $u_1 = x_2 x_3 \cdots x_n$, $u_2 = x_1 x_3 \cdots x_n$, ..., $u_n = x_1 x_2 \cdots x_{n-1}$.

(xi) Suppose that, for instance, $x_n < 0$; then with the help of (48) reach a contradiction to the given inequality:

$$x_1 + x_2 + \cdots + x_n < x_1 + x_2 + \cdots + x_{n-1}$$
$$\leq \sqrt{(n-1)(x_1^2 + x_2^2 + \cdots + x_{n-1}^2)} < \sqrt{(n-1)(x_1^2 + x_2^2 + \cdots + x_n^2)}.$$

(xii) Add the inequalities $a_k b_k c_k \leq (a_k^3 + b_k^3 + c_k^3)/3$, $1 \leq k \leq n$, under the assumption $a_1^3 + a_2^3 + \cdots + a_n^3 = b_1^3 + b_2^3 + \cdots + b_n^3 = c_1^3 + c_2^3 + \cdots + c_n^3 = 1$.

(xiii) Use (xii) with $b_k = c_k = 1$, $1 \leq k \leq n$.

(xiv) Use (xii) with $b_k = a_k$ a $c_k = 1$, $1 \leq k \leq n$.

(xv) By (39) with $u_k = \sqrt{a_k/(a_{k+1} + a_{k+2})}$ and $v_k = \sqrt{a_k(a_{k+1} + a_{k+2})}$, where $a_{n+1} = a_1$, $a_{n+2} = a_2$, and using the inequality $a_i a_j \leq (a_i^2 + a_j^2)/2$, you get

$$(a_1 + a_2 + \cdots + a_n)^2 \leq \left(\frac{a_1}{a_2 + a_3} + \frac{a_2}{a_3 + a_4} + \cdots + \frac{a_n}{a_1 + a_2} \right)$$
$$\times (a_1 a_2 + a_1 a_3 + a_2 a_3 + a_2 a_4 + \cdots + a_n a_1 + a_n a_2)$$
$$\leq \left(\frac{a_1}{a_2 + a_3} + \cdots + \frac{a_n}{a_1 + a_2} \right)$$
$$\times \left(\frac{a_1^2 + a_2^2}{2} + \frac{a_1^2 + a_3^2}{2} + \cdots + \frac{a_n^2 + a_1^2}{2} + \frac{a_n^2 + a_2^2}{2} \right)$$
$$= 2(a_1^2 + a_2^2 + \cdots + a_n^2) \left(\frac{a_1}{a_2 + a_3} + \frac{a_2}{a_3 + a_4} + \cdots + \frac{a_n}{a_1 + a_2} \right).$$

Upon dividing by $2(a_1^2 + a_2^2 + \cdots + a_n^2)$ you obtain the desired inequality.

(xvi) Choosing $x_k = 1$ for all k, obtain for the sum S of all numbers a_k the estimate $S^2 \leq S$, or $S \leq 1$. Conversely, if $S \leq 1$, then

$$\left(\sum_{k=1}^{n} a_k x_k^2 \right)^2 \leq \sum_{k=1}^{n} a_k \cdot \sum_{k=1}^{n} a_k x_k^4 \leq \sum_{k=1}^{n} a_k x_k^4,$$

where the left inequality is (39) with $u_k = \sqrt{a_k}$ and $v_k = \sqrt{a_k} \cdot x_k^2$.

6.3 (i) $\frac{L(n+1)}{L(n)} < \frac{R(n+1)}{R(n)} \Longleftrightarrow 1 - a < \frac{1+na}{1+(n+1)a}$. The last inequality can be verified with an equivalent transformation.

(ii) Write the given inequality symbolically as $L(n) < P(n) < Q(n)$. Then

$$\frac{L(n+1)}{L(n)} < \frac{P(n+1)}{P(n)} < \frac{Q(n+1)}{Q(n)} \Longleftrightarrow \frac{4(n+1)}{n+2} < \frac{(2n+1)(2n+2)}{(n+1)^2} < 4.$$

The inequalities on the right can be verified by equivalent transformations.

(iii) $L(n+1) - L(n) = \frac{1}{3n+2} + \frac{1}{3n+3} + \frac{1}{3n+4} - \frac{1}{n+1} = \frac{2}{(3n+2)(3n+3)(3n+4)}$.

(iv) $\frac{L(n+1)}{L(n)} > \frac{R(n+1)}{R(n)} \iff 2^n > n+1$. This last inequality can be verified by induction.

(v) For $n = 1$: $x^2 + y^2 + 1 \geq xy + x + y$, which is inequality (27). Next, find that $L(n+1) - L(n) \geq R(n+1) - R(n) \iff x^{2^{n+1}} + y^{2^{n+1}} + 1 - x^{2^n} - y^{2^n} \geq (xy)^{2^n}$. The last inequality is again (27).

(vi) For the induction proof, use the estimates

$$\frac{1}{2} < \frac{1}{2^n} + \frac{1}{2^n + 1} + \cdots + \frac{1}{2^{n+1} - 1} < 1,$$

which can be verified using the approach in 2.5.(ii).

(vii) $\frac{L(n+1)}{L(n)} > \frac{R(n+1)}{R(n)} \iff (1 + a^{2n+1} + a^{2n+2} + a^{4n+3}) < (1+a)(1+a^{n+1})$. Since $1 + a > 1$, the last inequality is satisfied as long as $1 + a^{2n+1} + a^{2n+2} + a^{4n+3} < 1 + a^{n+1}$, which for $a > 0$ is equivalent to $a^n + a^{n+1} + a^{3n+2} < 1$. The left-hand side of the last inequality is largest for $n = 1$ and $a = \frac{7}{12}$, in which case it is, however, less than 0.992.

(viii) $\frac{L(n+1)}{L(n)} > \frac{R(n+1)}{R(n)} \iff (2n+2)! > (n+1)!(n+2)^{n+1}$. This is equivalent to $(n+3)(n+4)\ldots(2n+2) > (n+2)^n$, which is the product of the inequalities $n + k > n + 2$, $3 \leq k \leq n + 2$.

(ix) Use finite induction on k. For $k = 1$, (52) holds. If (52) holds for some $k < n$, then

$$\left(1 + \frac{1}{n}\right)^{k+1} = \left(1 + \frac{1}{n}\right)^k \left(1 + \frac{1}{n}\right) \geq \left(1 + \frac{k}{n}\right)\left(1 + \frac{1}{n}\right)$$
$$= 1 + \frac{k+1}{n} + \frac{k}{n^2} > 1 + \frac{k+1}{n}$$

and

$$\left(1 + \frac{1}{n}\right)^{k+1} = \left(1 + \frac{1}{n}\right)^k \left(1 + \frac{1}{n}\right) < \left(1 + \frac{k}{n} + \frac{k^2}{n^2}\right)\left(1 + \frac{1}{n}\right)$$
$$= 1 + \frac{k+1}{n} + \frac{(k+1)^2}{n^2} - \frac{n(k+1) - k^2}{n^3}$$
$$< 1 + \frac{k+1}{n} + \frac{(k+1)^2}{n^2},$$

since $n(k+1) - k^2 > 0$. (This last inequality is obtained by multiplying $n > k$ and $k + 1 > k$.)

Remark. This approach shows that the left-hand part of (52) holds for all $k > 0$; this also follows from Bernoulli's inequality (17) or directly from the binomial theorem.

(x) For $n = 6$: $729 > 720 > 64$. Further, it suffices to show that

$$\frac{\left(\frac{n+1}{2}\right)^{n+1}}{\left(\frac{n}{2}\right)^n} \geq n + 1 \geq \frac{\left(\frac{n+1}{3}\right)^{n+1}}{\left(\frac{n}{3}\right)^n},$$

which, upon dividing by $(n+1)$, can be rewritten as

$$\frac{1}{2}\left(1+\frac{1}{n}\right)^n \geq 1 \geq \frac{1}{3}\left(1+\frac{1}{n}\right)^n.$$

These last inequalities follow from (53).

(xi) If, for instance, $k \leq n$, then $\sqrt[n]{k} \leq \sqrt[n]{n}$. Hence it suffices to show that $\sqrt[n]{n} < 2$ for all $n \geq 2$. Rewrite this as $n < 2^n$ and use induction.

(xii) For $n = p$ you have equality in (54). Further, it suffices to verify that

$$\frac{\sqrt{4n+1}}{\sqrt{4n+5}} \leq \frac{Q(n+1)}{Q(n)} = \frac{2n+1}{2n+2} \leq \frac{\sqrt{3n+1}}{\sqrt{3n+4}}$$

(use an equivalent transformation). In comparison with 2.6.(x), show that

$$\frac{p}{n} < \frac{4p+1}{4n+1} < \frac{3p+1}{3n+1} < \frac{2p+1}{2n+1} \qquad (n > p).$$

(xiii) Since $L(1) = 1 < 2 = R(1)$ and $L(n+1) = \sqrt{n+1+L(n)}$, it suffices to show that $\sqrt{n+1+(\sqrt{n}+1)} < \sqrt{n+1}+1$. This becomes clear upon squaring.

6.5 (i) The power with the eight fives is greater. Denote by A_n, resp. B_n, the powers with n fives, resp. n eights, and show by induction that $A_n > 2B_{n-3}$, $n \geq 4$.

(ii) Rewrite the inequality as $(a+b-ab)^n \geq a^n+b^n-a^nb^n$. For $n = 1$ there is equality; show that $(a^n+b^n-a^nb^n)(a+b-ab) > a^{n+1}+b^{n+1}-a^{n+1}b^{n+1}$ for all $n \geq 1$, rewriting it as $L-R = (b^n-b^{n+1})(a-a^{n+1})+(a^n-a^{n+1})(b-b^{n+1}) > 0$.

(iii) Rewrite the induction condition

$$\left(A-\frac{B}{n+C}\right)\left[1+\frac{1}{(n+1)(n+2)}\right] \leq A - \frac{B}{n+1+C}$$

in the equivalent form $A[n^2 + (2C+1)n + C^2 + C] \leq B[n^2 + 4n + C + 3]$, which holds if $A = B$ and $2C+1 = 4$, that is, $C = \frac{3}{2}$ (then $C^2+C < C+3$). Considering the inequality for $n = 1$, obtain then $A = B = \frac{5}{2}$.

(iv) A more exact bound is of the form

$$Q(n) \leq \frac{Q(p) \cdot (2p+1)}{\sqrt{(4p+3)n+p+1}} \qquad (n \geq p).$$

(v) Use induction to verify the more exact bound

$$\left(1+\frac{1}{2}\right)\left(1+\frac{1}{4}\right)\cdots\left(1+\frac{1}{2^n}\right) \leq 3\left(1-\frac{1}{2^n}\right).$$

(vi) For a proof by induction (on k) of the stronger inequality

$$\frac{1}{n^n} > \frac{1}{S(n)} + \frac{1}{S(n+1)} + \cdots + \frac{1}{S(n+k)} + \frac{1}{(n+k+1)^{n+k+1}}$$

it suffices to show that for all $n \geq 2$ you have $\frac{1}{n^n} > \frac{1}{S(n)} + \frac{1}{(n+1)^{n+1}}$. Show that this inequality is equivalent to $\frac{S(n)}{S(n-1)} < \frac{(n+1)^{n+1}}{n^n}$; this last one can be proved by induction. If this inequality holds for some n (which is clearly the case for $n = 2$), then

$$\frac{S(n+1)}{S(n)} = \frac{S(n) + (n+1)^{n+1}}{S(n-1) + n^n} < \frac{(n+1)^{n+1}}{n^n} < \frac{(n+2)^{n+2}}{(n+1)^{n+1}}.$$

(The first inequality follows from the rule that when the positive numbers p, q, r and s satisfy $\frac{p}{q} < \frac{r}{s}$, then also $\frac{p+r}{q+s} < \frac{r}{s}$; the second inequality is a consequence of $(1 + \frac{1}{n})^n < (1 + \frac{1}{n+1})^{n+1}$, which was given as a hint.)

6.7 (i) By 6.6.(i),

$$\left(1 - \frac{1}{2^2}\right)\left(1 - \frac{1}{2^3}\right)\cdots\left(1 - \frac{1}{2^n}\right) \geq 1 - \frac{1}{2^2} - \frac{1}{2^3} - \cdots - \frac{1}{2^n} = \frac{1}{2} + \frac{1}{2^n}.$$

(ii) The inequality $L(n+1) - L(n) \geq R(n+1) - R(n)$ is obtained as follows:

$$(n+1)x_{n+1} \geq \frac{1}{2}(x_1 + x_2 + \cdots + x_n) + \frac{n+2}{2}x_{n+1},$$

which is equivalent to $x_{n+1} \geq (x_1 + x_2 + \cdots + x_n)/n$.

(iii) The inequality $L(n+1) - L(n) \geq R(n+1) - R(n)$ is obtained as follows: $(n+1)2^n x_{n+1} + x_1 + 2x_2 + \cdots + 2^{n-1}x_n \geq (2^{n+1} - 1)x_{n+1} + 2^n(x_1 + x_2 + \cdots + x_n)$, which is equivalent to

$$(n2^n - 2^n + 1)x_{n+1} \geq (2^n - 1)x_1 + (2^n - 2)x_2 + \cdots + (2^n - 2^{n-1})x_n.$$

This is the sum of the n inequalities $(2^n - 2^{k-1})x_{n+1} \geq (2^n - 2^{k-1})x_k$ for $k = 1, 2, \ldots, n$.

(iv) Show that the inequality $b_{n+1}b_{n-1} \geq b_n^2$ is equivalent to

$$(a_1 a_2 \cdots a_{n-1})^2 (a_{n+1})^{n^2 - n} \geq (a_n)^{n^2 + n - 2};$$

denote this by $L(n) \geq R(n)$. Show that the sequence of inequalities $L(n) \geq R(n)$ satisfies condition (i) of the principle 6.2.

(v) The inequality for fixed $k \in \mathbb{N}$ follows from the chain $S(0) \geq S(1) \geq \cdots \geq S(k)$, where $S(j) = a_1^{k-j}b_1^j + a_2^{k-j}b_2^j + \cdots + a_n^{k-j}b_n^j$, $0 \leq j \leq k$. The inequality $S(j) \geq S(j+1)$ follows from the statement of Example 6.6.(iv), where $u_i = a_i$, $v_i = b_i$ and $x_i = a_i^{k-j-1}b_i^j$, $1 \leq i \leq n$.

(vi) For $n = 1$ the statement is clear. Assume that the statement holds for some $n \geq 1$ and consider an arbitrary set of $n+1$ positive numbers x_1, x_2, $\ldots$, x_{n+1} such that $x_1 + x_2 + \cdots + x_{n+1} = n + 1$. Order them such that $x_n \leq 1 \leq x_{n+1}$. Then for n positive numbers $y_1 = x_1$, $y_2 = x_2$, $\ldots$, $y_{n-1} = x_{n-1}$ and $y_n = x_n + x_{n+1} - 1$ you have $y_1 + y_2 + \cdots + y_n = n$, from which, by the hypothesis, $x_1 x_2 \cdots x_{n-1}(x_n + x_{n+1} - 1) \leq 1$. It now suffices to add that $x_n + x_{n+1} - 1 \geq x_n x_{n+1}$.

(vii) $L(4) - R(4) = (x_1 - x_2 + x_3 - x_4)^2 \geq 0$,

$$L(n+1) - L(n) \geq R(n+1) - R(n)$$
$$\iff 2x_{n+1}(x_2 + \cdots + x_{n-1}) + (x_{n+1} - 2x_1)(x_{n+1} - 2x_n) \geq 0.$$

The inequality on the right holds as long as $x_{n+1} \leq x_1$ and $x_{n+1} \leq x_n$. By a cyclic permutation of the given $(n+1)$-tuple $x_1, x_2, \ldots, x_{n+1}$, arrange $x_{n+1} \leq x_j$ $(1 \leq j \leq n)$; doing this, the inequality $L(n+1) \geq R(n+1)$ does not change.

(viii) For $n = 1$ the inequality is equivalent to $a_1^5(a_1 - 1)^2 \geq 0$. The induction step from $n = k$ to $n = k+1$ can be carried out with a "difference method," as long as

$$a_{k+1}^7 + a_{k+1}^5 \geq 2a_{k+1}^6 + 4a_{k+1}^3(a_1^3 + a_2^3 + \cdots + a_k^3).$$

You may clearly assume that $a_1 < a_2 < \cdots < a_{k+1}$. Then the expression in parentheses on the right-hand side of the last inequality is not greater than $1^3 + 2^3 + \cdots + (a_{k+1} - 1)^3$. According to 2.15.(i) of Chapter 1, this sum is equal to $(a_{k+1} - 1)^2 a_{k+1}^2/4$. Hence

$$2a_{k+1}^6 + 4a_{k+1}^3(a_1^3 + a_2^3 + \cdots + a_k^3) \leq 2a_{k+1}^6 + a_{k+1}^3(a_{k+1} - 1)^2 a_{k+1}^2$$
$$= a_{k+1}^7 + a_{k+1}^5.$$

7.5 (i) Use the fact that the n-tuples $a_1^p, a_2^p, \ldots, a_n^p$ and $a_1^{-1}, a_2^{-1}, \ldots, a_n^{-1}$ have opposite orderings.

(ii) Use (74) with $p = 1$, $a_1 = x_1 x_2 \cdots x_n = 1$, $a_2 = x_2 x_3 \cdots x_n, \ldots,$ $a_{n-1} = x_{n-1} x_n$, and $a_n = x_n$.

(iii) By Theorem 7.3, for two triples (a^3, b^3, c^3) and $\left(\frac{a}{abc}, \frac{b}{abc}, \frac{c}{abc}\right)$ having the same ordering, the inequalities

$$a^3 \cdot \frac{1}{bc} + b^3 \cdot \frac{1}{ac} + c^3 \cdot \frac{1}{ab} \geq a^3 \cdot \frac{1}{ac} + b^3 \cdot \frac{1}{ab} + c^3 \cdot \frac{1}{bc},$$
$$a^3 \cdot \frac{1}{bc} + b^3 \cdot \frac{1}{ac} + c^3 \cdot \frac{1}{ab} \geq a^3 \cdot \frac{1}{ab} + b^3 \cdot \frac{1}{bc} + c^3 \cdot \frac{1}{ac}$$

hold. Add them and divide the result by 2.

7.8 (i) Since $s_{ij} = x_i y_i - x_i y_j + x_j y_j - y_i x_j$, you will find that the sum $s_{1j} + s_{2j} + \cdots + s_{nj}$ for each j is equal to

$$(x_1 y_1 + \cdots + x_n y_n) - (x_1 + \cdots + x_n)y_j + nx_j y_j - (y_1 + \cdots + y_n)x_j.$$

Hence the sum of all n^2 numbers s_{ij} is equal to

$$2n(x_1 y_1 + \cdots + x_n y_n) - 2(x_1 + \cdots + x_n)(y_1 + \cdots + y_n).$$

(ii) Assume that $a_1 \leq a_2 \leq \cdots \leq a_n$, and use the left-hand part of (76) for $x_k = a_k$ and $y_k = a_{n+1-k}^{-1}$, $1 \leq k \leq n$.

(iii) If $a_1, a_2, \ldots, a_n \in \mathbb{R}^+$, then for $r > 0$ and $s > 0$,

$$(a_1^r + a_2^r + \cdots + a_n^r)(a_1^s + a_2^s + \cdots + a_n^s) \leq n(a_1^{r+s} + a_2^{r+s} + \cdots + a_n^{r+s}),$$

while for $r > 0$ and $s < 0$ the opposite inequality holds. Both inequalities are strict, unless $a_1 = a_2 = \cdots = a_n$.

(iv) Use the right-hand part of (76) for the ordered n-tuples $(1, 2, \ldots, n)$ and $(\frac{1}{n}, \frac{1}{n-1}, \ldots, \frac{1}{1})$.

(v) Use induction on k and the following consequence of Chebyshev's inequality:

$$(a_1^k + a_2^k + \cdots + a_n^k)(a_1 + a_2 + \cdots + a_n) \leq n(a_1^{k+1} + a_2^{k+1} + \cdots + a_n^{k+1}).$$

(vi) Between the values of the left and the right sides there is the number $n(a_1^4 + a_2^4 + \cdots + a_n^4)$.

(vii) Use the right-hand part of (76) with $y_k = 2^{k-1}$, $1 \leq k \leq n$.

(viii) Rewrite the desired inequality as

$$n[(n-1)xy + 1] \geq [(n-1)x + 1][(n-1)y + 1].$$

This is the right-hand part of (76) for the n-tuples $x_1 = x_2 = \cdots = x_{n-1} = x$, $x_n = 1$ and $y_1 = y_2 = \cdots = y_{n-1} = y$, $y_n = 1$. (Since by assumption $(x-1)(y-1) \geq 0$ holds, these n-tuples have the same ordering.)

(ix) By Chebyshev's inequality, for each $j = 1, 2, \ldots, k-1$,

$$n \cdot \sum_{i=1}^{n} (a_{1i} a_{2i} \cdots a_{ji}) a_{j+1,i} \geq \left(\sum_{i=1}^{n} a_{1i} a_{2i} \cdots a_{ji} \right) \left(\sum_{i=1}^{n} a_{j+1,i} \right);$$

hence the desired assertion can be proved by induction.

(x) Choose (if they exist) indices i and j such that $x_i < A < x_j$. Check how both sides of the inequality in question change if in the n-tuple $x_1, \ldots, x_n$ you replace the pair (x_i, x_j) by the pair (x_i', x_j'), where $x_i' = \min\{A, x_i + x_j - A\}$ and $x_j' = \max\{A, x_i + x_j - A\}$. This change increases the number of those terms x_k that are equal to the (fixed) mean A of the whole n-tuple $x_1, \ldots, x_n$. If you repeat this change several times, you finally obtain the n-tuple $\bar{x}_1, \ldots, \bar{x}_n$ consisting of n numbers A.

8.3 The inequalities (i)–(vii) follow from (84) with the following sets of numbers:

(i) a^4, a^3, $4a$, 4.

(ii) $a^2 b$, $a^2 c$, $b^2 c$, $b^2 a$, $c^2 a$, $c^2 b$.

(iii) a^4, $a^3 b$, $a^3 b$, ab^3, ab^3, b^4.

(iv) $\sqrt{a}$, $\sqrt{a}$, $\sqrt[3]{b}$, $\sqrt[3]{b}$, $\sqrt[3]{b}$.

(v) a^3, b^6, 8.

(vi) $a/2$, $a/2$, $4/a^2$.

(vii) $1, a, a^2, \ldots, a^{n-1}$.

The inequalities (viii)–(xv) follow from (85) with the following sets of numbers:

(viii) $a, b, b, c, c, c, d, d, d, d$.

(ix) $A_1 = k^2 + 3k$, $A_2 = A_3 = \cdots = A_{k+3} = k^2 + 5k + 6$.

(x) $a/2, a/2, b$.

(xi) $a, b/2, b/2, c/3, c/3, c/3$.

(xii) $A_1 = A_2 = \cdots = A_5 = 1 - a$, $A_6 = 1 + a$, $A_7 = A_8 = 1 + 2a$ (for $a > 1$ trival).

(xiii) $A_1 = A_2 = \cdots = A_k = b - a$, $A_{k+1} = b + ka$.

(xiv) $1, 2, 2, 3, 3, 3, \ldots, k, k, \ldots, k$.

(xv) $A_1 = A_2 = \cdots = A_m = k$, $A_{m+1} = A_{m+2} = \cdots = A_{m+k} = m$.

(xvi) If you set $S = a_1 + a_2 + \cdots + a_k$, then by (85),

$$L \le \left[\frac{(1 + a_1) + (1 + a_2) + \cdots + (1 + a_k)}{k} \right]^k = \left(1 + \frac{S}{k} \right)^k = \sum_{j=0}^{k} \binom{k}{j} \frac{S^j}{k^j}.$$

It suffices to add that $\binom{k}{j} \cdot \frac{1}{k^j} \le \frac{1}{j!}$ $(0 \le j \le k)$.

8.5 (i) $L \ge n^2 \mathcal{G}_n(a_1, \ldots, a_n) \mathcal{G}_n(a_1^{n-1}, \ldots, a_n^{n-1}) = n^2 a_1 a_2 \cdots a_n$.

(ii) $(a_1 + \cdots + a_n) \left(\frac{1}{a_1} + \cdots + \frac{1}{a_n} \right) \ge n^2 \mathcal{G}_n(a_1, \ldots, a_n) \cdot \mathcal{G}_n \left(\frac{1}{a_1}, \ldots, \frac{1}{a_n} \right)$
$= n^2$.

(iii) Multiply the two AM–GM inequalities

$$\frac{1}{n} \sum_{k=1}^{n} a_k^2 \ge \sqrt[n]{a_1^2 \cdots a_n^2} \quad \text{and} \quad \frac{1}{n} \sum_{k=1}^{n} \frac{1}{b_k^2} \ge \sqrt[n]{\frac{1}{b_1^2} \cdots \frac{1}{b_n^2}}$$

and take the square root of the result.

(iv) For a lower bound of the left-hand side use the three AM–GM inequalities

$$a^2 + b^2 + c^2 \ge 3 \cdot \sqrt[3]{a^2 b^2 c^2},$$
$$ab + bc + ca \ge 3 \cdot \sqrt[3]{a^2 b^2 c^2},$$
$$a^2 b^2 + b^2 c^2 + c^2 a^2 \ge 3 \cdot \sqrt[3]{a^4 b^4 c^4}.$$

(v) $L + n = \dfrac{(a_1 - a_3) + (a_2 + a_3)}{a_2 + a_3} + \dfrac{(a_2 - a_4) + (a_3 + a_4)}{a_3 + a_4} + \cdots$
$+ \dfrac{(a_n - a_2) + (a_1 + a_2)}{a_1 + a_2} = \dfrac{a_1 + a_2}{a_2 + a_3} + \dfrac{a_2 + a_3}{a_3 + a_4} + \cdots + \dfrac{a_n + a_1}{a_1 + a_2}$.

The sum of these n terms is at least n, since their product is 1.

(vi) $n + 1 - L_n = 1 + \frac{1}{2} + \frac{2}{3} + \cdots + \frac{n-1}{n} \ge n \cdot \sqrt[n]{1 \cdot \frac{1}{2} \cdots \frac{n-1}{n}} = n \sqrt[n]{\frac{1}{n}}$.

(vii)

$$L \geq n\mathcal{G}_n \left(\frac{a^{b_1-b_2}}{b_1+b_2}, \ldots, \frac{a^{b_n-b_1}}{b_n+b_1} \right) = n\mathcal{G}_n \left(\frac{1}{b_1+b_2}, \ldots, \frac{1}{b_n+b_1} \right)$$

$$= \frac{n}{\mathcal{G}_n(b_1+b_2, \ldots, b_n+b_1)} \geq \frac{n}{\mathcal{A}_n(b_1+b_2, \ldots, b_n+b_1)} = R.$$

(viii) Multiply the inequalities $(j = 1, 2, \ldots, k)$

$$n + a_j = (n+1)\mathcal{A}_{n+1}(a_j, 1, 1, \ldots, 1) \geq (n+1)\mathcal{G}_{n+1}(a_j, 1, 1, \ldots, 1).$$

(ix)

$$L + n = \frac{a_1 + (b-a_1)}{b-a_1} + \cdots + \frac{a_n + (b-a_n)}{b-a_n}$$

$$= nb \cdot \mathcal{A}_n \left(\frac{1}{b-a_1}, \ldots, \frac{1}{b-a_n} \right) \geq nb\mathcal{G}_n \left(\frac{1}{b-a_1}, \ldots, \frac{1}{b-a_n} \right)$$

$$= \frac{nb}{\mathcal{G}_n(b-a_1, \ldots, b-a_n)} \geq \frac{nb}{\mathcal{A}_n(b-a_1, \ldots, b-a_n)} = \frac{n^2 b}{nb-1}.$$

This implies $L \geq \frac{n^2 b}{nb-1} - n = R$.

(x) Multiply the inequalities $(k = 1, 2, \ldots, n)$

$$\frac{1}{a_k} - 1 = \sum_{j \neq k} \frac{a_j}{a_k} \geq (n-1)\mathcal{G}_{n-1} \left(\frac{a_1}{a_k}, \ldots, \frac{a_{k-1}}{a_k}, \frac{a_{k+1}}{a_k}, \ldots, \frac{a_n}{a_k} \right).$$

(xi) Use the AM–GM inequality for the following 2^{n+1} numbers: 2^n times a_1^2, 2^{n-1} times $a_2^4, \ldots, 2$ times $a_n^{2^n}$, and 2 times the number 1.

(xii) If $G = \sqrt[n]{a_1 a_2 \cdots a_n}$, then G^k is the geometric mean of the set of $\binom{n}{k}$ products $a_{i_1} a_{i_2} \cdots a_{i_k}$, where $1 \leq i_1 < i_2 < \cdots < i_k \leq n$, for each $k = 1, 2, \ldots, n-1$. Hence by the AM–GM inequality for these sets of numbers you get

$$(1+a_1)(1+a_2)\cdots(1+a_n) = 1 + \sum_{k=1}^{n-1} \left(\sum a_{i_1} a_{i_2} \cdots a_{i_k} \right) + a_1 a_2 \cdots a_n$$

$$\geq 1 + \sum_{k=1}^{n-1} \binom{n}{k} G^k + G^n = (1+G)^n .$$

By taking roots you obtain the desired inequality.

8.9 (i) Under the assumption $0 < m \leq x_j \leq M$, $1 \leq j \leq n$, add, resp. multiply, the inequalities $v_j m \leq v_j x_j \leq v_j M$, resp. $m^{v_j} \leq x_j^{v_j} \leq M^{v_j}$.

(ii) Use (94) with $n = 2$, $v_1 = 1 - p$, $a_1 = a$, $v_2 = p$, and $a_2 = a + b$ $(a_1 \neq a_2)$.

(iii) Use (94) with $n = 3$ and with v_j according to (98), $a_1 = d^{b-c}$, $a_2 = d^{c-a}$, $a_3 = d^{a-b}$.

(iv) By (94) with $n = 2$, $\frac{c}{c+d} \cdot \frac{a}{c} + \frac{d}{c+d} \cdot \frac{b}{d} \geq \left(\frac{a}{c}\right)^{\frac{c}{c+d}} \left(\frac{b}{d}\right)^{\frac{d}{c+d}}$. Raising this to the power $c + d$ and simplifying, you obtain the desired inequality.

(v) Rewrite the inequality as

$$(b+c)^{v_1} \cdot (c+a)^{v_2} \cdot (a+b)^{v_3} \leq \frac{2}{3}(a+b+c),$$

where the v_j are as in (98). By (94) the left-hand side is at most equal to

$$v_1(b+c) + v_2(c+a) + v_3(a+b) = 2 \cdot \frac{ab + bc + ca}{a+b+c}.$$

It therefore suffices to prove the inequality $3(ab + bc + ca) \leq (a+b+c)^2$.

(vi) By (94) with $n = 2$, $v_1 = a$, and $v_2 = b$ you have the following inequalities: $2ab = ab + ba \geq b^a \cdot a^b$, $a^2 + b^2 = a \cdot a + b \cdot b \geq a^a \cdot b^b$, and finally

$$2 = a \cdot \frac{1}{a} + b \cdot \frac{1}{b} \geq \left(\frac{1}{a}\right)^a \left(\frac{1}{b}\right)^b = \frac{1}{a^a \cdot b^b}.$$

Equality occurs only when $a = b = \frac{1}{2}$.

(vii) By (94) with $n = 2$, $v_1 = c$, and $v_2 = 1 - c$ you have the inequalities

$$a^c b^{1-c} \leq ca + (1-c)b \quad \text{and} \quad (1-a)^c (1-b)^{1-c} \leq c(1-a) + (1-c)(1-b),$$

and by adding them you obtain the desired inequality.

(viii) By raising this to the power $1/a$, you obtain (99) with the exponent $p = (a+1)/a > 1$.

(ix) By raising this to the power $1/b$, you obtain (100) with $p = a/b < 1$ and $x = 1/a$.

(x) The inequality (101) is homogeneous. Under the assumption $x_1^p + \cdots + x_n^p = y_1^q + \cdots + y_n^q = 1$ add n inequalities (96) for $x = x_k$ and $y = y_k$.

(xi) Use (101) with $n = 2$ and $p = \frac{5}{2}$, $q = \frac{5}{3}$, $x_1 = a^2$, $x_2 = b^2$, $y_1 = c^3$, $y_2 = d^3$.

(xii) Rewrite the sum of the two Hölder inequalities

$$\sum x_k (x_k + y_k)^{p-1} \leq \left(\sum x_k^p\right)^{\frac{1}{p}} \left(\sum (x_k + y_k)^p\right)^{\frac{p-1}{p}}$$

and

$$\sum y_k (x_k + y_k)^{p-1} \leq \left(\sum y_k^p\right)^{\frac{1}{p}} \left(\sum (x_k + y_k)^p\right)^{\frac{p-1}{p}}.$$

(xiii) Use (94) with $v_k = p_k/(p_1 + p_2 + \cdots + p_n)$ and $a_k = x_k/p_k$, $1 \leq k \leq n$.

3   Hints and Answers to Chapter 3

1.1 You have $a \cdot 0 = 0$, and thus $a \mid 0$ for any $a \in \mathbb{Z}$. Similarly, $a \cdot 1 = a$ gives $a \mid a$ for any $a \in \mathbb{Z}$. If $a \cdot x = b$, $b \cdot y = c$, then $a \cdot (xy) = c$. If $a \cdot x = b$, $a \cdot y = c$, then $a(x + y) = b + c$, $a(x - y) = b - c$. For $c \neq 0$ it follows from $ax = b$ that $acx = bc$ and conversely. If $ax = b$, $b > 0$, then in the case $a \leq 0$ clearly $a \leq b$, and in the case $a > 0$ you have $x > 0$, hence $x \geq 1$, since $x \in \mathbb{Z}$, which implies $b = ax \geq a$.

1.2 (i) If $3n + 4 \mid 7n + 1$, then $3n + 4 \mid (7 \cdot (3n + 4) - 3(7n + 1))$, i.e., $3n + 4 \in \{\pm 1, \pm 5, \pm 25\}$, which means that $n \in \{-3, -1, 7\}$.

(ii) For $n \in \{3, 14\}$. [Use $9(5n^2 + 2n + 4) - (15n - 4)(3n + 2) = 44$.]

(iii) Use the binomial theorem for $(3 + 1)^n$, or prove the assertion by induction.

(iv) Use the binomial theorem for $(4 - 1)^{2n+3}$, or induction.

(v) Use the binomial theorem for $(n + 1)^n$.

(vi) Substitute $2^n - 1$ for n in (v).

(vii) Use the identities $2a + 3b = 4(9a + 5b) - 17(2a + b)$, $9a + 5b = 81(2a + 3b) - 17(9a + 14b)$.

(viii) If $2^n - 2 = nt$, then for $m = 2^n - 1$ you have $2^m - 2 = 2(2^{nt} - 1) = 2m(2^{n(t-1)} + 2^{n(t-2)} + \cdots + 2^n + 1)$.

1.5 (i) Use the facts that $a^2 + (a + 1)^2 = 2a(a + 1) + 1$ and that $a(a + 1)$ is an even number for all $a \in \mathbb{Z}$.

(ii) Take the square $(2q + 1)^2 = 4q(q + 1) + 1$ and proceed as in (i).

(iii) Induction on n.

1.13 (i) 1.   (ii) 17, if $17 \mid n + 8$, and 1 otherwise.

(iii) 3.   (iv) 11, if $11 \mid n + 7$, and 1 otherwise.

(v) Proceed as in 1.12.(iii), choose $i_{k+1} = 6c + 1$.

(vi) For odd n choose 2 as one of the summands; for $n = 4k$ or $n = 4k+2$ ($k \in \mathbb{N}$) choose $2k - 1$.

(vii) Use $(2^{2^m} - 1) = (2^{2^{m-1}} + 1)(2^{2^{m-2}} + 1) \cdots (2^{2^n} + 1)(2^{2^n} - 1)$.

(viii) $d \mid a$, $d \mid b$ imply $d \mid (a, b)$. Conversely, $(a, b) \mid a$, $(a, b) \mid b$, and thus $(a, b) \mid ka + lb = d$.

(ix) $(2^n - 1, 2^m - 1) = 2^d - 1$, where $d = (n, m)$. To prove this, use (5) of Chapter 1 to verify that $2^d - 1 \mid 2^n - 1$, $2^d - 1 \mid 2^m - 1$. For $d = n$ or $d = m$ the proof is easy. Otherwise, by 1.8 you have $d = tn + sm$ for appropriate $s, t \in \mathbb{Z}$. Show that $t > 0$, $s < 0$ or $t < 0$, $s > 0$. In the first case set $k = 2^{n(t-1)} + 2^{n(t-2)} + \cdots + 2^n + 1$, $l = -2^d(2^{m(-s-1)} + 2^{m(-s-2)} + \cdots + 2^m + 1)$, show that $(2^n - 1)k + (2^m - 1)l = 2^d - 1$, and use (viii). The second case can be dealt with analogously.

1.15 (i) Use 1.8, or prove $((u, v), (x, y)) = (u, v, x, y)$ and use 1.14.(i) three times.

(ii) Use a proof by contradiction, apply 1.8 or 1.15.(i) for $d = a$.

(iii) Show that (ac, b) is a common divisor of ac, bc, from which it follows by 1.14.(i) that $(ac, b) \mid c$. Use this to show that the numbers (ac, b) and (c, b) divide each other.

(iv) Let $m = (ab, ac, bc)$, $n = \frac{abc}{m}$. Show that $n \in \mathbb{Z}$ and $n = [a, b, c] \cdot q$, where $q \in \mathbb{N}$. Show that q is a common divisor of the numbers $\frac{ab}{m}, \frac{ac}{m}, \frac{bc}{m}$ and derive from 1.14.(i) that $\left(\frac{ab}{m}, \frac{ac}{m}, \frac{bc}{m}\right) = 1$.

(v) Substitute ab, ac, bc for a, b, c in (iv) and divide both sides by abc.

(vi) By 1.14.(iii), $(x, y) = 30$ implies the existence of $u, v \in \mathbb{N}$ such that $x = 30u$, $y = 30v$, $(u, v) = 1$ and $u + v = 5$. Then $x \in \{30, 60, 90, 120\}$, $y = 150 - x$.

(vii) $x = 495$, $y = 315$.

(viii) $x \in \{20, 60, 140, 420\}$, $y = \frac{8400}{x}$.

(ix) $x = 2$, $y = 10$, or $x = 10$, $y = 2$.

2.4 (i) One of the primes has to be even, i.e., 2. Then the second one is $6n + 3$, which for $n \in \mathbb{N}$ is not a prime.

(ii) $n = 3$. [Of the numbers n, $n + 1$, $n + 2$, exactly one is divisible by 3; the same holds for n, $n + 1 + 9$, $n + 2 + 12$.]

(iii) $p = 3$. [It follows from 1.3 that any prime $p \neq 3$ is of the form $3k \pm 1$, where $k \in \mathbb{N}$, while $2(3k \pm 1)^2 + 1$ is divisible by 3.]

(iv) $p = 5$. [Proceed as in (iii), distinguish between the cases $p = 5k \pm 1$ and $p = 5k \pm 2$.]

(v) $n = 2$. [For $n = 1$, $2^n - 1 = 1$; for $n > 2$, 2^n is of the form $3k \pm 1$, where $k \in \mathbb{N}$, $k > 1$. Then $2^n \mp 1 = 3k$ is composite.]

(vi) If $p = 3k \pm 1$, $n \in \mathbb{N}$, then $8p^2 + 1$ is divisible by 3. If $p = 3k$, $k \in \mathbb{N}$, then $p = 3$ and $8p^2 + 2p + 1 = 79$ is a prime.

(vii) $n = 4$. [Consider divisibility by 5.]

(viii) Rewrite in the form $5^{20} + 2^{30} = (5^{10} + 2^{15})^2 - 2 \cdot 5^{10} \cdot 2^{15} = ((5^{10} + 2^{15}) + 5^5 \cdot 2^8)((5^{10} + 2^{15}) - 5^5 \cdot 2^8)$.

(ix) For $n = 4k^4$, where $k \in \mathbb{N}$, $k \geq 2$, you have
$$m^4 + 4k^4 = (m^2 + 2k^2)^2 - (2mk)^2 = ((m + k)^2 + k^2)((m - k)^2 + k^2).$$

(x) Use the fact that $p - 1$ is even, and add the first and the last summands, the second and the second-to-last, etc. Then rewrite it as $\frac{m}{n} = \frac{pq}{(p-1)!}$, which means that $m(p - 1)! = pqn$, and note that p does not divide $(p - 1)!$.

(xi) The condition is satisfied, for instance, by the numbers $a_i = i \cdot n! + 1$, where $i = 1, \ldots, n$.

(xii) If $x = (a, d)$ and $y = (b, c)$, the natural numbers $a_1 = \frac{a}{x}$, $d_1 = \frac{d}{x}$, $b_1 = \frac{b}{y}$, $c_1 = \frac{c}{y}$ satisfy $(a_1, d_1) = (b_1, c_1) = 1$, by 1.14.(iii). From $ab = cd$ it follows that $a_1 b_1 = c_1 d_1$, and by 1.14.(ii) you obtain $a_1 \mid c_1$ and $c_1 \mid a_1$. Hence $a_1 = c_1$, $b_1 = d_1$ and finally $a^n + b^n + c^n + d^n = (a_1^n + b_1^n)(x^n + y^n)$.

2.6 (i) For any divisor d of a decompose d and $\frac{a}{d}$ according to 2.5; consider the primes that can occur, and how the exponents of a particular prime in the decompositions of d, $\frac{a}{d}$, a are connected to each other.

(ii) This follows directly from (i).

(iii) Write a, b, c according to 2.5 and note that in the decomposition of a, b the same prime cannot occur.

2.8 (i) If $(a+b, ab) > 1$, there exists a prime p dividing $a+b$ and ab. By 2.2., p divides one of the numbers a, b, and from $p \mid a + b$ it follows that it also divides the other one; hence $p \mid (a, b)$, which is a contradiction. Therefore, $(a + b, ab) = 1$.

(ii) (a, b). [Apply (i) to $x = \frac{a}{(a,b)}$, $y = \frac{b}{(a,b)}$.]

(iii) If p is the smallest integer bigger than 1 dividing a, then in the case $\frac{a}{p} \neq p$ the numbers $\frac{a}{p}, p$ are factors of $(a - 1)!$. In the case $\frac{a}{p} = p$ you have $a = p^2$, and from $a > 4$ it follows that $p > 2$, and thus $p, 2p$ are factors of $(a - 1)!$.

(iv) Write $(p-1)! = p^m - 1 = (p-1)(p^{m-1} + p^{m-2} + \cdots + p + 1)$, which implies $m > 1$ and $(p - 2)! = p^{m-1} + \cdots + p + 1$. By (iii), $p - 1$ divides $(p - 2)!$; hence it also divides $p^{m-1} + \cdots + p + 1 = (p^{m-1} - 1) + (p^{m-2} - 1) + \cdots + (p - 1) + m$. This means that $p - 1 \mid m$, and so $m \geq p - 1$, thus $(p-1)! = p^m - 1 \geq p^{p-1} - 1$. However, by multiplying the inequalities $p > 2$, $p > 3, \ldots, p > p - 1$ together, you obtain $p^{p-2} > (p - 1)!$, a contradiction.

(v) Proceed as in 2.7.(ii). Assume that $p_1 = 3$, $p_2 = 7$, $p_3 = 11$, $\ldots$, p_n are all the primes of the given form, and consider the number $N = 4p_2 \cdots p_n + 3$.

(vi) If $2^{2^n} + 1 = m^5 - k^5$ is a prime, then it follows from the factorization $m^5 - k^5 = (m - k)(m^4 + m^3k + m^2k^2 + mk^3 + k^4)$ that $m = k + 1$, and thus $2^{2^n} = (k + 1)^5 - k^5 - 1 = 5k(k^3 + 2k^2 + 2k + 1)$, a contradiction.

(vii) Show that for any prime p and $r \in \mathbb{N}$ such that $p^r \mid n$, the number $m = 1 + a + \cdots + a^{n-1}$ is also divisible by p^r. Use the expression

$$m = \left(1 + a^{p^r} + \cdots + a^{n-p^r}\right) \cdot \prod_{i=1}^{r}\left(1 + a^{p^{i-1}} + a^{2p^{i-1}} + \cdots + a^{(p-1)p^{i-1}}\right),$$

which is clear for $a = 1$, and which for $a \neq 1$ can be shown by multiplying both sides by $a - 1$. Then any one of the r factors $1 + a^{p^{i-1}} + a^{2p^{i-1}} + \cdots + a^{(p-1)p^{i-1}} = p + (a^{p^{i-1}} - 1) + (a^{2p^{i-1}} - 1) + \cdots + (a^{(p-1)p^{i-1}} - 1)$ is divisible by p, since from $p^r \mid (a - 1)^s$ it follows that $p \mid a - 1$.

(viii) If the prime p divides $a + b$ and $a^2 + b^2$, then it also divides $2a^2 = a^2 + b^2 + (a + b)(a - b)$ and $2b^2$. For $p \neq 2$ you then have $p \mid a$, $p \mid b$, a contradiction. For $p = 2$ it follows from $2 \mid ab$, $2 \mid a + b$ that $2 \mid a$, $2 \mid b$, a contradiction.

2.11 (ii) If $v_p(a) \leq v_p(b)$, then $v_p(a) = 0$, and thus $v_p(ab) = v_p(b)$. Analogously in the case $v_p(a) \geq v_p(b)$.

(iii) Assume that $\sqrt[m]{n} = \frac{r}{s}$, where $s > 1$ and $(r, s) = 1$. Then $r^m = s^m \cdot n$, from which for any prime p dividing s it follows that $p \mid r^m$, hence $p \mid r$, which contradicts $(r, s) = 1$.

(iv) For any prime p you have $v_p(\sqrt[s]{n^r}) = \frac{r}{s} v_p(n) \in \mathbb{N}$ and thus $s \mid r v_p(n)$, which implies with 1.14.(ii) that $s \mid v_p(n)$. If therefore $n = p_1^{t_1} \cdots p_k^{t_k}$, where $p_1, \ldots, p_k$ are distinct primes, then $m = p_1^{t_1/s} \cdots p_k^{t_k/s} \in \mathbb{N}$ and $m^s = n$.

(v) Set $c = \sqrt{a} + \sqrt{b}$. Then the number $d = \frac{a-b}{c} = \sqrt{a} - \sqrt{b}$ is also rational, and therefore $\sqrt{a} = \frac{c+d}{2}$ and $\sqrt{b} = \frac{c-d}{2}$ are rational numbers as well.

(vi) The assertion is clearly true if one of the numbers a, b, c, d is zero. In the case $abcd \neq 0$ it is enough if for any prime p you show that $v_p(ac) \geq k$, $v_p(bd) \geq k$, $v_p(bc + ad) \geq k$, or $bc + ad = 0$ implies $v_p(bc) \geq k$, $v_p(ad) \geq k$. In the case $v_p(bc) = v_p(ad)$ use (8); otherwise, use (10).

3.5 (i) Use the facts that $2^4 \equiv 3 \pmod{13}$ and $7^2 \equiv -3 \pmod{13}$.

(ii) Use $72^{2n+2} - 47^{2n} + 28^{2n-1} \equiv (-3)^{2n+2} - (-3)^{2n} + 3^{2n-1} \equiv 3^{2n-1}(3^3 - 3 + 1) \pmod{25}$.

(iii) This follows from $5^5 \equiv 4^5 \equiv 3^5 \equiv 1 \pmod{11}$.

(iv) By 1.3, any $a \in \mathbb{Z}$ is congruent to 0, 1, or 2 modulo 3. If $a \equiv 0 \pmod 3$, then $b^2 \equiv 0 \pmod 3$, and by 2.2 you have $b \equiv 0 \pmod 3$. If $a \equiv 1 \pmod 3$ or $a \equiv 2 \pmod 3$, then you always have $a^2 \equiv 1 \pmod 3$; this implies $b^2 \equiv 2 \pmod 3$, which is impossible.

(v) Proceed as in (iv): If $a \not\equiv 0 \pmod 7$, then a^2 is congruent to 1, 2, or 4 modulo 7, which implies that b^2 is congruent to 6, 5, or 3 modulo 7, but this is impossible.

(vi) It suffices to choose $a = 1$, $b = 2$.

(vii) Verify that for $abc \not\equiv 0 \pmod 3$ each of the numbers a^3, b^3, c^3 is congruent to 1 or -1 modulo 9, and thus $a^3 + b^3 + c^3 \not\equiv 0 \pmod 9$.

(viii) Proceed as in (vii).

(ix) Since $121 = 11^2$ and $a^2 + 3a + 5 \equiv (a - 4)^2 \pmod{11}$, it follows from $a^2 + 3a + 5 \equiv 0 \pmod{121}$ that $11 \mid (a-4)^2$, and thus $11 \mid a - 4$, i.e., there exists a $t \in \mathbb{Z}$ such that $a = 11t + 4$. But then $0 \equiv a^2 + 3a + 5 \equiv 121t^2 + 121t + 33 \equiv 33 \pmod{121}$, a contradiction.

(x) The solution is given by all n congruent to 1 or 2 modulo 6. [Consider all six possibilities, i.e., n congruent to 0, 1, 2, 3, 4, 5 modulo 6.]

(xi) Such a number does not exist. For, if $2^{n-1}(2^{n+1} - 1)$ were a third power of a natural number, then by 2.6.(iii), both factors would be third powers of natural numbers. First of all, this means that $n - 1$ is divisible by 3, i.e., $n = 3k + 1$ for $k \in \mathbb{Z}$, $k \geq 0$. Then $2^{n+1} - 1 = 4 \cdot 8^k - 1 \equiv 3 \pmod 7$; however, as one can easily see, a third power of a natural number can only have remainders 0, 1, or 6 upon division by 7.

(xii) Write down consecutively the remainders of $2^1, 2^2, 2^3, \ldots, 2^{12}$ upon division by 13.

(xiii) For $n = 2k$ ($k \in \mathbb{N}_0$) show that $3 \mid 19 \cdot 8^n + 17$, for $n = 4k + 1$ ($k \in \mathbb{N}_0$) show that $13 \mid 19 \cdot 8^n + 17$, and for $n = 4k + 3$ ($k \in \mathbb{N}_0$) show that $5 \mid 19 \cdot 8^n + 17$.

(xiv) For $n = 1$ and $n = 2$ this is clearly true. For $n > 2$ use equation (5) from Chapter 1: $n^n - n^2 + n - 1 = (n^{n-2} - 1)n^2 + (n - 1) = (n - 1)(n^{n-1} + n^{n-2} + \cdots + n^3 + n^2 + 1)$. Derive from $n \equiv 1 \pmod{n - 1}$ that

$n^{n-1} + \cdots + n^2 + 1 \equiv 0 \pmod{n-1}$.

(xv) For $e \in \{-1, 1\}$ set $b_e = a^2 + ae + 1$. Use the fact that $(a^2 + 1)^3 = (b_e - ae)^3 \equiv -a^3 e = -ae(b_e - ae - 1) \equiv a^2 + ae \equiv -1 \pmod{b_e}$. Both assertions are therefore true.

3.8 (i) 400, 504, 48, 96, 720.

(ii) For $5 \mid n$ you have $\varphi(5n) = 5\varphi(n)$; otherwise, $\varphi(5n) = \varphi(5) \cdot \varphi(n) = 4\varphi(n)$.

(iii) $p = 2$, $x \geq 1$. [$p^{x-1} \in \mathbb{N}$ implies $x \geq 1$, and thus $\varphi(p^x) = (p-1)p^{x-1}$.]

(iv) $x = 2$, $y = 3$. [From $3 \mid 600$ it follows that $x \geq 1$, and similarly $y \geq 1$; hence $\varphi(3^x 5^y) = 2 \cdot 3^{x-1} \cdot 4 \cdot 5^{y-1}$.]

(v) $p = 3$, $x \geq 1$. [From $2 \cdot 3p^{x-2} \in \mathbb{Z}$ it follows that $x \geq 1$, and thus $\varphi(p^x) = (p-1)p^{x-1} = 2 \cdot 3 \cdot p^{x-2}$.]

(vi) $(p,q) \in \{(3,61), (5,31), (11,13), (13,11), (31,5), (61,3)\}$. [Distinguish between $p = q$, where $p(p-1) = 120$, and $p \neq q$, where $(p-1)(q-1) = 120$, and consider all decompositions of the number 120 into two factors.]

(vii) $p = 2$, $m \in \mathbb{N}$ arbitrary odd. [If $p \mid m$, then $\varphi(pm) = p\varphi(m)$; otherwise, $\varphi(pm) = (p-1)\varphi(m)$.]

(viii) p any prime, $m \in \mathbb{N}$ any number divisible by p.

(ix) $p = q$, $m \in \mathbb{N}$ arbitrary, or $p = 2$, $q = 3$, $m \in \mathbb{N}$ even and not divisible by 3, or $p = 3$, $q = 2$, $m \in \mathbb{N}$ even and not divisible by 3. [See the hint to (vii).]

(x) No solution. [From $\varphi(m) = 14$, and since neither 8 nor 15 is prime, it follows that $7^2 \mid m$, hence $6 \cdot 7 \mid \varphi(m)$, a contradiction.]

(xi) $m \in \{17, 32, 34, 40, 48, 60\}$. [$\varphi(m) = 16$ implies $m = 2^a \cdot 3^b \cdot 5^c \cdot 17^d$, where $a, b, c, d \in \mathbb{N}_0$, $b \leq 1$, $c \leq 1$, $d \leq 1$. So $\varphi(m) = 2^b \cdot 4^c \cdot 16^d$ if $a = 0$ and $\varphi(m) = 2^{a-1} \cdot 2^b \cdot 4^c \cdot 16^d$ if $a > 0$. Hence either $a = 0$ and $b + 2c + 4d = 4$, or $a > 0$ and $(a-1) + b + 2c + 4d = 4$.]

(xii) $m = 2^a$, where $a \in \mathbb{N}$. [$\frac{m}{2} \in \mathbb{Z}$ implies $m = 2^a \cdot t$, where $a \in \mathbb{N}$ and $t \in \mathbb{N}$ is odd. Then $\varphi(m) = 2^{a-1}\varphi(t)$; hence $t = \varphi(t)$, i.e., $t = 1$.]

(xiii) $m = 2^a \cdot 3^b$, where $a, b \in \mathbb{N}$. [$\frac{m}{3} \in \mathbb{Z}$ implies $m = 3^b \cdot t$, where $b \in \mathbb{N}$, $t \in \mathbb{N}$, $\frac{t}{3} \notin \mathbb{N}$. Then $\varphi(m) = 2 \cdot 3^{b-1} \cdot \varphi(t)$, so $\varphi(t) = \frac{t}{2}$. Use (xii).]

(xiv) No solution. [$\frac{m}{4} \in \mathbb{Z}$ implies $m = 2^a \cdot t$, where $a \in \mathbb{N}$, $a \geq 2$ and $t \in \mathbb{N}$ is odd. Then $\varphi(m) = 2^{a-1}\varphi(t)$, so $\varphi(t) = \frac{t}{2}$. Use (xii).]

(xv) Proceed as in 3.7.(iv), or substitute (m, n) and $[m, n]$ for m and n in 3.7.(iv) and use 1.9.

(xvi) p^n. [With the exception of the first summand, factor out $(p-1)$ and use (5) of Chapter 1.]

(xvii) Write $m = p_1^{n_1} \cdots p_k^{n_k}$ and note that by 2.6.(i) the positive divisors of m are of the form $p_1^{m_1} \cdots p_k^{m_k}$, where $0 \leq m_1 \leq n_1, \ldots, 0 \leq m_k \leq n_k$, and that you have $\varphi(p_1^{m_1} \cdots p_k^{m_k}) = \varphi(p_1^{m_1}) \cdots \varphi(p_k^{m_k})$. Consider the product $(\varphi(p_1^0) + \varphi(p_1^1) + \varphi(p_1^2) + \cdots + \varphi(p_1^{n_1})) \cdots (\varphi(p_k^0) + \varphi(p_k^1) + \varphi(p_k^2) + \cdots + \varphi(p_k^{n_k}))$, which by (xvi) is equal to $p_1^{n_1} \cdots p_k^{n_k} = m$. By multiplying out you obtain exactly the sum of the products $\varphi(p_1^{m_1}) \cdots \varphi(p_k^{m_k})$, where $0 \leq m_1 \leq n_1$, $\ldots, 0 \leq m_k \leq n_k$.

3.15 (i) 0 for integers divisible by 5, and 1 for all others. [Use Euler's theorem for $m = 5^3$.]

(ii) Distinguish between the cases where a is or is not divisible by the prime p, and use Fermat's theorem.

(iii) Decompose $2730 = 2 \cdot 3 \cdot 5 \cdot 7 \cdot 13$ and use (ii) and 3.3.(vi).

(iv) Decompose $341 = 31 \cdot 11$. Using (ii) with $341 = (11-1) \cdot 34 + 1$ you obtain $2^{341} \equiv 2 \pmod{11}$, and from $341 = (31-1) \cdot 11 + 1 + 10$ it follows that $2^{341} \equiv 2 \cdot 2^{10} \pmod{31}$. Use $2^{10} = 32^2 \equiv 1 \pmod{31}$. Similarly, $3^{341} \equiv 3 \cdot 3^{10} \not\equiv 3 \pmod{31}$, since $3^{10} = 243^2 \equiv (-5)^2 \equiv -6 \pmod{31}$.

(v) Consider divisibility by 11, and use $2^{4n+1} = 2 \cdot 2^{4n} \equiv 2 \pmod{10}$.

(vi) Consider divisibility by 29, and $2^{6n+2} \equiv 4 \pmod{28}$.

(vii) Use $2^{10n+1} \equiv 2 \pmod{22}$ for divisibility by 23.

(viii) Use $2^{6n+2} \equiv 4 \pmod{36}$ for divisibility by 37.

(ix) First convince yourself that the given number is an integer. Consider divisibility by 7, using the fact that for $k \in \mathbb{N}$, $2^k \equiv 2 \pmod 6$ if k is odd, and $2^k \equiv 4 \pmod 6$ if k is even.

(x) Show that $2 \cdot 3^{6n+3} + 4 \equiv 2 \pmod{28}$ and that the given numbers are divisible by 29.

(xi) For $n = 6k + 4$ ($k \in \mathbb{N}_0$) the given numbers are divisible by 7, for $n = 12k + 1$ ($k \in \mathbb{N}_0$) they are divisible by 13, and so on.

(xii) For $n = 1$ this is clearly true. For $n > 1$ write $n = p_1^{a_1} \cdots p_k^{a_k}$, where $p_1 < p_2 < \cdots < p_k$ are primes and $a_1, \ldots, a_k \in \mathbb{N}$. Note that from $1 < p_1 - 1 < p_2 - 1 < \cdots < p_k - 1 \leq n - 1$ it follows that $(p_1 - 1)(p_2 - 1) \cdots (p_k - 1) \mid (n-1)!$. Also, $p_1^{a_1-1} \cdots p_k^{a_k-1} \mid n$, and together $\varphi(n) \mid n!$, which by Euler's theorem implies that $2^{n!} \equiv 1 \pmod n$.

(xiii) For $p = 2$ consider $n = 2k$ ($k \in \mathbb{N}$), and for $p > 2$ consider $n = (kp - 1)(p - 1)$ ($k \in \mathbb{N}$).

4.4. (i) $x = 6 + 9t$ $(t \in \mathbb{Z})$.   (ii) $x = 7 + 10t$ $(t \in \mathbb{Z})$.

(iii) $x = -5 + 31t$ $(t \in \mathbb{Z})$.   (iv) $x = 1 + 5t$ $(t \in \mathbb{Z})$.

(v) $x = 1 + 3t$ $(t \in \mathbb{Z})$.   (vi) No solution.

(vii) Use induction on n. For $n = 1$ it suffices to choose $k_1 = 1$. If $k_1 < k_2 < \cdots < k_{n-1}$ are chosen such that $ak_1 + b, \ldots, ak_{n-1} + b$ are pairwise relatively prime, denote by m their product and show that $(a, m) = 1$. Let $k_n > k_{n-1}$ denote some solution of the congruence $ax \equiv 1 - b \pmod m$ (existence is guaranteed by Theorem 4.2); then $ak_n + b \equiv 1 \pmod m$ and the numbers $k_1, \ldots, k_n$ solve the problem.

4.8. (i) $x \equiv -3 \pmod{55}$.   (ii) No solution.

(iii) $x \equiv 20 \pmod{42}$.

(iv) $x \equiv 13 \pmod{36}$. [The first congruence follows from the second.]

(v) $x \equiv 11 \pmod{30}$.   (vi) $x \equiv 4 \pmod{105}$.

(vii) For even a no solution; for odd a the solution is $x \equiv 4a - 3 \pmod{24}$.

(viii) No solution when a is not divisible by 4; when a is divisible by 4, then $x \equiv -2 \pmod{10}$ is the solution. [The second congruence gives $x = -2 + 10t$, where $t \in \mathbb{Z}$, and substituting this into the first one, $-4 + 20t \equiv a$

(mod 4). For a not divisible by 4, this has no solution, while for $4 \mid a$ it is satisfied by any $t \in \mathbb{Z}$.]

(ix) If the integer c satisfies the system (17), then this system is equivalent to the system obtained from (17) by substituting all the right sides by c. Use 3.3.(vi).

(x) Use induction on k with the help of Theorem 4.6.

(xi) Choose k distinct primes $p_1, p_2, \ldots, p_k$. The numbers $p_1^n, p_2^n, \ldots, p_k^n$ are pairwise relatively prime, and by (x) there exists an $m \in \mathbb{Z}$ such that $m \equiv -i \pmod{p_i^n}$ for $i = 1, 2, \ldots, k$. The numbers $m+1, m+2, \ldots, m+k$ have the required property.

(xii) To prove the more difficult part of the problem, set $m_i = p_1^{n_{i,1}} \cdots p_s^{n_{i,s}}$, $i = 1, \ldots, k$, where $p_1, \ldots, p_s$ are all the distinct primes dividing $m = [m_1, \ldots, m_k]$, $n_{i,j} \in \mathbb{N}_0$ for $i = 1, \ldots, k$; $j = 1, \ldots, s$. Note that the system (17) is equivalent to the the the system

$$x \equiv c_i \pmod{p_j^{n_{i,j}}} \qquad (i = 1, \ldots, k; j = 1, \ldots, s).$$

For each $j = 1, \ldots, s$ choose $i(j) \in \{1, \ldots, k\}$ such that for all $i \in \{1, \ldots, k\}$ you have $n_{i,j} \leq n_{i(j),j}$. If the condition of the problem is satisfied, this last system is equivalent to the system

$$x \equiv c_{i(j)} \pmod{p_j^{n_{i(j),j}}} \qquad (j = 1, \ldots, s),$$

which, according to (x), always has solutions.

4.13 (i) $x \equiv 1 \pmod 7$.

(ii) $x \equiv 1 \pmod 5$ or $x \equiv 3 \pmod 5$. [It is convenient to write the given congruence in the form $2x^2 + 2x + 1 \equiv 0 \pmod 5$, respectively $x^2 + x + 3 \equiv 0 \pmod 5$.]

(iii) No solution.   (iv) $x \equiv 1 \pmod 5$.

(v) $x \equiv 1 \pmod 6$ or $x \equiv 3 \pmod 6$. [It is convenient to first solve the congruence $x^2 - 4x + 3 \equiv 0 \pmod 2$.]

(vi) $x \equiv 0 \pmod 5$. [By 3.11 you have $x^5 \equiv x \pmod 5$; rewrite the given congruence as $-x^2 \equiv 0 \pmod 5$.]

(vii) $x \equiv 1 \pmod{11}$. [Using 3.11, rewrite as $x - 1 \equiv 0 \pmod{11}$.]

(viii) $x \equiv 1 \pmod 5$ or $x \equiv 2 \pmod 5$. [Using 3.11, rewrite as $x^2 - 3x + 2 \equiv 0 \pmod 5$; the trinomial on the left can be factored: $x^2 - 3x + 2 = (x-2)(x-1)$.]

(ix) $x \equiv 3 \pmod{35}$.

(x) $x \equiv 24 \pmod{45}$. [Solve separately for the moduli 5 and 9. For modulus 5 rewrite as $3(x+1)^2 \equiv 0 \pmod 5$. In the case of modulus 9 it is convenient to first solve modulo 3, which gives $x = 3s + 1$ or $x = 3s$; the first case gives a contradiction, while the second one leads to $x = 6 + 9t$.]

(xi) No solution. [First solve modulo 11 and substitute the solution $x = 11t - 5$.]

(xii) All $x \equiv a \pmod{35}$ are solutions, where $a \in \{1, 4, 6, 9, 11, 16, 19, 24, 26, 29, 31, 34\}$.

(xiii) This condition is satisfied if and only if m is a power of a prime. [For $m = p^n$, where p is a prime, use Euler's Theorem 3.13 to show that the choice $F(x) = x^{\varphi(m)}$ gives $F(a) \equiv 1 \pmod{m}$ for any integer a that is not divisible by p, while the inequality $\varphi(m) = p^{n-1}(p-1) \geq p^{n-1} \geq n$ shows that $F(a) \equiv 0 \pmod{m}$ for any integer a divisible by p.

On the other hand, suppose that an integer m is divisible by two different primes p and q and that a polynomial $F(x)$ with integer coefficients satisfies the given condition. Then there are integers a, b such that $F(a) \equiv 0 \pmod{m}$ and $F(b) \equiv 1 \pmod{m}$. Due to Theorem 4.6 there is an integer c such that $c \equiv a \pmod{p}$ and $c \equiv b \pmod{q}$, so Theorem 4.10 gives $F(c) \equiv F(a) \equiv 0 \pmod{p}$ and $F(c) \equiv F(b) \equiv 1 \pmod{q}$. But in this case you have neither $F(c) \equiv 0 \pmod{m}$ nor $F(c) \equiv 1 \pmod{m}$, a contradiction.]

4.15 (i) If a is a solution of the congruence $x^2 \equiv t \pmod{p}$, then $a^2 \equiv t \pmod{p}$, so $(a, p) = 1$ and by 3.11 you have $t^{(p-1)/2} \equiv a^{p-1} \equiv 1 \pmod{p}$.

(ii) Evaluate $(\pm t^{k+1})^2 = t^{2k+2} = t^{2k+1} \cdot t \equiv t \pmod{p}$.

5.4 (i) $x = 3t + 2$, $y = 2t + 1$, $t \in \mathbb{Z}$ arbitrary. [Other forms of the solution are also possible.]

(ii) $x = 1 + 15t$, $y = -1 - 17t$, $t \in \mathbb{Z}$ arbitrary.

(iii) No solution, since $(16, 48) = 16$, which does not divide 40.

(iv) $x = 2t$, $y = 4 - 5t$, $t \in \mathbb{Z}$ arbitrary.

(v) $x = -3 + 13t$, $y = 4 - 17t$, $t \in \mathbb{Z}$ arbitrary.

(vi) $x = 9 + 77t$, $y = 7 + 60t$, $t \in \mathbb{Z}$ arbitrary.

(vii) $x = -7 - 5s - 5t$, $y = 7 + 3s + 6t$, $z = 1 + 2s$, s, t are arbitrary integers. [This expression is not unique; the solution can, for instance, be written as $x = 3 + 5s$, $y = -5 - 6s - 3t$, $z = 1 + 2t$, $s, t \in \mathbb{Z}$, and in many other ways. This is similar in (viii) and (x)–(xii) below.]

(viii) $x = -3 + 5s + 10t$, $y = 4 - 12s - 9t$, $z = 7s$, $s, t \in \mathbb{Z}$ arbitrary.

(ix) No solution, since $(105, 119, 161) = 7$, and 83 is not divisible by 7.

(x) $x = -3 + s + 2r - 11t$, $y = 4 - 2s - 3r + 10t$, $z = s$, $u = r$, $s, r, t \in \mathbb{Z}$ arbitrary.

(xi) $x = 4 - 5s - 4r - 10t$, $y = 4s - r + t$, $z = -1 + r + 3t$, $u = r$, $s, r, t \in \mathbb{Z}$ arbitrary.

(xii) $x = 2t$, $y = 2s - t$, $z = 2r - s$, $u = 2 - r$, $r, s, t \in \mathbb{Z}$ arbitrary. Or more easily, simplifying to the form $x = -2y - 4z - 8u + 16$, we immediately see the solution $x = -2r - 4s - 8t + 16$, $y = r$, $z = s$, $u = t$.

5.8 (i) $x = 2 + 11t$, $y = 4t + 11t^2$ or $x = 9 + 11t$, $y = 7 + 18t + 11t^2$, where $t \in \mathbb{Z}$ is arbitrary.

(ii) No solution.

(iii) $x = 2 + 21t$, $y = 441t^3 + 168t^2 + 20t + 1$, where $t \in \mathbb{Z}$ is arbitrary.

(iv) $x = 125t - 12$, $y = 15625t^3 - 4500t^2 + 434t - 14$, where $t \in \mathbb{Z}$ is arbitrary. [First solve the congruence $x^3 + 2x + 2 \equiv 0 \pmod{5}$, obtaining the solutions $x = 1 + 5r$, $x = 3 + 5r$, $r \in \mathbb{Z}$; substitute them back into $x^3 + 2x + 2$ and solve the same congruence modulo 25. The case $x = 1 + 5r$

leads to the contradiction $5 \equiv 0 \pmod{25}$, while the second case gives $x = -12 + 25s$, $s \in \mathbb{Z}$. Substitute this again into $x^3 + 2x + 2$ and solve the congruence $(25s - 12)^3 + 2(25s - 12) + 2 \equiv 0 \pmod{125}$, which leads to $s = 5t$, $t \in \mathbb{Z}$, i.e., $x = -12 + 125t$.]

(v) $x = 11t - 4$, $y = 121t^3 - 253t^2 + 136t - 22$, $t \in \mathbb{Z}$ arbitrary.

(vi) No solution.

(vii) $x = 3 + 30t$, $y = -1 + 6t + 30t^2$ or $x = -3 + 30t$, $y = -1 - 6t + 30t^2$, where $t \in \mathbb{Z}$ is arbitrary. [The congruence $x^2 - 39 \equiv 0 \pmod{30}$ is easiest to solve using a system of three congruences modulo $2, 3, 5$ (see the paragraph before 4.12).]

(viii) $x = 2t$, $y = 2s + 1$, $z = -\frac{1}{2}(t + s + 1)(t + s)$ or $x = 2t + 1$, $y = 2s$, $z = -\frac{1}{2}(t + s + 1)(t + s)$, where t, s are arbitrary integers.

(ix) $x = t$, $y = t + 3s$, $z = t^2 + 3ts + 3s^2 - 2t - 3s$, where $t, s \in \mathbb{Z}$ are arbitrary. [Show that $x \equiv y \pmod 3$.]

(x) No solution. [Use the fact that for arbitrary $a \in \mathbb{N}$ you have $a^2 \equiv 0 \pmod 8$ or $a^2 \equiv 1 \pmod 8$ or $a^2 \equiv 4 \pmod 8$. Hence $x^2 + (2y)^2 + (3z)^2 \equiv 7 \pmod 8$ has no solution.]

(xi) $x = -600 - 2t + t^2$, $y = 300 + \frac{t(3-t)}{2}$, $t \in \mathbb{Z}$ arbitrary. [Choose the substitution $x + 2y = z$.]

(xii) $x = -1 - 5s - 7s^2$, $y = 2 + 7s - (1 + 5s + 7s^2)^2$ or $x = -3 - 9s - 7s^2$, $y = 4 + 7s - (3 + 9s + 7s^2)^2$. [Choose the substitution $t = x^2 + y$ and solve the equation $t^2 + t + 1 = -7x$.]

5.10 (i) $x = 2k + 1$, $y = 2s$, $z = \frac{1}{8}(5 \cdot 25^k + 9^s + 2)$, where $k, s \in \mathbb{N}_0$ are arbitrary. [Note that $5^{2k} \equiv 1 \equiv 3^{2s} \pmod 8$, $5^{2k+1} \equiv 5 \pmod 8$, $3^{2s+1} \equiv 3 \pmod 8$).]

(ii) $x = 4 + 12k$, $y = \frac{1}{13}(2^{4+12k} - 3)$, where $k \in \mathbb{N}_0$ is arbitrary. [Use the fact that the smallest natural number n for which $2^n \equiv 1 \pmod{13}$ is $n = 12$, and therefore by 3.14.(v) for any $x \in \mathbb{N}_0$ you have $2^x \equiv 2^4 \pmod{13}$ if and only if $x \equiv 4 \pmod{12}$.]

(iii) $(x, y) \in \{(-2, -11), (-4, 3), (-1, -6), (-5, -2), (1, -2), (-7, -6), (5, 3), (-11, -11)\}$. [Express $y = x - 1 - \frac{8}{x+3}$ and consider the cases where $x + 3$ divides 8.]

(iv) $x = 2$, $y = 8$ or $x = -2$, $y = 0$. [Express $y = 2x + 1 + \frac{33}{x^2+7}$ and consider the cases where $x^2 + 7$ divides 33.]

(v) $x = t^5 + 2$, $y = t^4 - 3$, $t \in \mathbb{Z}$ arbitrary. [Express $y = -3 + \sqrt[5]{(x - 2)^4}$, which, by 2.11.(iv), implies $x - 2 = t^5$ for appropriate $t \in \mathbb{Z}$. (Consider the cases $x < 2$, $x = 2$, $x > 2$; 2.11.(iv) is stated only for natural numbers).]

(vi) $x = (t + 1)^2$, $y = t^2$, $t \in \mathbb{N}_0$ arbitrary. [Use $2\sqrt{y} = x - 1 - y$ and 2.11.(iii) to obtain $y = t^2$, where $t \in \mathbb{N}_0$.]

(vii) $(x, y) \in \{(3, 2), (-3, -2), (-3, 4), (3, -4), (-7, 6), (7, -6), (-17, 12), (17, -12)\}$. [It follows directly from the problem that y is even. If you solve the equation for x, you find that y has to be a divisor of 12; hence y can only be among the numbers $\pm 2, \pm 4, \pm 6, \pm 12$.]

(viii) $(x, y, z) \in \{(1, 1, 5), (1, -1, -1), (-1, 1, -1), (-1, -1, 1)\}$. [By solving for z, you find that $xy \mid 1$, which is possible only when $|x| = |y| = 1$.]

5.13 (i) $(x, y) \in \{(3, 2), (6, 3), (12, 4), (30, 5)\}$. [Rewrite as $(x + 6)(y - 6) = -36$, which implies $y < 6$.]

(ii) $(x, y, z) \in \{(2, 4, 4), (4, 2, 4), (4, 4, 2), (2, 3, 6), (2, 6, 3), (3, 2, 6), (3, 6, 2), (6, 2, 3), (6, 3, 2), (3, 3, 3)\}$. [Assume that $x \leq y \leq z$ and show that $1 < x < 4$ (if $4 \leq x \leq y \leq z$, you would have $\frac{1}{x} + \frac{1}{y} + \frac{1}{z} \leq \frac{3}{4}$). Distinguish between the cases $x = 2$, from which $(y - 2)(z - 2) = 4$, and $x = 3$, which gives $(2y - 3)(2z - 3) = 9$.]

(iii) $x = 2$, $y = 1$. [If $y = 1$, then $x(x^{x-1} - 2) = 0$, which implies $x = 2$. If $y \geq 2$, then $y(y^{y-1} - 1) \geq 2$, which gives $x(x^{x-1} - 2) \leq -2$. Of course, for $x \geq 2$ you have $x(x^{x-1} - 2) \geq 0$ and for $x = 1$, $x^x - 2x = -1$.]

(iv) $(x, y, z) \in \{(2, 3, 4), (2, 4, 3), (3, 2, 4), (3, 4, 2), (4, 2, 3), (4, 3, 2)\}$. [Suppose $x \leq y \leq z$ and show that $z \geq 5$ or $z \leq 3$ leads to a contradiction.]

(v) $x = y = 1$ or $x = 3$, $y = 2$. [As in 5.12.(iii), show that $x \geq 0$, $y \geq 1$. If y is odd, then $2^x = 3^y - 1 \equiv 2 \pmod 4$, which means $x \leq 1$. For $y = 2k$ you have $2^x = 3^{2k} - 1 = (3^k - 1)(3^k + 1)$; hence $3^k - 1$ and $3^k + 1$ are powers of 2. Therefore, from $3^k - 1 < 3^k + 1$ it follows that $2(3^k - 1) \leq 3^k + 1$, which implies $3^k \leq 3$, i.e., $k \leq 1$.]

(vi) $(x, y) \in \{(0, 0), (1, 0), (0, 1), (2, 1), (1, 2), (2, 2)\}$. [Rewrite as $(x-1)^2 + (y - 1)^2 + (x - y)^2 = 2$, which implies $0 \leq x \leq 2$, $0 \leq y \leq 2$.]

5.15 (i) $x = 11$, $y = 3$. [If x, y is a solution of the given equation, then $4x^2 = 4 + 4y + 4y^2 + 4y^3 + 4y^4$, and thus $(2y^2 + y)^2 = 4y^4 + 4y^3 + y^2 < (2x)^2 < 4y^4 + 4y^3 + 9y^2 + 4y + 4 = (2y^2 + y + 2)^2$, which implies $4x^2 = (2x)^2 = (2y^2 + y + 1)^2 = 4y^4 + 4y^3 + 5y^2 + 2y + 1$. Equating this with the given equation, you get $y^2 - 2y - 3 = 0$, i.e., $y = 3$ ($y = -1 \notin \mathbb{N}$), and thus $x = 11$.]

(ii) $(x, y) \in \{(0, -1), (-1, -1), (0, 0), (-1, 0), (5, 2), (-6, 2)\}$. [A solution x, y of the given equation satisfies $(2x + 1)^2 = 4y^4 + 4y^3 + 4y^2 + 4y + 1 = (2y^2 + y)^2 + 3y^2 + 4y + 1 = (2y^2 + y + 1)^2 - (y^2 - 2y)$. Show that for any $y \in \mathbb{Z}$, $y \neq -1$, you have $3y^2 + 4y + 1 = (3y + 1)(y + 1) > 0$, and that for any $y \in \mathbb{Z}$, $y \neq 0$, $y \neq 1$, $y \neq 2$, you have $y^2 - 2y = (y - 2)y > 0$. Hence for arbitrary $y \in \mathbb{Z}$, $y \notin \{-1, 0, 1, 2\}$, you get $(2y^2 + y)^2 < (2x + 1)^2 < (2y^2 + y + 1)^2$, which is a contradiction. It remains to consider $y \in \{-1, 0, 1, 2\}$.]

(iii) $x = 0$, $y = 1$ or $x = 0$, $y = -1$. [Show that a solution x, y of the given equation satisfies the following: If $x > 0$, then $(x^3 + 1)^2 = x^6 + 2x^3 + 1 < x^6 + 3x^3 + 1 = y^4 < x^6 + 4x^3 + 4 = (x^3 + 2)^2$; if $x \leq -2$, then $x^3 + 3 < 0$; hence $(x^3 + 2)^2 = x^6 + 4x^3 + 4 < x^6 + 3x^3 + 1 = y^4 < x^6 + 2x^3 + 1 = (x^3 + 1)^2$, and in both cases you get contradictions. Therefore, $x = -1$ or $x = 0$; the case $x = -1$ leads to $y^4 = -1$, a contradiction.]

(iv) $x = 0$, $y = 1$ or $x = 0$, $y = -1$. [For $x^2 > 0$ you have $(x^4 + x^2)^2 = x^8 + 2x^6 + x^4 < y^2 = x^8 + 2x^6 + 2x^4 + 2x^2 + 1 < x^8 + 2x^6 + 3x^4 + 2x^2 + 1 = (x^4 + x^2 + 1)^2$, a contradiction.]

(v) For an appropriate $k \in \mathbb{N}$ you have $2n^2 = kd$. Suppose there exists an $x \in \mathbb{N}$ such that $x^2 = n^2 + d$. Then $k^2x^2 = k^2n^2 + k^2d = n^2(k^2 + 2k)$, which implies $k^2 + 2k > 0$ and $\sqrt{k^2 + 2k} = \frac{kx}{n}$. By 2.11.(iii) you have $\sqrt{k^2 + 2k} = m \in \mathbb{N}$. But then $k^2 < k^2 + 2k = m^2 < k^2 + 2k + 1 = (k+1)^2$, a contradiction.

5.17 (i) $x = y = z$, where $z \in \mathbb{N}$ is arbitrary. [Use the AM–GM inequality for x^3, y^3, z^3.]

(ii) No solutions.

(iii) $y = 2x$, $z = 3x$, where $x \in \mathbb{N}$ is arbitrary. [Use Cauchy's inequality for the triples x, y, z and $1, 2, 3$; see Chapter 2, Section 5.4.]

(iv) No solutions. [By symmetry you may assume that the natural numbers $x \le y \le z \le u \le v$ satisfy the equation. Then clearly $x \ge 2$, but at the same time $u < 3$, since for $u \ge 3$ you would obtain

$$\frac{1}{x^2} + \frac{1}{y^2} + \frac{1}{z^2} + \frac{1}{u^2} + \frac{1}{v^2} \le \frac{1}{4} + \frac{1}{4} + \frac{1}{4} + \frac{1}{9} + \frac{1}{9} = \frac{35}{36} < 1.$$

Then $x = y = z = u = 2$, from which upon substitution you get $1/v^2 = 0$, a contradiction.]

5.20 (i) $x = y = 0$ or $x = y = 2$. [Rewrite as $(x - 1)(y - 1) = 1$.]

(ii) $(x, y) \in \{(0,0), (0,-1), (-1,0), (-1,-1)\}$. [Rewrite as $(2(2y+1))^2 - (2x+1)^2 = 3$ and use the identity $A^2 - B^2 = (A+B)(A-B)$. The number 3 can be written as a product of two integers in only four ways.]

(iii) $(x, y) \in \{(11,4), (11,-4), (-11,4), (-11,-4), (13,8), (13,-8), (-13,8), (-13,-8), (19,16), (19,-16), (-19,16), (-19,-16), (53,52), (53,-52), (-53,52), (-53,-52)\}$. [Note that if (x_0, y_0) is a solution, then $(x_0, -y_0)$, $(-x_0, y_0)$, $(-x_0, -y_0)$ are also solutions, and therefore it suffices to solve the equation in $\mathbb{N}_0$. If you rewrite the equation as $(x+y)(x-y) = 105$, then the conditions $x \ge 0$, $y \ge 0$ imply $x + y \ge 0$, and therefore also $x - y \ge 0$, and since $x + y \ge x \ge x - y$, it suffices to solve only the four systems of equations.]

(iv) For $p = 3$ the equation has the solution $x = 5$, for $p = 17$ it has the solution $x = 13$, and for all remaining primes it has no solution. [Decompose $2x^2 - x - 36 = (2x - 9)(x + 4)$ and consider the six possible decompositions of p^2 as a product of two integers. For each of these decompositions solve a system of two equations in the two unknowns x, p.]

(v) $x = 8$, $y = 5$. [Decompose $2x^2 + 5xy - 12y^2 = (2x - 3y)(x + 4y)$ and use the fact that $x + 4y \ge 5$ holds for $x, y \in \mathbb{N}$, and solve the three systems of equations corresponding to the decompositions $28 = 4 \cdot 7 = 2 \cdot 14 = 1 \cdot 28$.]

(vi) $x = 3$. [Rewrite the equation as $2^{x^2-4}(2^{x+2} - 1) = 2^5 \cdot 31$. Since for $x \in \mathbb{N}$ the number $2^{x+2} - 1 \in \mathbb{N}$ is odd, it follows that $2^{x+2} - 1 = 31$, i.e., $x = 3$. Convince yourself by substituting that $x = 3$ satisfies the equation.]

(vii) $x = 4$, $y = 3$, $z = 1$. [Rewrite as $2^{x-1} = (y-1)(y^z + y^{z-1} + \cdots + y + 1)$. Then $y - 1$ is a power of 2, which means that $y = 2$ or $y \ge 3$ is odd. From $y = 2$ it would follow that 2^{x-1} is odd; thus $x = 1$ and $z = 0 \notin \mathbb{N}$. If $y \ge 3$ is odd, then z is odd, since $y^z + y^{z-1} + \cdots + y + 1 \ge y + 1 \ge 4$ is a power of 2. Hence $z = 2v - 1$ for an appropriate $v \in \mathbb{N}$, and $2^{x-1} = y^{2v} - 1 = (y^v - 1)(y^v + 1)$. This implies $y^v - 1 = 2^s$, $y^v + 1 = 2^t$, where $s, t \in \mathbb{N}_0$ satisfy $s + t = x - 1$, and thus $2 = (y^v + 1) - (y^v - 1) = 2^t - 2^s$. Therefore, $t > s$ and $2 = 2^s(2^{t-s} - 1)$, i.e., $s = 1$, $t = 2$. Then $x = s + t + 1 = 4$ and $y^v = 3$, so $y = 3$, $v = 1$, and finally $z = 2v - 1 = 1$.]

(viii) No solution. [Rewrite as $2^{x+1} = y^{z+1} + 1$, which implies that y is odd. If z is odd or $y = 1$, then $2^{x+1} = y^{z+1} + 1 \equiv 1 + 1 \pmod 4$, which would imply $x = 0 \notin \mathbb{N}$. Therefore, z is even and $y > 1$. Then $2^{x+1} = (y + 1)(y^z - y^{z-1} + \cdots - y + 1)$. Now $y^z - y^{z-1} + \cdots - y + 1 = (y - 1)(y^{z-1} + y^{z-3} + \cdots + y) + 1 > 1$ is an odd divisor of 2^{x+1}, which is a contradiction.]

(ix) Assume to the contrary that for appropriate $k, r \in \mathbb{N}$ you have $k + (k+1) + \cdots + (k+r) = 2^n$, which implies $(2k+r)(r+1) = 2^{n+1}$. Since $r+1 > 1$ and $2k+r > 1$, both numbers are even. Then also $(2k+r) - (r+1) = 2k - 1$ has to be even, which is a contradiction.

(x) Any natural number N that is not of the form 2^n, $n \in \mathbb{N}_0$, can be expressed as $N = 2^n(2m + 1)$, where $n \in \mathbb{N}_0$, $m \in \mathbb{N}$. You have to show that there exist $k, r \in \mathbb{N}$ such that $k + (k + 1) + \cdots + (k+r) = N$, i.e., $(2k+r)(r+1) = 2^{n+1} \cdot (2m+1)$. In the case $2^n > m$ set $r = 2m$, $k = 2^n - m$, and in the case $2^n \leq m$ set $r = 2^{n+1} - 1$, $k = m + 1 - 2^n$.

(xi) $(x, y, z) \in \{(6, 6, 2), (6, 2, 6), (18, -4, 0), (18, 0, -4)\}$. [Subtracting the two equations from each other, rewrite them as $(y-1)(z-1) = 5$ and solve four systems of equations in terms of y, z.]

(xii) $(x, y, z) \in \{(1, 1, 1), (-5, 4, 4), (4, -5, 4), (4, 4, -5)\}$. [From Table 4.12 in Chapter 1 (for $x_1 = x$, $x_2 = y$, $x_3 = z$) you get with $\sigma_1 = 3$, $s_3 = 3$,

$$(3 - x)(3 - y)(3 - z) = 27 - 9\sigma_1 + 3\sigma_2 - \sigma_3$$

$$= \sigma_1\sigma_2 - \sigma_3 = \frac{1}{3}(\sigma_1^3 - s_3) = 8,$$

$$(3 - x) + (3 - y) + (3 - z) = 9 - \sigma_1 = 6.$$

Therefore, either all of the numbers $3 - x$, $3 - y$, $3 - z$ are even, or exactly one is even. In the first case $3 - x = 3 - y = 3 - z = 2$, and in the second case two of the numbers $3-x$, $3-y$, $3-z$ are equal to -1 and the remaining one is 8.]

(xiii) Use the result of Example 5.19.(v) to show that there exist $a, b \in \mathbb{N}$, $a > b$ such that $xy = 2ab(a^2 - b^2)$, $z^2 = a^2 + b^2$. If 7 does not divide xy, then it divides neither a nor b. It is easy to verify that the square of a number not divisible by 7 is congruent to 1, 2 or 4 modulo 7. Since none of the sums $1 + 2$, $1 + 4$, $2 + 4$ is divisible by 7, nor congruent to 1, 2, 4 modulo 7, z is not divisible by 7 and $a^2 \equiv b^2 \pmod 7$ must hold. Then $7 \mid a^2 - b^2$ and thus $7 \mid xy$, which is a contradiction. The condition $(x, y) = 1$ is necessary, since, for example, $15^2 + 20^2 = 5^4$, while 7 does not divide $15 \cdot 20$.

6.3 (i) No solution. [Solve as a congruence modulo 3 and use the fact that the congruence $-y^2 \equiv 1 \pmod 3$ does not hold for any $y \in \mathbb{Z}$.]

(ii) No solution. [The left-hand side is congruent to 3 or 5 modulo 8, since from the equation it follows that y cannot be even.]

(iii) No solution. [$y^2 \equiv 2 \pmod 3$ holds for no $y \in \mathbb{Z}$.]

(iv) No solution. [For $x \equiv 1 \pmod 3$ you have $3x^2 + 3x + 7 \equiv 4 \pmod 9$, and for $x \not\equiv 1 \pmod 3$, $3x^2 + 3x + 7 \equiv 7 \pmod 9$ holds. On the other hand, y^3 can be congruent only to 0, -1, or 1 modulo 9.]

(v) No solution. [By Fermat's Theorem 3.11 you have $(x+1)^3 + (x+2)^3 + (x+3)^3 + (x+4)^3 - (x+5)^3 \equiv x+1+x+2+x+3+x+4-x-5 = 3x+5 \equiv 2 \pmod 3$.]

(vi) No solution. [You have $19^z \equiv 1 \pmod 3$, while $2^x + 7^y \equiv (-1)^x + 1 \pmod 3$.]

(vii) No solution. [When x is even, then $2^x + 5^y \equiv 1 + (-1)^y \not\equiv 1 \pmod 3$, while $19^z \equiv 1 \pmod 3$. For odd $x = 2n + 1$, $n \in \mathbb{N}_0$, you have $2^x + 5^y \equiv 2 \cdot 4^n \equiv 2 \cdot (-1)^n \pmod 5$, while $19^z \equiv (-1)^z \pmod 5$.]

(viii) $x = y = 1$, $z \in \mathbb{N}$ arbitrary, or $x = y = 3$, $z = 1$. [The case $z = 1$ was dealt with in 6.2.(iii). For $z > 1$, first deal with $x \leq 7$: $1! + 2! + \cdots + 7! = 5913$ is divisible by 73, but not by 73^2, and for $x = 2, 3, \ldots, 6$, $1! + \cdots + x!$ is divisible by 3 but not by 27. Since for $n \geq 9$, $n! \equiv 0 \pmod{27}$ holds, for $x \geq 8$ you have $1! + 2! + \cdots + x! \equiv 1! + 2! + \cdots + 8! = 46233 \equiv 9 \pmod{27}$, and thus $1! + 2! + \cdots + x!$ is divisible by 9 but not by 27.]

(ix) For arbitrary $x \in \mathbb{Z}$ you have $x^{10} + rx^7 + s \equiv x + x + 1 \equiv 1 \pmod 2$, and thus $x^{10} + rx^7 + s \neq 0$.

(x) Suppose that x, y, z is a solution, and set $d = (x, y)$, $x_1 = \frac{x}{d}$, $y_1 = \frac{y}{d}$. Use 2.5. to conclude from $d^2 \mid pz^2$ that $d \mid z$ holds and set $z_1 = \frac{z}{d}$. Then $x_1^2 + y_1^2 = pz_1^2 \equiv 0 \pmod p$, and from 3.14.(vi) it follows that $x_1 \equiv y_1 \equiv 0 \pmod p$, which is a contradiction to $(x_1, y_1) = 1$.

(xi) $x = y = z = u = 0$. [Proceed as in 6.2.(ii).]

(xii) $x = y = z = u = 0$. [Proceed as in 6.2.(ii) and use 3.14.(vi) for $p = 7$.]

(xiii) No solution. [Assume that $x, y, z \in \mathbb{Z}$ is a solution of the given equation. Then y is odd, since otherwise $x^2 = y^3 + (4z+2)^3 - 1 \equiv 7 \pmod 8$, which is impossible. Rewriting, $x^2 + 1 = (4z + 2)^3 + y^3 = ((4z+2) + y) \cdot A$, where $A = (4z + 2)^2 - y(4z + 2) + y^2 = (2z + 1 - y)^2 + 3(2z + 1)^2$. Hence $A \geq 3$ and $A \equiv 3 \pmod 4$. As in 6.2.(iv) show that there exists a prime p dividing A such that $p \equiv 3 \pmod 4$. For such a p you have $x^2 + 1 \equiv 0 \pmod p$, and by 3.14.(vi) it follows that $x \equiv 1 \equiv 0 \pmod p$, which is a contradiction.]

(xiv) No solution. [Assume that $x, y \in \mathbb{Z}$ is a solution of the given equation. Then x is even, since otherwise $y^3 = x^2 + 5 \equiv 6 \pmod 8$, which is impossible. Hence $y^3 = x^2 + 5 \equiv 1 \pmod 4$, and therefore y is odd, $y \equiv y \cdot y^2 = y^3 \equiv 1 \pmod 4$. So for appropriate $m, n \in \mathbb{Z}$ you have $x = 2n$, $y = 4m+1$. Rewriting, $n^2 + 1 = \frac{1}{4}(x^2 + 4) = \frac{1}{4}(y^3 - 1) = \frac{1}{4}(y-1)(y^2 + y + 1) = m \cdot A$, where $A = y^2 + y + 1$. You have $A = (y^2 + y + 1) \equiv 3 \pmod 4$, $4A = (2y + 1)^2 + 3 > 0$, and thus $A \geq 3$. As in 6.2.(iv) prove the existence of a prime p dividing a such that $p \equiv 3 \pmod 4$. Then $n^2 + 1 = m \cdot A \equiv 0$ $\pmod p$, and by 3.14.(vi), $n \equiv 1 \equiv 0 \pmod p$, which is a contradiction.]

(xv) Assume that $x, y, z \in \mathbb{N}$ satisfy the given equation. Then $(4x - 1)(4y-1) = 4(4xy-x-y)+1 = 4z^2+1$, while $4x-1 \geq 3$, $4x-1 \equiv 3 \pmod 4$. As in the preceding exercises show that there exists a prime p dividing $4x-1$ such that $p \equiv 3 \pmod 4$. Then $(2z)^2 + 1 = (4x - 1)(4y - 1) \equiv 0 \pmod p$, and by 3.14.(vi) you have $2z \equiv 1 \equiv 0 \pmod p$, a contradiction.

6.6 (i) $x = y = z = 0$. [Show that if $x, y, z \in \mathbb{Z}$ is a solution, then $\frac{x}{2}, \frac{y}{2}, \frac{z}{2} \in \mathbb{Z}$ is also a solution.]

(ii) $x = y = z = u = 0$. [Show that if $x, y, z, u \in \mathbb{Z}$ is a solution, then $\frac{x}{2}, \frac{y}{2}, \frac{z}{2}, \frac{u}{2} \in \mathbb{Z}$ is also a solution.]

(iii) $x = y = z = u = 0$. [Proceed as in 6.5.(iii), but modulo 8.]

(iv) $x = y = z = 0$. [Show that if $x, y, z \in \mathbb{Z}$ is a solution, then $\frac{x}{3}, \frac{y}{3}, \frac{z}{3} \in \mathbb{Z}$ is also a solution.]

(v) $x = y = z = 0$, $u \in \mathbb{N}$ arbitrary. [Show that if $x, y, z, u \in \mathbb{Z}$ is a solution, then xy is even (otherwise, $1+1+z^2$ would be congruent to 1 or 0 modulo 4, which is impossible), and moreover, both x, y must be even (if exactly one of the numbers x, y were even, then $1+z^2 \equiv 0 \pmod 4$, which is impossible). Then, of course, z must also be even. Set $x_1 = \frac{x}{2}$, $y_1 = \frac{y}{2}$, $z_1 = \frac{z}{2}$, $u_1 = 2u$, $d = x^2 + y^2 + z^2$ and show that x_1, y_1, z_1, u_1 satisfy the given equation and $x_1^2 + y_1^2 + z_1^2 < d$ if $d > 0$.]

(vi) $x = 0$, $y = 11^t$, $z = t$ or $x = 11^t$, $y = 0$, $z = t$, where $t \in \mathbb{N}_0$ is arbitrary. [First show that from $x, y \in \mathbb{Z}$ it follows that $z \geq 0$. Set $d = x^2 y^2 z$ and for an arbitrary solution x, y, z with the property $d > 0$ show with the help of 3.14.(vi) that $x \equiv y \equiv 0 \pmod{11}$. Hence $z > 1$ and $x_1 = \frac{x}{11} \in \mathbb{Z}$, $y_1 = \frac{y}{11} \in \mathbb{Z}$, $z_1 = z - 1 \in \mathbb{N}$, where

$$0 < x_1^2 y_1^2 z_1 = \frac{x^2 y^2 z_1}{11^4} < x^2 y^2 z_1 < x^2 y^2 z = d.$$

By the principle 6.4 the equation has no solution satisfying $d > 0$.]

(vii) $x = y = z = u = 0$. [Proceed as in 6.5.(v) and solve the auxiliary equation $x^2 + y^2 + z^2 + u^2 = 2^{2v-1} xyzu$, where $x, y, z, u \in \mathbb{Z}$, $v \in \mathbb{N}$. Assume that x, y, z, u, v is a solution of this equation such that $d = x^2 + y^2 + z^2 + u^2 > 0$. Then $x^2 + y^2 + z^2 + u^2$ is even. For all numbers x, y, z, u to be odd, you would need $x^2 + y^2 + z^2 + u^2 \equiv 1+1+1+1 = 4 \pmod 8$, while $2^{2v-1} xyzu$ is congruent to 2 or 6 modulo 8 for $v = 1$, and is congruent to 0 modulo 8 when $v \geq 2$. For exactly two of the numbers x, y, z, u to be even, you would need $x^2 + y^2 + z^2 + u^2 \equiv 0+0+1+1 = 2 \pmod 4$, while $2^{2v-1} xyzu$ would be divisible by 8. Hence all the numbers x, y, z, u are even, and thus $x_1 = \frac{x}{2}$, $y_1 = \frac{y}{2}$, $z_1 = \frac{z}{2}$ and $u_1 = \frac{u}{2}$ are integers. Furthermore, x_1, y_1, z_1, u_1, v_1, where $v_1 = v + 1$, is a solution of the auxiliary equation, and $0 < x_1^2 + y_1^2 + z_1^2 + u_1^2 < d$.]

(viii) $x = y = z = 0$. [Show that if $x, y, z \in \mathbb{Z}$ is a solution such that $d = x^2 + y^2 + z^2 > 0$, then yz must be even (otherwise, $2x^2 = 7z^2 - y^2 \equiv$

$7 - 1 = 6 \pmod 8$, which would imply $x^2 \equiv 3 \pmod 4$, but this cannot occur for any $x \in \mathbb{Z}$). Hence one of the numbers y, z is even, and from the equation it follows that the other one is also even. But then $2x^2 = 7z^2 - y^2$ is divisible by 4, and thus x is even. Setting $x_1 = \frac{x}{2}$, $y_1 = \frac{y}{2}$, $z_1 = \frac{z}{2}$, you obtain an integer solution satisfying $0 < x_1^2 + y_1^2 + z_1^2 < d$.]

6.10 (i) For instance, $x = -50t^3 - 75t^2 - 36t - 4$, $y = 5t + 2$, $z = 1$, where $t \in \mathbb{Z}$ is arbitrary. [Choose $z = 1$ and use the method of 5.5.]

(ii) Rewrite as $-1 = (2x+1)^2 - 2y^2 = ((2x+1) + \sqrt{2}y)((2x+1) - \sqrt{2}y)$. The equation is clearly satisfied by $x = 3$, $y = 5$. As in 6.9.(ii) use the fact that the equation $u^2 - 2v^2 = 1$ has as a solution $u = 3$, $v = 2$, and multiply out as $((2x+1) + \sqrt{2}y)(3 + 2\sqrt{2}) = (6x + 3 + 4y) + \sqrt{2}(4x + 2 + 3y)$. If you set $x_1 = 3$, $y_1 = 5$ and $x_{n+1} = 3x_n + 2y_n + 1$, $y_{n+1} = 4x_n + 2 + 3y_n$ for any $n \in \mathbb{N}$, then x_n, y_n for all $n \in \mathbb{N}$ are solutions of the given equation, and $0 < x_1 < x_2 < \cdots, 0 < y_1 < y_2 < \cdots$.

(iii) Rewrite as $1 = (2y)^2 - 3(2x+1)^2 = (2y + \sqrt{3}(2x+1))(2y - \sqrt{3}(2x+1))$. The equation is satisfied by $x = 7$, $y = 13$. Use the fact that the equation $u^2 - 3v^2 = 1$ has the solution $u = 2$, $v = 1$. Multiplying $(2y + \sqrt{3}(2x + 1))$ by $2 + \sqrt{3}$ will not lead to a convenient form (the term not belonging to $\sqrt{3}$ is odd), hence you have to multiply by $(2 + \sqrt{3})^2 = 7 + 4\sqrt{3}$. You get $(2y + \sqrt{3}(2x + 1))(7 + 4\sqrt{3}) = (14y + 24x + 12) + \sqrt{3}(14x + 7 + 8y)$. If you set $x_1 = 7$, $y_1 = 13$ and $x_{n+1} = 7x_n + 4y_n + 3$, $y_{n+1} = 12x_n + 7y_n + 6$ for any $n \in \mathbb{N}$, then x_n, y_n are solutions of the given equation for all $n \in \mathbb{N}$, and $0 < x_1 < x_2 < \cdots, 0 < y_1 < \cdots$.

(iv) For fixed y, z the equation $x^2 + y^2 + z^2 = 3xyz$ is quadratic in x, and its roots x_1, x_2 satisfy, by Vieta's relations, $x_1 + x_2 = 3yz$. Hence, if x, y, z are a solution of the given equation, then so are $y, z, 3yz - x$. The equation is certainly satisfied by $x = 1$, $y = 1$, $z = 1$. Set $u_1 = u_2 = u_3 = 1$ and $u_{n+3} = 3u_{n+1}u_{n+2} - u_n$ for $n \in \mathbb{N}$. Prove by induction that $u_1 = u_2 = u_3 < u_4 < u_5 < \cdots$ (for an arbitrary $n \in \mathbb{N}$ assume that $1 \le u_n \le u_{n+1} \le u_{n+2}$ and show $u_{n+2} < u_{n+3}$) and that for all $n \in \mathbb{N}$, u_n, u_{n+1}, u_{n+2} is a solution of the given equation (assume, for an arbitrary n, that u_n, u_{n+1}, u_{n+2} is a solution and show that $u_{n+1}, u_{n+2}, u_{n+3}$ is also a solution).

(v) For instance, $x = n - 1 + 2t$, $y = 2t(n-1) + 2t^2 - 1 + \frac{1}{2}(n-1)(n-2)$, $z = 2t(n-1) + 2t^2 + \frac{1}{2}(n-1)(n-2)$. [Choose $z = y + 1$ and use the method of 5.5.]

(vi) Rewrite the equation as $(2x+n)^2 + (2y+n)^2 + (2z+n)^2 + n^2 = 4n^6$ and show that an arbitrary number triple x, y, z solving this equation is such that none of these numbers exceeds $|n^3| - \frac{n}{2}$, and none is less than $-|n^3| - \frac{n}{2}$.

6.12 (i) Proceed as in 6.11.(ii). [If $x_0, x_1, \ldots, x_s$ is a solution for $s \in \mathbb{N}$, set $y_0 = 5x_0$, $y_i = 3x_i$ for $i = 1, 2, \ldots, s$, $y_{s+1} = 4x_0$ and show that $y_0, y_1, \ldots, y_{s+1}$ is a solution for $s + 1$.]

(ii) Proceed as in 6.11.(i). [Use the fact that for $s = 1$ the equation can be written as a quadratic, and has therefore no more than two solutions.

Further, use the fact that natural numbers a, b with $a > b$ satisfy $a + \frac{1}{a} > a \geq b + 1 \geq b + \frac{1}{b}$, which implies $\frac{a}{1+a^2} < \frac{b}{1+b^2}$. From $u \leq (s+1)x_1/(1+x_1^2)$ it again follows that $x_1 < x_1 + (1/x_1) \leq (s+1)/u$.]

(iii) Proceed as in 6.11.(ii). [Use the fact that $(1/12)^3 + (1/15)^3 + (1/20)^3 = (1/10)^3$ and $(1/(5 \cdot 7 \cdot 13))^3 + (1/(5 \cdot 12 \cdot 13))^3 + (1/(7 \cdot 12 \cdot 13))^3 + (1/(5 \cdot 7 \cdot 12 \cdot 13))^3 = (1/(5 \cdot 7 \cdot 12))^3$; these identities can be obtained from the identities $3^3 + 4^3 + 5^3 = 6^3$, $1^3 + 5^3 + 7^3 + 12^3 = 13^3$. If $x_0, x_1, \ldots, x_s \in \mathbb{N}$ are such that $(1/x_0)^3 = (1/x_1)^3 + \cdots + (1/x_s)^3$, set $y_0 = 10x_0$, $y_i = 12x_i$ for $i = 1, 2, \ldots, s$, $y_{s+1} = 15x_0$, $y_{s+2} = 20x_0$ and show that $(1/y_0)^3 = (1/y_1)^3 + \cdots + (1/y_{s+2})^3$.]

7.3 (i) By 7.2.(iv) the number of such m is exactly $[(10^7 - 1)/786] - [10^6/786] = 11450$.

(ii) By 7.2.(iv) the number of such m is exactly $999 - [999/5] - [999/7] + [999/35] = 686$ (the numbers divisible by 35 were subtracted once as divisible by five, and a second time as divisible by 7; hence they must be added in again).

(iii) In a similar fashion as in (ii) you see that there are exactly $99 - [99/2] - [99/3] + [99/6] = 33$ such numbers.

(iv) By 7.2.(iv) there are exactly $[1000/5] - [1000/25] = 160$ such numbers.

(v) Use 7.2.(ii) and show that

$$\left[\frac{[x]}{n}\right] \leq \frac{[x]}{n} \leq \frac{x}{n} < \frac{[x]+1}{n} \leq \left[\frac{[x]}{n}\right] + 1.$$

(vi) Distinguish between even and odd n.

(vii) By (35) you have $[-x] \leq -x < [-x] + 1$, and thus $-[-x] \geq x > -[-x] - 1$.

(viii) If both numbers $\langle x \rangle$, $\langle y \rangle$ are less than $\frac{1}{2}$, then both sides of the inequality are equal to $2([x]+[y])$. If, for instance, $\langle x \rangle \geq \frac{1}{2}$, then $x \geq [x] + \frac{1}{2}$, which means that $[2x] \geq 2[x] + 1$. From this last inequality and from the inequalities $[2y] \geq 2[y]$ and $[x + y] \leq [x] + [y] + 1$ (see 7.2.(iii)) the desired inequalities follow also in that case.

(ix) By 7.2.(i) you have $[xy] = [([x]+\langle x \rangle)([y]+\langle y \rangle)] = [x] \cdot [y] + [\langle x \rangle \cdot [y] + [x]\langle y \rangle + \langle x \rangle \cdot \langle y \rangle] \geq [x] \cdot [y]$. Since $[\langle x \rangle [y] + [x]\langle y \rangle + \langle x \rangle \langle y \rangle] \leq [[y] + [x] + \langle x \rangle \langle y \rangle] = [y] + [x] + [\langle x \rangle \langle y \rangle] = [y] + [x]$, it follows that $[xy] \leq [x][y] + [x] + [y]$.

(x) If $x > y \geq 1$, the assertion follows from 7.2.(iv) for $n = 1$. Reduce the remaining cases to this one with the help of 7.2.(i).

(xi) By 7.2.(i) you have $[x]+[-x] = [[x]-x]$ and by (35), $-1 < [x]-x \leq 0$.

(xii) This follows from (xi), since $\langle x \rangle + \langle -x \rangle = x - [x] + (-x) - [-x] = -([x] + [-x])$.

(xiii) This follows from 7.2.(iii) with $y = x$.

(xiv) This follows from (xiii), since $\langle 2x \rangle - 2\langle x \rangle = 2x - [2x] - 2(x - [x]) = -[2x] + 2[x]$.

(xv) By (35) you have for $m = [\sqrt[n]{[x]}]$ that $m \leq \sqrt[n]{[x]} < m + 1$, and thus $m^n \leq [x] < (m + 1)^n$, which implies $[x] + 1 \leq (m + 1)^n$. You obtain then $m^n \leq [x] \leq x < [x] + 1 \leq (m + 1)^n$, and so $m \leq \sqrt[n]{x} < m + 1$, i.e., $m = [\sqrt[n]{x}]$.

(xvi) Since $4n(n + 1) < (2n + 1)^2$, you have $2\sqrt{n(n + 1)} < 2n + 1$, and thus $(\sqrt{n} + \sqrt{n + 1})^2 < 4n + 2$, which implies $[\sqrt{n} + \sqrt{n + 1}] \leq [\sqrt{4n + 2}]$. If $[\sqrt{n} + \sqrt{n + 1}] < [\sqrt{4n + 2}]$, there would exist an $m \in \mathbb{N}$ such that $\sqrt{n} + \sqrt{n + 1} < m \leq \sqrt{4n + 2}$. But then $2\sqrt{n(n + 1)} < m^2 - (2n + 1) \leq 2n + 1$. Squaring the left inequality and keeping in mind that m has to be an integer, you would obtain $m^2 = 2(2n + 1)$, which, however, is not possible: The square of a natural number is either odd or divisible by 4.

(xvii) This follows from 7.2.(v) by setting $\frac{x}{n}$ for x.

(xviii) Similar as in (xvii).

(xix) By 7.2.(v) you have $\sum_{k=0}^{m-1}[x + \frac{k}{m} + \frac{j}{n}] = [mx + \frac{mj}{n}]$ and $\sum_{j=0}^{n-1}[x + \frac{k}{m} + \frac{j}{n}] = [nx + \frac{nk}{m}]$, and thus both sides of the desired identity are equal to $\sum_{k=0}^{m-1} \sum_{j=0}^{n-1}[x + \frac{k}{m} + \frac{j}{n}]$.

(xx) This follows from (xix); however, it is also possible to use the second identity in 7.2.(v) in the same way as the first one was used in identity (xix).

(xxi) Show that $\tau^2 = \tau + 1$. Then for $d = \tau n - [\tau n]$ you have $d = (\tau + 1)n - [(\tau + 1)n] = \tau^2 n - [\tau^2 n]$, which implies that $-\frac{d}{\tau} = d(1 - \tau) = (\tau^2 n - [\tau^2 n]) - (\tau^2 n - \tau[\tau n]) = \tau[\tau n] - [\tau^2 n]$. Then $[\tau^2 n] = \tau[\tau n] + \frac{d}{\tau}$. From this you obtain the result if you note that $0 < d < 1 < \tau$.

7.5 (i) You have to determine the highest power of 10 that divides $\binom{125}{62} = \frac{125!}{62!63!}$. Since by 7.4 you have $v_5(125!) = 25+5+1 = 31$, $v_5(62!) = v_5(63!) = 12 + 2 = 14$, then $v_5(\binom{125}{62}) = 31 - 14 - 14 = 3$ (using the notation of 2.9). Similarly, $v_2(125!) = 62 + 31 + 15 + 7 + 3 + 1 = 119$, $v_2(62!) = v_2(63!) = 31 + 15 + 7 + 3 + 1 = 57$, and thus $v_2(\binom{125}{62}) = 119 - 57 - 57 = 5$. The given binomial coefficient is therefore divisible by 2^5 and 5^3, but not by 5^4. Hence its digit representation (to base 10) ends in three zeros.

(ii) These are exactly the numbers of the form $n = 2^t$, $t \in \mathbb{N}_0$. For by 7.4 you have $v_2((2^t)!) = 2^{t-1} + 2^{t-2} + \cdots + 2 + 1 = 2^t - 1$, while all other numbers do not satisfy this condition. If $n = 2m + 1$, $m \in \mathbb{N}$, then by 2.9 and 7.4, $v_2(n!) = v_2(2m+1)+v_2((2m)!) = v_2((2m)!) = m+[m/2]+[m/4]+ \cdots + [m/2^k] \leq m + (m/2) + (m/4) + \cdots + (m/2^k) = (m/2^k) \cdot (2^k + 2^{k+1} + \cdots + 2 + 1) = (m/2^k)(2^{k+1} - 1) < (m/2^k) \cdot 2^{k+1} = 2m = n-1$, and thus the number $n!$ is not divisible by 2^{n-1}. If $n = 2^t(2m + 1)$, $m \in \mathbb{N}$, $t \in \mathbb{N}$, then $v_2(n!) = v_2(2^t(2m+1)!) = 2^{t-1}(2m+1)+2^{t-2}(2m+1)+\cdots+2(2m+1)+ (2m+1)+v_2((2m+1)!)$ holds by 7.4. Now, $v_2((2m+1)!) < 2m$ by what you have already shown, so $v_2(n!) < (2m+1)(2^{t-1}+2^{t-2}+\cdots+2+1)+2m = (2m+1)(2^t-1)+2m = n-1$. Again, the number $n!$ is not divisible by 2^{n-1}.

(iii) To prove the assertion it suffices to show that for every prime p the inequality $v_p(\binom{2m}{m})+v_p(\binom{2n}{n}) \geq v_p(\binom{m+n}{m})$ holds. But this follows from 7.4 with the use of 7.3.(viii), since for sufficiently large k you have

$$v_p\left(\binom{2m}{m}\right) + v_p\left(\binom{2n}{n}\right) - v_p\left(\binom{m+n}{m}\right)$$

$$= \sum_{i=1}^{k}\left(\left[\tfrac{2m}{p^i}\right] - 2\left[\tfrac{m}{p^i}\right]\right) + \sum_{i=1}^{k}\left(\left[\tfrac{2n}{p^i}\right] - 2\left[\tfrac{n}{p^i}\right]\right)$$

$$- \sum_{i=1}^{k}\left(\left[\tfrac{m+n}{p^i}\right] - \left[\tfrac{m}{p^i}\right] - \left[\tfrac{n}{p^i}\right]\right)$$

$$= \sum_{i=1}^{k}\left(\left[\tfrac{2m}{p^i}\right] + \left[\tfrac{2n}{p^i}\right] - \left[\tfrac{m}{p^i}\right] - \left[\tfrac{n}{p^i}\right] - \left[\tfrac{m+n}{p^i}\right]\right) \geq 0.$$

(iv) Set $y_n = \langle 2^n\sqrt{2}\rangle$. If $y_n < \frac{1}{2}$, then $x_{n+1} = 2x_n$ is even. If, on the other hand, $y_n \geq \frac{1}{2}$, then $1 - y_{n+1} = 2(1 - y_n)$, and thus for appropriate $r \in \mathbb{N}$ you have $1 - y_{n+r} > \frac{1}{2}$, i.e., $y_{n+r} < \frac{1}{2}$.

(v) It is clear that $1 < x_0 < x_1 < x_2 < \cdots$. If for some $n \in \mathbb{N}$ the number x_n is even, then $x_n = 2^r(2k-1)$ for appropriate $r, k \in \mathbb{N}$, and the number $x_{n+r} = 3^r(2k-1)$ is odd. If, on the other hand, x_n is odd, then $x_n = 2^r(2k-1) + 1$ for appropriate $r, k \in \mathbb{N}$ and $x_{n+r} = 3^r(2k-1) + 1$ is even.

(vi) Set $k = [\sqrt{m}]$. Then $m = k^2 + j$ for an appropriate j, $0 \leq j \leq 2k$. In the case $0 \leq j \leq k$, set $f^r(m) = f(f^{r-1}(m))$ and prove by induction that $f^{2r}(m) = (k+r)^2 + (j-r)$ holds for $r = 0, 1, \ldots, j$. Then $f^{2j}(m)$ is the square of the integer $k+j$. The case $k < j \leq 2k$ can be reduced to the first case by considering $f(m)$ instead of m.

(vii) From the statement of the problem it follows that each of the numbers $1, 2, \ldots, 2\,000$ is divisible by at most one of the given numbers $a_1, \ldots, a_n$. Using 7.2.(iv), you find that $[\frac{2000}{a_1}] + \cdots + [\frac{2000}{a_n}] \leq 2000$. Hence $(\frac{2000}{a_1} - 1) + \cdots + (\frac{2000}{a_n} - 1) < 2000$, which means that $\frac{1}{a_1} + \cdots + \frac{1}{a_n} < 1 + \frac{n}{2000} \leq 2$.

7.7 (i) Proceed as in the special case $m = 10$ in 7.6.(iii).

(ii) $x = 0$ or $x = 1$. [Use the fact that $x = [x^2] \in \mathbb{Z}$.]

(iii) The equation is solved by exactly those x satisfying $-\sqrt{3} < x \leq -\sqrt{2}$ or $\sqrt{2} \leq x < \sqrt{3}$. [Rewrite the equation as $2 \leq x^2 < 3$ and solve the inequality.]

(iv) $x = 1$. [Use the fact that $x = [3x^2 - x] - 1 \in \mathbb{Z}$, and thus $[3x^2 - x] = 3x^2 - x$.]

(v) $x \in \{0, \frac{3}{2}, 3\}$. [Use the fact that $2\langle x\rangle \in \mathbb{Z}$, and thus $\langle x\rangle = 0$ or $\langle x\rangle = \frac{1}{2}$.]

(vi) $x = 5$, $y = 7$. [Set $k = [\sqrt{y-1}]$ and from the equation $x = 1 + k^2$ derive the inequalities $k + 1 < \sqrt{y + 2\sqrt{x}} < k + 2$. The second equation then implies $y = 2k + 3$. Show that from the condition $k = [\sqrt{2k+2}]$ it follows that $(k-1)^2 \leq 3$. Verify by substituting that of the values 0, 1, 2 only $k = 2$ works.]

(vii) $x \in \{-\frac{2}{5}, \frac{1}{5}, \frac{4}{5}, \frac{7}{5}, 2\}$. [Using 7.2.(v) for $n = 2$ write the left-hand side of the equation as $m = [\frac{4x-2}{3}] \in \mathbb{Z}$. Then $x = \frac{3m+4}{5}$, so $m = [\frac{4m+2}{5}]$. Solve the pair of inequalities $m \leq \frac{4m+2}{5} < m+1$ in integers and verify that each solution $m \in \{-2, -1, 0, 1, 2\}$ leads to a solution of the given equation.]

(viii) $x = 5$. [The condition $[\sqrt[3]{m}] = k$ is for $m, k \in \mathbb{N}$ equivalent to the pair of inequalities $k^3 \leq m \leq (k+1)^3 - 1$. For a fixed k, the number of such m is exactly $(k+1)^3 - k^3 = 3k^2 + 3k + 1$. Then the left-hand side of the equation is of the form $\sum_{k=1}^{x-1} S_k$, where $S_k = k(3k^2 + 3k + 1)$. Since for $k \in \mathbb{N}$ you have $S_k > 0$ and $S_1 = 7$, $S_2 = 38$, $S_3 = 111$, $S_4 = 244$, the equation has the unique solution $x = 5$.]

(ix) For all $k \in \mathbb{N}$ it follows from (35) that $k[x] \leq kx < k[x] + k$, which means that $k[x] \leq [kx] \leq k[x] + k - 1$. Therefore, $S = [x] + [2x] + [4x] + [8x] + [16x] + [32x] \leq 63[x] + (1 + 3 + 7 + 15 + 31) = 63[x] + 57$ and $S \geq 63[x]$. Hence for some $m = [x] \in \mathbb{Z}$ you have $63m \leq S \leq 63m + 57$, and therefore S can be congruent only to $0, 1, 2, \ldots, 57$ modulo 63. On the other side, $12345 \equiv 60 \pmod{63}$, and thus the equation cannot have solutions.

Remark. Show that the sum S, upon division by 63 with different values of x, gives only 32 different remainders. To do this, use the expression

$$\langle x \rangle = \frac{a_1}{2} + \frac{a_2}{4} + \frac{a_3}{8} + \frac{a_4}{16} + \frac{a_5}{32} + r,$$

where $a_k \in \{0, 1\}$ for $k = 1, 2, \ldots, 5$ and $0 \leq r < \frac{1}{32}$.

(x) $n^2 - n + 1$ solutions. [If $n = 1$, then $x = 1$ is a solution. Now assume that $n > 1$. By rewriting you obtain $([x] + \langle x \rangle)^2 - [[x]^2 + 2[x]\langle x \rangle + \langle x \rangle^2] = \langle x \rangle^2$, and thus, in view of $[x]^2 \in \mathbb{Z}$, you get $2[x]\langle x \rangle = [2[x]\langle x \rangle + \langle x \rangle^2]$. Since $0 \leq \langle x \rangle^2 < 1$, this last equation is equivalent to the condition $2[x]\langle x \rangle \in \mathbb{Z}$, i.e., $\langle x \rangle = \frac{k}{2[x]}$, where $k \in \{0, 1, \ldots, 2[x] - 1\}$. For each of the values $[x] = 1, 2, \ldots, n-1$ the term $\langle x \rangle$ must take on exactly $2[x]$ different values, and for $[x] = n$ you have $\langle x \rangle = 0$ (since $x \leq n$), and thus the number of all solutions is $2 + 4 + \cdots + 2(n-1) + 1 = n(n-1) + 1 = n^2 - n + 1$.]

(xi) Answer: Exactly the squares of natural numbers cannot be expressed in this form. Denote $f(n) = [n + \sqrt{n} + \frac{1}{2}]$. The difference $f(n+1) - f(n) = 1 + [\sqrt{n+1} + \frac{1}{2}] - [\sqrt{n} + \frac{1}{2}] \geq 1$ is greater than 1 if and only if there exists a natural number m satisfying $\sqrt{n} + \frac{1}{2} < m \leq \sqrt{n+1} + \frac{1}{2}$, which is equivalent to $n < m^2 - m + \frac{1}{4} \leq n + 1$, i.e., $n = m^2 - m$. On the other hand it follows from $\sqrt{n} + 1 > \sqrt{n+1}$ that $f(n+1) - f(n) \leq 2$. Hence

$$f(n+1) - f(n) = \begin{cases} 2 & \text{if } n = m^2 - m \text{ for some } m \in \mathbb{N}, \ m > 1, \\ 1 & \text{otherwise.} \end{cases}$$

Therefore, the values $f(n)$ for $n \in \mathbb{N}$ are all natural numbers in succession, with the exception of the integer 1 (since $f(1) = 2$) and the numbers of the form $f(m^2 - m) + 1 = m^2 - m + 1 + [\sqrt{m^2 - m} + \frac{1}{2}] = m^2$. The last equation follows from the bounds $m - \frac{3}{2} < \sqrt{m^2 - m} < m - \frac{1}{2}$, valid for all $m \in \mathbb{N}$.

8.2 (i) 89. [Find a number $10a + b$ satisfying the equation $10a + b = a + b^2$, which means $9a = b(b-1)$. Use the fact that the numbers b and $b - 1$ cannot be both divisible by 3.]

(ii) 8281. [Find a number $1000a + 100(b+1) + 10a + b = x^2$. Rewrite it as $101(10a + b) = (x+10)(x-10)$ and use the fact that 101 is a prime.]

(iii) $55^2 = 3025$. [Find an $x \in \mathbb{N}$ such that $(10x+5)^2 = 3005+100a+10b$. Then $b = 2$, $x^2 + x = 30 + a$, which means that $30 \le x^2 + x < 40$, and thus $x = 5$.]

(iv) $19^2 = 361$. [Show that n is odd and is not divisible by 5 and check all such n, $11 \le n \le 31$, among which only 19 satisfies the condition of the problem.]

(v) 432. [Find a number $x = 100a + 10b + c$ for which $2x = 3a!b!c!$. Since $100 \le x \le 999$, you have $67 \le a!b!c! \le 666$. This implies that none of the numbers a, b, c exceeds 5, since $6! = 720 > 666$. Furthermore, if all the numbers a, b, c are less than 4, then two of them must be 3, and the remaining one must be 2. Then $a!b!c!$ is always divisible by 8, which means that x is divisible by 4. Now, x is clearly divisible by 3. So altogether $1 \le a \le 5$, $0 \le b \le 5$, $0 \le c \le 5$, $3 \mid a+b+c$, $4 \mid 2b + c$, $67 \le a!b!c! \le 666$. If $c = 0$, then $b = 4$, $a = 5$; if $c = 2$, then $b = 3$, $a = 4$ or $b = 5$, $a = 2$; if $c = 4$, then $b = 0$, $a = 5$ or $b = 2$, $a = 3$ or $b = 4$, $a = 1$. Of the numbers 540, 432, 252, 504, 324, 144, only 432 satisfies the given problem.]

(vi) $x = 112$, $y = 896$. [Set $x = 100a+10b+c$. Since $y = 800a+80b+8c < 1000$ and $x \ge 100$, you have $a = 1$. Then $y - x = 700 + 70b + 7c = n^2$ for an appropriate $n \in \mathbb{N}$, which means that $700 \le n^2 < 1000$, or $26 < n < 32$. Since $7 \mid n$, you have $n = 28$, which means that $x = n^2/7 = 112$, $y = 896$. In this case $112896 = 336^2$.]

(vii) The number n is divisible by 5 and is odd, so $n = 10x + 5$, where $x \in \mathbb{N}$. Then $n^2 = (10x+5)^2 = 100x(x+1) + 25$. Since $x(x+1)$ is an even number, its final digit is even, and this is the hundreds digit of n^2.

(viii) $n = 1$ or $n = 9$. [Assume that n has k digits, i.e., $10^{k-1} \le n < 10^k$. Then $n^2 < 10^{2k}$, and thus the digit sum of n^2 is at most $9 \cdot 2k$, so $10^{k-1} \le n \le 18k$. For $k \ge 3$ it is easy to show by induction that $10^{k-1} > 18k$, hence $k \le 2$. Therefore, n^2 has at most four digits, i.e., $n^2 = 1000a+100b+10c+d$, where at least one of the digits a, b, c, d is nonzero, and $n = a+b+c+d$. This implies $n(n-1) = 9(111a+11b+c)$. Since n, $n-1$ cannot both be divisible by 3, either n or $n-1$ is divisible by 9. Since $1 \le n = a+b+c+d \le 36$, n can only be among the numbers 1, 9, 10, 18, 19, 27, 28, 36, among which only 1 and 9 work out.]

(ix) 153846. [Write the desired number x in the form $x = 10n + 6$ and assume that it has k digits. Then $4x = 6 \cdot 10^{k-1} + n$, which implies $13n = 2(10^{k-1} - 4)$. The smallest k such that $10^{k-1} \equiv 4 \pmod{13}$ is $k = 6$, which means that $x = 153846$.]

(x) n can be an arbitrary number divisible by 10, or any of the numbers 11, 22, 33, 44, 55, 66, 77 88, 99, 12, 24, 36, 48, 13, 26, 39, 14, 28, 15, 16, 17, 18, 19. [It is easy to see that upon removing the final digit, the number n

will be decreased by at least a factor 10, and that the factor 10 is achieved exactly for all multiples of 10. Write n in the form $10m + a$, where $m \in \mathbb{N}$, $0 \leq a \leq 9$, and assume that for some $x \geq 1$ you have $10m + a = m(10 + x)$, which means that $a = mx$, so $x \leq 9$. For $x = 1$ you have $1 \leq m = a \leq 9$; for $x = 2$ it is $2 \leq 2m = a \leq 8$; for $x = 3$, $3 \leq 3m = a \leq 9$; for $x = 4$, $4 \leq 4m = a \leq 8$; for $5 \leq x \leq 9$, $m = 1$, $a = x$.]

(xi) Show by induction that for each $m \in \mathbb{N}$ there exists an $n \in \mathbb{N}$ in the decimal representation of which there occur only ones and zeros and such that $s(n) = m$ and $s(n^2) = m^2$. For $m = 1$ one can choose $n = 1$. If $n \in \mathbb{N}$ satisfies $s(n) = m$ and $s(n^2) = m^2$ for some $m \in \mathbb{N}$, and if $r \in \mathbb{N}$ is determined by the condition $10^r > n \geq 10^{r-1}$, then set $n' = 10^r n + 1$. It is easy to see that the decimal representation of n' again consists only of zeros and ones, and that $s(n') = s(n) + 1 = m + 1$. Since $10^{2r} > 2n \cdot 10^r$, you also have $s((n')^2) = s(n^2) + s(2n) + 1 = (m + 1)^2$.

(xii) It is clear that for any $m, n \in \mathbb{N}$ you have $s(m + n) \leq s(m) + s(n)$. If $a_k, \ldots, a_1, a_0$ are the digits of n, i.e., if $n = \sum_{i=0}^{k} 10^i a_i$, then $s(mn) = s(\sum_{i=0}^{k} 10^i m a_i) \leq \sum_{i=0}^{k} s(m a_i) \leq \sum_{i=0}^{k} a_i s(m) = s(n) s(m)$. Hence $s(n) = s(1000n) \leq s(125) s(8n) = 8 s(8n)$. The example $n = 125$ shows that $s(n) < 8 s(8n)$ does not generally hold.

(xiii) The case $m \leq 2$ is clear, so assume $m > 2$. Set $S_n = n + s(n)$. If n does not end in a 9, then $S_{n+1} = S_n + 2$. If, on the other hand, n does end in a 9, then $S_{n+1} < S_n$. Let k be the largest natural number satisfying $S_k < m$. Then $S_{k+1} > S_k$; hence $m \leq S_{k+1} = S_k + 2 < m + 2$.

8.4 (i) Use the fact that $2^4 \equiv 1 \pmod 5$, and thus $2^{2^n} \equiv 1 \pmod 5$ for every integer $n \geq 2$.

(ii) 07. [Use the congruences $7^4 \equiv 1 \pmod{100}$ and $9^9 \equiv 1 \pmod 4$.]

(iii) 6667. [Since $\varphi(2^4) = 8$ and $\varphi(5^4) = 500$ divide $1\,000$, you have $3^{1000} \equiv 1 \pmod{10^4}$, and thus for $x = 3^{999}$ you have $3x \equiv 1 \equiv 1 - 10^4 \pmod{10^4}$, which implies $x \equiv -3333 \pmod{10^4}$.]

(iv) 36. [Note that $14^{14^{14}} = 7^{14^{14}} \cdot 2^{14^{14}}$, where $7^{14^{14}} \equiv 1 \pmod{100}$, because $7^4 \equiv 1 \pmod{100}$ and $4 \mid 14^{14}$. Since $14^{14} \equiv 1 \pmod 5$ and $14^{14} \equiv 0 \pmod 4$, you have $14^{14} \equiv 16 \pmod{20}$, and since $\varphi(25) = 20$, it follows from Euler's Theorem 3.13 (or from 3.14.(v)) that $2^{14^{14}} \equiv 2^{16} = 16^4 \equiv 9^4 = 81^2 \equiv 6^2 = 36 \pmod{25}$. Since also $2^{14^{14}} \equiv 0 \equiv 36 \pmod 4$, you get $2^{14^{14}} \equiv 36 \pmod{100}$, and altogether $14^{14^{14}} \equiv 36 \pmod{100}$.]

(v) Since $n \equiv k \pmod 4$ and $7^4 \equiv 1 \pmod{100}$, it follows from 3.14.(v) that $7^n \equiv 7^k \pmod{100}$. Since $\varphi(5^3) = 100$ and $\varphi(2^3) = 4$ divide 100, it follows from 3.14.(v) that $7^{7^n} \equiv 7^{7^k} \pmod{10^3}$. In complete analogy: Since $\varphi(5^4)$ and $\varphi(2^4)$ divide 10^3, you get $7^{7^{7^n}} \equiv 7^{7^{7^k}} \pmod{10^4}$, and similarly $7^{7^{7^{7^n}}} \equiv 7^{7^{7^{7^k}}} \pmod{10^5}$ and $7^{7^{7^{7^{7^n}}}} \equiv 7^{7^{7^{7^{7^k}}}} \pmod{10^6}$.]

(vi) For $n = 100$ both numbers end in 001. [Find the smallest $n \in \mathbb{N}$ for which $19^n \equiv 313^n \pmod{1000}$. Since $\varphi(5^3) = 100$ and $\varphi(2^3) = 4$

divide the number 100, by Euler's theorem, $19^{100} \equiv 1 \equiv 313^{100}$ (mod 125), $19^{100} \equiv 1 \equiv 313^{100}$ (mod 8), so that $19^{100} \equiv 1 \equiv 313^{100}$ (mod 1000). Show that therefore the desired n has to divide 100 (divide the number 100 by n with remainder: $100 = nq + r$, where $0 \le r < n$, then $(19^n)^q \cdot 19^r \equiv (313^n)^q \cdot 313^r$ (mod 1000), which implies $19^r \equiv 313^r$ (mod 1000), and thus $r = 0$). If $n \ne 100$, then $n \mid 50$ or $n \mid 20$ must hold. Now neither $19^{50} \equiv 313^{50}$ (mod 1000) nor $19^{20} \equiv 313^{20}$ (mod 1000) will hold (you have $19^2 \equiv 1$ (mod 5), $313^2 \equiv -1$ (mod 5), which with 3.4.(vi) implies $19^{50} \equiv 1$ (mod 125), $313^{50} \equiv -1$ (mod 125); similarly, from $19^4 \equiv 21$ (mod 25), $313^4 \equiv 11$ (mod 25) it follows that $19^{20} \equiv 21^5 \equiv 101$ (mod 125), $313^{20} \equiv 11^5 \equiv 51$ (mod 125)).]

8.6 (i) Note that $\frac{1}{9}(10^{2n} - 1) - \frac{2}{9}(10^n - 1) = \frac{1}{9}(10^{2n} - 2 \cdot 10^n + 1) = (\frac{1}{3}(10^n - 1))^2$, where $(10^n - 1)/3 \in \mathbb{N}$. Furthermore, $(10^n - 1)/3 = \underbrace{33\ldots3}_{n}$.

(ii) Note that $m_1 m_2 = \frac{2}{9}(10^{2n} - 2 \cdot 10^n + 1) = \frac{2}{9}(10^{2n} - 1) - \frac{4}{9}(10^n - 1)$, and so $m_1 m_2$ is the difference between the number consisting of $2n$ twos and the number consisting of n fours. Therefore, $m_1 m_2 = 222\ldots21777\ldots78$, with $n - 1$ twos and as many sevens.

(iii) Divide n by k with remainder: $n = qk + r$, where $0 \le r < k$. Then $10^n - 1 = 10^{qk+r} - 1 = 10^r(10^{qk} - 1) + 10^r - 1 = 10^r(10^{(q-1)k} + 10^{(q-2)k} + \cdots + 10^k + 1)(10^k - 1) + (10^r - 1)$. Since $0 \le 10^r - 1 < 10^k - 1$, the remainder in the division of $10^n - 1$ by $10^k - 1$ is $10^r - 1$. Hence $10^k - 1 \mid 10^n - 1$ if and only if $k \mid n$. Now of course, $10^k - 1 \mid 10^n - 1$ if and only if $\frac{1}{9}(10^k - 1) \mid \frac{1}{9}(10^n - 1)$, i.e., $\underbrace{111\ldots1}_{k} \mid \underbrace{111\ldots1}_{n}$.

(iv) Assume that the desired number is written as m ones. The condition of the problem means that $m > k$, and furthermore (iii) implies $k \mid m$. Since for $m = 2k$ you have $\frac{1}{9}(10^{2k} - 1) = \frac{1}{3}(10^k - 1) \cdot \frac{1}{3}(10^k + 1)$ and the number $10^k + 1$ is not divisible by 3, then $(10^k + 1)/3$ is not an integer, and therefore the number consisting of $2k$ ones is not divisible by the number consisting of k threes. Hence $m \ge 3k$. However, for $m = 3k$ you already have $\frac{1}{9}(10^{3k} - 1) = \frac{1}{3}(10^k - 1) \cdot \frac{1}{3}(10^{2k} + 10^k + 1)$, where $\frac{1}{3}(10^{2k} + 10^k + 1)$ is an integer. The number consisting of $3k$ ones is therefore the desired smallest number.

(v) $a = 5$ (for $n = 10$) or $a = 6$ (for $n = 11$ or $n = 36$). [Clearly, $a \ne 0$. The given sum is $\frac{1}{2}n(n+1)$. For $a = 1$ it follows from $\frac{1}{9}(10^k - 1) = \frac{1}{2}n(n+1)$ that $(3n + 1)(3n + 2) = 2 \cdot 10^k = 2^{k+1} \cdot 5^k$, which is a contradiction, since $(3n + 1, 3n + 2) = 1$ and $2^{k+1} < 2^{2k} = 4^k < 5^k$. Furthermore, $8(\frac{1}{2}n(n + 1)) + 1 = (2n + 1)^2$, and so the digit a is not among the digits 2, 4, 7, 9, since otherwise $(2n + 1)^2$ would be congruent to 3 or 7 modulo 10, which is impossible. If the digit a were equal to 3 or 8, then $(2n+1)^2$ would be congruent to 65 or 5 modulo 100, which is impossible. Hence $a = 5$ or $a = 6$.]

(vi) Prove by contradiction that no such n exists: Assume to the contrary that such numbers exist, and denote by n_0 the smallest one among them. Let $r \in \mathbb{N}$ be determined by the condition $10^r > n_0 \geq 10^{r-1}$. Since $M, 2M, \ldots, 9M$ do not have a digit sum less than m, it follows from $M \mid n_0$ that $n_0 \geq 10M$ and thus $r > m$. But then the number $n_1 = n_0 - 10^{r-1} + 10^{r-m-1} = n_0 - 9M \cdot 10^{r-m-1}$ is again divisible by M and its digit sum does not exceed the digit sum of n_0. Since $n_1 < n_0$, you have obtained a contradiction to the definition of n_0.

8.8 Proceed as in 8.7, but choose a k such that $3\sqrt[3]{m^2} \leq 10^{k-2}$ and $n = [10^k \cdot \sqrt[3]{m}] + 1$. Show that then $10^{3k}m < n^3 < 10^{3k}m + (10^{3k} - 1)$.

9.3 (i) Set $a_i = \frac{1}{9}(10^i - 1)$ for $i = 1, 2, \ldots, n+1$ and proceed as in 9.2.(iii). The difference $a_i - a_j$ for $i > j$ has the decimal representation of the desired form.

(ii) The natural number m divides a_n for an appropriate $n \in \mathbb{N}$ if and only if $(a, m) = 1$. [If $m \mid a_n$, then (a, m) must divide (a, a_n), but of course $(a, a_n) = (a, 1 + a \cdot a_{n-1}) = 1$. If, on the other hand, m is an arbitrary natural number relatively prime to a, use the same argument as in 9.2.(iii) to show that m divides $a_i - a_j$ for appropriate $1 \leq j < i \leq m + 1$. Then $a_i - a_j = a^j \cdot a_{i-j}$, and since $(m, a) = 1$, then m divides a_{i-j}.]

(iii) Prove this by induction on n, using the arguments of 9.2.(iv).

(iv) Use induction on n, as in (iii). To do this, use the fact that from among any five natural numbers divisible by 3^t one can choose a triple of numbers such that their sum is divisible by 3^{t+1}. [Distribute these five numbers over three sets according to the remainder upon division by 3^{t+1}. If it is possible, choose one number from each set. If one of the sets is empty, then by Dirichlet's principle one of the remaining sets must contain at least three numbers, which you then choose.]

9.5 (i) Proceed as in 9.4.(ii), but consider only nonempty subsets with even numbers of elements. Use the fact that the symmetric difference of sets with even numbers of elements has an even number of elements.

(ii) Proceed as in 9.4.(iii), but consider only nonempty subsets with one or two elements, the number of which is $\binom{45}{1} + \binom{45}{2} = 45 + \frac{45 \cdot 44}{2} = 1035$.

(iii) If the given numbers are $a_1, a_2, \ldots, a_{29}$, set $b_i = a_i \cdot a_1^2$ for $i = 1, 2, \ldots, 29$ and continue to work with the numbers $b_1, b_2, \ldots, b_{29}$. (Show that the product $a_k a_m a_n$ is the third power of a natural number if and only if $b_k b_m b_n$ is a third power.) Introduce the same triple $(\varepsilon_1, \varepsilon_2, \varepsilon_3)$ as in 9.4.(iv), but divide the numbers $b_1, \ldots, b_{29}$ into 14 groups: Keep the group that in 9.4.(iv) corresponds to the triple $(0, 0, 0)$, while of the remaining 26 you combine in one group always those groups corresponding to $(\varepsilon_1, \varepsilon_2, \varepsilon_3)$ and $(\varepsilon_1', \varepsilon_2', \varepsilon_3')$, where $\varepsilon_1 + \varepsilon_1'$, $\varepsilon_2 + \varepsilon_2'$ and $\varepsilon_3 + \varepsilon_3'$ are divisible by 3; thus you obtain 13 groups. If in the group corresponding to $(0, 0, 0)$ there are at least three numbers, choose three numbers from this group. If, on the other hand, there are no more than two numbers in this group, then the remaining 27 are distributed over 13 groups, and by Dirichlet's principle at least one of them contains at least three numbers. If you divide this group into the two original groups from 9.4.(iv), then either one of them contains at least three numbers, in which case choose these three numbers; or both contain at least one number, in which case choose one number from each group, and take b_1 as the third number.

(iv) Proceed as in (iii) and divide the numbers into the same 14 groups. If the group corresponding to $(0,0,0)$ contains some number, then choose it and set $k = 1$; otherwise, there are again 27 numbers distributed over 13 groups; hence at least one of them contains at least three numbers. Divide them again into two groups and either choose one from each $(k = 2)$, or choose three numbers from one group $(k = 3)$.

(v) Proceed as in 9.4.(v). Consider all n-tuples $(\varepsilon_1, \varepsilon_2, \ldots, \varepsilon_n)$, where $0 \le \varepsilon_i < s$ for $i = 1, 2, \ldots, n$, of which there are s^n. The rest is almost the same.

9.7 Note that if x is an arbitrary integer, then x^2 is congruent to one of the numbers $0, 1, 2^2, 3^2, \ldots, (\frac{p-1}{2})^2$, of which there are $\frac{p+1}{2} < p$.

10.3 (i) Since $19 \equiv 62 \pmod{43}$, a polynomial $F(x)$ with integer coefficients must satisfy $F(19) \equiv F(62) \pmod{43}$.

(ii) Proceed as in 10.2.(iii).

(iii) Assume that for some $n \in \mathbb{Z}$ you have $F(n) = 0$ and show that for each $i = 1, \ldots, 5$ you have $a_i = n \pm 1$ or $a_i = n \pm 73$, and thus the numbers $a_1, a_2, \ldots, a_5$ are not all different.

(iv) Divide the five numbers $a_1, \ldots, a_5$ into two groups according to which value of the polynomial $F(x)$ they take on. If none of these groups is empty, then by Dirichlet's principle one of them must contain at least three numbers, and the other one at least one, which by (ii) is a contradiction.

(v) Find r, s such that $F(r) = 0$ and $F(s) = 1$. Then by (ii) you have $s = r + 1$ or $s = r - 1$. In the case $s = r + 1$ prove by induction that $F(r + m) = m$ holds for each $m \in \mathbb{N}_0$. (This is true for $m = 0$ and $m = 1$. For integers $m > 2$ assume that $F(r + m - 2) = m - 2$, $F(r + m - 1) = m - 1$. Given an m, there exists a $k \in \mathbb{Z}$ such that $F(k) = m$, and by (ii) you have $k = r + m - 2$ or $k = r + m$. The first case does not occur, since $F(r + m - 2) = m - 2 \ne m = F(k)$. Hence $k = r + m$, and so $F(r + m) = F(k) = m$.) By 3.15 in Chapter 1 you have $F(x) = x - r$. Similarly, in the case $s = r - 1$ show that $F(x) = -x + r$. The solution of the problem is therefore given exactly by all linear polynomials $a_1 x + a_0$, where $a_1 = 1$ or $a_1 = -1$, and $a_0 \in \mathbb{Z}$ is arbitrary.

10.6 (i) Exactly when $6a, 2b, a+b+c, d \in \mathbb{Z}$. [Rewrite $ax^3 + bx^2 + cx + d = 6a\binom{x}{3} + 2(b + 3a)\binom{x}{2} + (a + b + c)\binom{x}{1} + d$.]

(ii) Yes, since $F(0), F(1), \ldots, F(7)$ are integers.

(iii) For $n \ge 2$ it exists, for $n = 1$ it does not exist. [Proceed as in 10.5.(iii), obtain the conditions $a_0, \ldots, a_{n-1} \in \mathbb{Z}$, $a_n \notin \mathbb{Z}$, $(n + 1)a_n \in \mathbb{Z}$, $\frac{1}{2}(n + 2)(n + 1)a_n \in \mathbb{Z}$. For $n \ge 2$ choose $a_n = \frac{2}{n+1}$; for $n = 1$ it follows from the conditions $2a_1 \in \mathbb{Z}$, $3a_1 \in \mathbb{Z}$ that also $a_1 \in \mathbb{Z}$; for $n = 1$ it is therefore not possible to satisfy all conditions.

(iv) Such a polynomial exists for $n = 3$, $n = 4$, and $n \geq 6$, but does not exist for $n \leq 2$ and $n = 5$. [Proceed as in 10.5.(iii), and see whether the conditions $a_n \notin \mathbb{Z}$, $(n+1)a_n \in \mathbb{Z}$, $\frac{1}{2}(n+2)(n+1)a_n \in \mathbb{Z}$, $\frac{1}{6}(n+3)(n+2)(n+1)a_n \in \mathbb{Z}$ are satisfied. For $n = 3$, $n = 4$, or $n \geq 6$, $a_n = \frac{6}{n+1}$ satisfies them. For $n = 1$ it follows from $2a_1 \in \mathbb{Z}$ and $3a_1 \in \mathbb{Z}$ that $a_1 \in \mathbb{Z}$. For $n = 2$ it follows from $3a_2 \in \mathbb{Z}$ and $10a_2 \in \mathbb{Z}$ that $a_2 = 10a_2 - 3 \cdot 3a_2 \in \mathbb{Z}$. For $n = 5$ it follows from $6a_5 \in \mathbb{Z}$, $21a_5 \in \mathbb{Z}$ and $56a_5 \in \mathbb{Z}$ that also $a_5 = -57 \cdot 6a_5 + 19 \cdot 21a_5 - 56a_5 \in \mathbb{Z}$ (this last construction can be carried out by use of Bézout's equality; it follows from $(6, 21, 56) = 1$).]

(v) Let $F(x) = b_k x^k + \cdots + b_1 x + b_0$. By Theorems 2 and 3 in 10.4 there exist integers $a_0, \ldots, a_k$ such that

$$\frac{F(x)}{p} = a_0 + a_1 \binom{x}{1} + \cdots + a_k \binom{x}{k},$$

which implies that the polynomial $k!F(x)/p$ has integer coefficients, so $p \mid b_i \cdot k!$ for all $i = 0, 1, \ldots, k$. Since $(p, k!) = 1$, this means that $p \mid b_i$.

10.10 (i)–(iv) Use 10.8 in (i) for $p = 5$, in (ii) for $p = 3$ (it does not work for $p = 2$), in (iii) for $p = 7$ ($p = 3$ not possible). In (iv), first divide by 5, then use 10.8 for $p = 5$, and finally show that the original polynomial cannot be decomposed either.

(v) $x^{125} - 75$. [Proceed as in 10.9.(ii), use 10.8 for $p = 3$.]

Bibliography

A. Books cited in text

[1] E. Beckenbach and R. Bellman, *Inequalities*, Springer-Verlag, Berlin-Göttingen-Heidelberg, 1961.

[2] A. I. Borevich and I. R. Shafarevich, *Number Theory*, Academic Press, New York-London, 1966.

[3] P. S. Bullen, D. S. Mitrinović, and P. M. Vasić, *Means and Their Inequalities*, Reidel, Dordrecht, 1987.

[4] L. Childs, *A Concrete Introduction to Higher Algebra*, second edition, Springer-Verlag, New York, 1995.

[5] P. Franklin, *A Treatise on Advanced Calculus*, Dover Publications, New York, 1964.

[6] H. W. Gould, *Combinatorial Identities; a Standardized Set of Tables Listing 500 Binomial Coefficient Summations*, revised edition, Morgantown, W.Va., 1972.

[7] S. L. Greitzer, *International Mathematical Olympiads 1959-1977*, Mathematical Association of America, Washington, D.C., 1978.

[8] S. Mac Lane and G. Birkhoff, *Algebra*, third edition, Chelsea Publishing Co., New York, 1988.

[9] J. Riordan, *Combinatorial Identities*, Wiley, New York, 1968.

[10] G. E. Shilov, *Linear Algebra*, Dover Publications, New York, 1977.

[11] W. Sierpiński, *Elementary Theory of Numbers*, Państwowe Wydawnictwo Naukowe, Warszawa, 1964.

[12] I. M. Vinogradov, *Elements of Number Theory*, Dover Publications, New York, 1954.

[13] B. L. van der Waerden, *Algebra, Vol. I*, Springer-Verlag, New York, 1991.

B. Additional English-language books

[14] E. J. Barbeau, M. S. Klamkin, and W. O. J. Moser, *Five Hundred Mathematical Challenges*, Mathematical Association of America, Washington, DC, 1995.

[15] E. Beckenbach and R. Bellman, *An Introduction to Inequalities*, Mathematical Association of America, Washington, DC, 1961.

[16] G. Birkhoff and S. Mac Lane, *A Survey of Modern Algebra*, third edition, Collier-Macmillan, London, 1965.

[17] M. Doob (ed.), *The Canadian Mathematical Olympiad, 1969–1993*, Canadian Mathematical Society, Ottawa, 1993.

[18] H. Dörrie, *100 Great Problems of Elementary Mathematics*, Dover Publications, 1965. Reprinted 1989.

[19] M. J. Erickson and J. Flowers, *Principles of Mathematical Problem Solving*, Prentice-Hall, 1999.

[20] A. Gardiner, *The Mathematical Olympiad Handbook. An Introduction to Problem Solving Based on the First 32 British Mathematical Olympiads 1965–1996*, Oxford University Press, 1997.

[21] G. T. Gilbert, M. I. Krusemeyer, and L. C. Larson, *The Wohascum County Problem Book*, Mathematical Association of America, Washington, DC, 1993.

[22] G. H. Hardy, J. E. Littlewood, and G. Pólya, *Inequalities*, second etition, Cambridge University Press, Cambridge, 1952.

[23] K. Hardy and K. S. Williams, *The Red Book of Mathematical Problems*, Dover Publications, Mineola, NY, 1996.

[24] K. Hardy and K. S. Williams, *The Green Book of Mathematical Problems*, Dover Publications, Mineola, NY, 1997.

[25] R. Honsberger, *From Erdős to Kiev. Problems of Olympiad Caliber*, Mathematical Association of America, Washington, DC, 1996.

[26] M. S. Klamkin, *USA Mathematical Olympiads 1972–1986*, Mathematical Association of America, Washington, DC, 1988.

[27] P. P. Korovkin, *Inequalities*, Little Mathematics Library, Mir, Moscow; distributed by Imported Publications, Chicago, Ill., 1986.

[28] J. Kürschák, *Hungarian Problem Book I, Based on the Eötvös Competitions, 1894–1905*, Mathematical Association of America, Washington, DC, 1963.

[29] J. Kürschák, *Hungarian Problem Book II, Based on the Eötvös Competitions, 1906–1928*, Mathematical Association of America, Washington, DC, 1963.

[30] L. C. Larson, *Problem-Solving Through Problems*, Springer-Verlag, New York, 1983.

[31] D. J. Newman, *A Problem Seminar*, Springer-Verlag, New York, 1982.

[32] G. Pólya, *Mathematical Discovery. On Understanding, Learning, and Teaching Problem Solving*, reprint in one volume, John Wiley & Sons, Inc., New York, 1981.

[33] G. Pólya, *How to Solve It. A New Aspect of Mathematical Method*, Princeton University Press, Princeton, NJ, 1988.

[34] J. Roberts, *Elementary Number Theory. A Problem Oriented Approach*, MIT Press, Cambridge, Mass.-London, 1977.

[35] D. O. Shklarsky, N. N. Chentzov, and I. M. Yaglom, *The USSR Olympiad Problem Book*, W. H. Freeman, San Francisco, 1962. Reprinted by Dover Publications, New York, 1993.

[36] W. Sierpiński, *250 Problems in Elementary Number Theory*, American Elsevier Publishing Co., Inc., New York; PWN Polish Scientific Publishers, Warsaw, 1970.

[37] H. Steinhaus, *One Hundred Problems in Elementary Mathematics*, Basic Books, Inc., New York, 1964. Reprinted by Dover Publications, New York, 1979.

[38] Ch. W. Trigg, *Mathematical Quickies*, Dover Publications, New York, 1985.

[39] W. A. Wickelgren, *How to Solve Mathematical Problems*, Dover Publications, Mineola, NY, 1995.

[40] P. Zeitz, *The Art and Craft of Problem Solving*, Wiley, New York, 1999.

C. East European sources

[41] I. L. Babinskaya, *Zadachi matematicheskikh olimpiad*, Nauka, Moskva, 1975.

[42] V. I. Bernik, I. K. Zhuk, and O. V. Melnikov, *Sbornik olimpiadnykh zadach po matematike*, Narodnaya osveta, Minsk, 1980.

[43] J. Fecenko, *O nerovnostiach, ktoré súvisia s aritmetickým a geometrickým priemerom*, Rozhledy mat.-fyz. **66** (1987/88), no. 2, 46–49.

[44] M. Fiedler and J. Zemánek, *Vybrané úlohy matematické olympiády (kategorie A + MMO)*, SPN, Praha, 1976.

[45] G. A. Galperin and A. K. Tolpygo, *Moskovskiye matematicheskiye olimpiady*, Prosveshcheniye, Moskva, 1986.

[46] V. M. Govorov, *Sbornik konkursnykh zadach po matematike*, Nauka, Moskva, 1986.

[47] J. Herman and J. Šimša, *O jednom užití Dirichletova principu v teorii čísel*, Rozhledy mat.-fyz. **66** (1987/88), no. 9, 353–356.

[48] K. Horák et al., *Úlohy mezinárodních matematických olympiád*, SPN, Praha, 1986.

[49] S. V. Konyagin et al., *Zarubezhnyye matematicheskiye olimpiady*, Nauka, Moskva, 1987.

[50] I. Korec, *Úlohy o velkých číslach*, Mladá fronta, edice ŠMM, Praha, 1988.

[51] V. A. Krechmar, *Zadachnik po algebre*, Nauka, Moskva, 1972.

[52] G. A. Kudrevatov, *Sbornik zadach po teorii chisel*, Prosveshcheniye, Moskva, 1970.

[53] V. B. Lidskiy et al., *Zadachi po elementarnoy matematike*, Gos. izd. fiz.-mat. lit., Moskva, 1960.

[54] P. S. Modenov, *Sbornik zadach po spetsialnom kurse elementarnoy matematiky*, Vysshaya shkola, Moskva, 1960.

[55] *N-tý ročník matematické olympiády*, SPN, Praha, (from 1953 to 1993).

[56] O. Odvárko et al., *Metody řešení matematických úloh I*, učební text MFF UK, Praha, 1977.

[57] I. Kh. Sivashinskiy, *Neravenstva v zadachakh*, Nauka, Moskva, 1967.

[58] I. Kh. Sivashinskiy, *Teoremy i zadachi po algebre i elementarnym funktsiyam*, Nauka, Moskva, 1971.

[59] S. Straszewicz, *Zadania z olimpiad matematicznych*, I, II, Warszawa, 1956, 1961.

[60] S. Straszewicz and J. Browkin, *Polskiye matematicheskiye olimipady*, Mir, Moskva, 1978.

[61] J. Šedivý et al., *Metody řešení metematických úloh II*, učební text MFF UK, Praha, 1978.

[62] V. S. Shevelyov, *Tri formuly Ramanudzhana*, Kvant, 1988, no. 6, 52–55

[63] J. Šimša, *Dolní odhady rozdílu průměrů*, Rozhledy mat.-fyz. **65** (1986/87), no. 10 (June), 403–407.

[64] N. V. Vasilyev and A. A. Yegorov, *Zadachi vsesoyuznykh matematicheskikh olimpiad*, Nauka, Moskva, 1988.

[65] N. V. Vasilyev et al., *Zaochnyye matematicheskiye olimpiady*, Nauka, Moskva, 1986.

[66] V. V. Vavilov et al., *Zadachi po matematike: Algebra*, Nauka, Moskva, 1987.

[67] V. V. Vavilov et al., *Zadachi po matematike: Uravneniya i neravenstva*, Nauka, Moskva, 1987.

[68] A. Vrba and K. Horák, *Vybrané úlohy MO kategorie A*, SPN, Praha, 1988.

[69] V. A. Vyshenskiy et al., *Sbornik zadach kiyevskikh matematicheskikh olimpiad*, Vishcha shkola, Kiyev, 1984.

Index